MATHEMATICS RESEARCH DEVELOPMENTS

ADVANCED ENGINEERING MATHEMATICS AND ANALYSIS

VOLUME 1

MATHEMATICS RESEARCH DEVELOPMENTS

MATHEMATICS RESEARCH DEVELOPMENTS

ADVANCED ENGINEERING MATHEMATICS AND ANALYSIS

VOLUME 1

RAMI A. MAHER

nova science publishers
New York

https://doi.org/10.52305/THXL5007

Library of Congress Cataloging-in-Publication Data

ISBN: 978-1-53619-869-0

Published by Nova Science Publishers, Inc. † New York

To the memory of my parents

Contents

PREFACE

Advanced engineering mathematics and analysis is a topic that deals with the implementation of mathematical theories, corollaries, lemmas, propositions, and algorithms to solve numerous engineering problems. These tools serve in two directions to solve a considered engineering problem directly or indirectly by setting up a solution framework. For example, linear algebraic systems theories instantly solve the loop currents evaluation problem in a dc electric circuit. An indirect example is the modeling of construction project networks by a matrix whose entry values represent activities durations. The book is intended for engineering students as well as for engineers at work.

Separated volumes are written to cover a significant part of this topic. It considers most of the questions that are raised in the engineering classrooms about where this specific mathematical tool is utilized in practice. Besides, for engineers at work, it represents a benefit reference. Most of the pure mathematical derivations are omitted without sacrificing the quality of the considered materials. To reduce the book size and clarify the content simultaneously, balancing between pure mathematical examples and physical problems is considered. Moreover, the book elucidates the subjects in a self-contain style. The reader can select what he wants to read without following a particular sequence of reading. The target material is arranged to be in three units that are split into separated printed volumes. All chapters in these volumes have their own sets of exercises. The chapters have appeared in a sequence, which is often used for teaching these materials.

In this volume, the first chapter deals with linear algebra concepts and matrix theory. It covers from very basics to advanced topics such as the concept and use of singular values and matrices' factorizations. In chapters two and three, ordinary differential equations are classified and solved besides modeling some engineering problems. In chapter two, the most popular solution methods are used to solve first-order nonlinear homogenous equations, first-order linear nonhomogeneous, second-order and high-order constant-coefficients equations, and first-order systems. Chapter three considers the series solution, based on the Frobenius method, to solve second-order variable-coefficients homogenous equations, including Bessel's equation and other particular equations. Since the differential equation is a large topic, the second part contains selected additional methods for nonlinear equations, such as the method of a parameter and the approach based on Jacobi Elliptic functions. To complete studying the differential equations topic, a set of numerical methods to solve differential equations is given in chapter five.

Chapter four is intended for the Laplace transforms and their applications. The importance of solving particular types of differential equations and their use in control systems makes the Laplace transform a usual topic in engineering analysis. The last chapter in this volume is devoted to reviewing the most fundamental topics in numerical computations. The chapter content subjects are selected to be helpful tools for the problems in the other chapters whenever the exact solution is impossible to obtain. For instance, the numerical integration methods can be utilized if the line integral of a vector function does not have an exact solution.

ACKNOWLEDGMENTS

I would like to thank my kind and patient family for their continuous encouragement to finish this manuscript. My dear colleague and friend Eng. Hikmat N. Georges deserves a lot of thanks and appreciation for his efforts in many helpful suggestions besides the proofreading notes. I would like to acknowledge the contribution of my students for their continuous puzzling questions. Finally, thanks are conveyed to colleagues in the engineering faculty at Isra University for their continuous support.

Chapter 1

LINEAR ALGEBRA AND MATRICES

1.1. INTRODUCTION

Perhaps linear algebra and matrices are some of the most usable topics in most scientific fields. For engineering students, this topic is essential in understanding and manipulating several other subjects and solving problems in other different areas. This fact is the reason behind studying the material in most engineering courses at the beginning. This importance guides the manipulation of the subject to be in a broad scope. However, the reader has to refer to specialist textbooks for more in-depth investigation and applications.

The concepts from linear algebra and matrices are related to each other, and therefore, can be treated as one subject. For instance, as we try to explain the concepts from linear algebra, we will need to know the definition of a matrix, and inversely. On the other hand, there are points of similarity between the nature of an algebraic system solution based on matrix theory and the solution of a first-order differential equation. These points of similarity motivate the introduction to linear algebra. Therefore, in the next section, we will introduce these topics' basics for a more straightforward explanation. There is a lot to be considered for using vectors and matrices in engineering and other scientific fields. However, we will consider only some of these applications throughout this chapter due to the book size limitation.

In our world, specifying the coordinates of a mass point represented by a three dimensions vector is a clear example of using vector and matrix representation. Graphic software uses a matrix transformation to process linear transformation to render images. Therefore, computers have embedded matrix arithmetic in graphic processing algorithms, especially to cause reflection and refraction. A pixel matrix can represent a monochrome image with a grayscale or intensity value (1 for the black color, 0 for white). Hence, the image can be reproduced. In the world of information security, many systems are designed to deal with matrices. Matrices are well-known tools to be used in the compression of digital information and are usually used in encryption or massages coding. Furthermore, the theory developed for simultaneous algebraic equations has a wide field of engineering applications, such as determining currents and voltages in passive electric networks or the stability problem of a fixed truss.

1.2. Motivation

Throughout the text, a commonly used mathematical expression is the definition of the polynomial. A polynomial of *n*-order, where *n* is an integer number, is given by

$$p_n(x) = a_n x^n + a_{n-1}x^{n-1} + a_{n-2}x^{n-2} + \cdots + a_1 x + a_0 \tag{1.1}$$

The coefficients a_i are real or complex numbers, including zero. Thus, it is a weighted combination of argument *x* and its powers. Polynomials can exist as a result of various operations of vectors and matrices, as will be seen. Characteristic polynomials and characteristic equations are fundamental terms in the theory of matrices.

The nth order polynomial can be factorized into a product of several smaller order polynomials; for instance, it can be factorized to *n* first-order polynomials, i.e.

$$p_n(x) = (x + q_1)(x + q_2) \ldots . (x + q_n); \; q_1 q_2 \ldots q_n = a_0 \tag{1.2}$$

Another form of writing a polynomial is to use the notation of summing, i.e.,

$$p_n(x) = \sum_{i=0}^{n} a_{n-i} x^{n-i} \tag{1.3}$$

A vector of *n* length, where *n* is an integer number, is a single row or column representation of *n* elements that could be variables or constants, or even functions, as illustrated in the following examples.

$$R_n = [r_1 \quad r_2 \quad r_3 \quad \ldots \quad r_n]$$
$$G_n(x, y, \ldots) = [g_1(x, y, \ldots) \quad g_2(x, y, \ldots) \quad \ldots \quad g_n(x, y, \ldots)]$$

The vectors in the above examples are written in row representation; also, vectors can be represented in column form such as

$$C_n = \begin{bmatrix} c_1 \\ c_2 \\ \vdots \\ c_n \end{bmatrix}, F_n = \begin{bmatrix} f_1(x) \\ f_2(x) \\ \vdots \\ f_1(x) \end{bmatrix}, \text{etc.}$$

If we have an equal length of vectors, then the algebraic sum (addition or subtraction) of vectors is the algebraic sum of their elements in the same position. The vectors obey the associative and distributive laws. The vector's product by a scalar value scales the vector correspondingly, i.e., multiply each entry of the vector by that scalar value. Furthermore, two vectors are said to be equal if and only if their elements in the same positions are equal.

A matrix is rectangular of a finite number *m* of rows and *n* columns with elements (entries) of thinks; only entries of real or complex numbers will be considered here. The uppercase letter is used to denote a matrix, say *A*. The matrix entries are indicated by the corresponding lowercase letter and their indices of the row and column, for example

$$A = \begin{bmatrix} a_{11} & a_{12} & \cdots & a_{1n} \\ a_{21} & a_{22} & \cdots & a_{2n} \\ \vdots & \vdots & \vdots & \vdots \\ a_{m1} & a_{m2} & \cdots & a_{mn} \end{bmatrix}, \mathrm{ent}_{ij}(A) = a_{ij}$$

To denote the *i*th row and the *j*th column, we write $\mathrm{Row}_i(A)$, and $\mathrm{Col}_j(A)$, respectively.

In manipulating matrices, we denote the matrix with its dimensions (usually, we say of order); for instance, $A_{3\times2}$ indicates a matrix of three rows and two columns. Vectors are a particular case of a matrix; every row or column of a matrix can be considered a specific vector. Most of the vector properties are applied equally to matrices. For instance, in particular, matrix addition of the same orders (conformable) has the following properties:

1. Commutative, i.e., $A + B = B + A$, and $A - B = -B + A$
2. Associative, i.e., $A + (B + C) = (A + B) + C$, and $A(BC) = (AB)C$
3. Cancelation law, i.e., $A + B = A + C$, then $B = C$
4. $A + M = A$, if and only if the matrix M has all its entries equal to zero, or a zero matrix $M = O$, and consequently, it holds $A - A = O$

Further, the multiplication of a matrix by a scalar constant value gives a matrix whose entries are multiplied by that constant value, i.e.,

$$cA \Longrightarrow \ \mathrm{c} \times \mathrm{ent}_{ij}(A) = \mathrm{ent}_{ij}(cA) = ca_{ij}, \forall\, i = 1,2 \ldots m, j = 1,2 \ldots n \qquad (1.4)$$

The properties of the scalar multiplication of conformable two matrices and two constants are:

1. $(c_1 + c_2)A = c_1A + c_2A$
2. $c(A + B) = cA + cB$
3. $c_1(c_2A) = (c_1c_2)A$

The multiplication may be the most usable arithmetic operation on matrices. To find the product of two conformable matrices, it means not to multiply each entry of one matrix by the corresponding placed entry of the second one but is more workable, as will be seen next. Unlike the algebraic sum of matrices, two matrices are conformable for multiplication if they have only the same orders or also when the number of rows in the preceding matrix is equal to the number of columns in the following one. Later, we have to explain and define matrices, arithmetic operations on matrices, and systems.

Matrices are a unique representation of several problems, even for small dimensions. One famous example is the network problem. Figure 1.1 shows a schematic road network, where the nodes represent cities, and the branches represent the roads connecting cities.

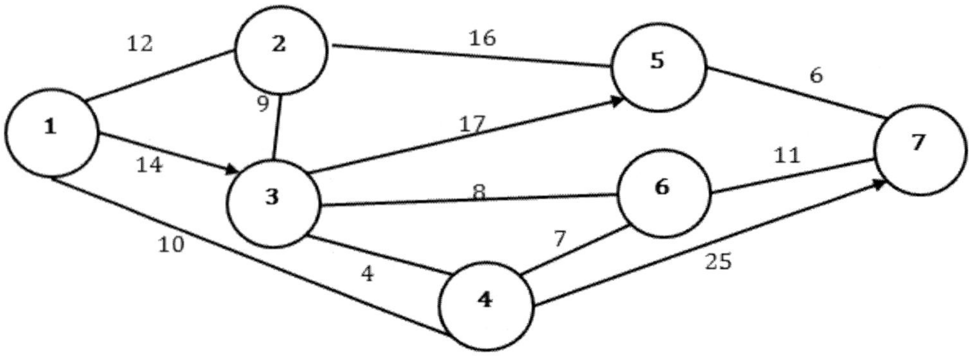

Figure 1.1. A road network.

The values assigned to the branches represent the distances between the cities. Branches with arrows indicate a one-direction road, while branches without arrows indicate a two-way road. One well-known problem is to find the minimum (or the maximum) distance between cities and the path of routing. It may be easy for a small number of nodes to solve the problem by trying a few trials. However, when the number of nodes is large, finding the economic path becomes a challenging task.

The road network can be represented by a square seven-dimension matrix, where elements are the distances between the nodes and zero diagonal elements.

$$R = \begin{bmatrix} 0 & 12 & 14 & 10 & \infty & \infty & \infty \\ 12 & 0 & 9 & \infty & 16 & \infty & \infty \\ \infty & 9 & 0 & 4 & 17 & 8 & \infty \\ 10 & \infty & 4 & 0 & \infty & 7 & 25 \\ \infty & 16 & \infty & \infty & 0 & \infty & 6 \\ \infty & \infty & 8 & 7 & \infty & 0 & 11 \\ \infty & \infty & \infty & \infty & 6 & 11 & 0 \end{bmatrix}$$

The assigned infinity means there is no road connecting the corresponding cities. It is evident that if all branches indicate two-way directions, then the matrix is symmetrical. There are several other applications. For instance, the network can represent the activity-on arrow for project planning to find the project's completion time. The communication network is another widely used in practice.

1.3. BASICS AND DEFINITIONS OF MATRICES

In previous sections, we introduced the definition of a matrix and some of its properties of addition and scalar multiplication only. This section will continue deliberating the types of matrices and related definitions, including the scalar-valued functions. These types of matrices will be used throughout the text of the chapter in different places. Along with the mentioned definitions, properties of some types of matrices are also introduced without proofing them. These properties can be framed as theorems with their proofs. The idea is to be familiar with the used terminology and the same simple computations before considering a rich elaboration. For explicit explanation purposes, we will use, for most of the definitions, one single matrix symbol A.

To complete the discussion in section 1.2, let us begin with the multiplication operation between matrices. It is already said that multiplication can only be carried out when the number of columns in the preceding matrix is equal to the number of rows in the following matrix. Thus, the multiplication operation is not limited to equaling order matrices.

Definition 1.1. Square matrix

A matrix of an equal number of rows and columns is called a square matrix, written with a single index, says A_n instead of writing $A_{n\times n}$.

Definition 1.2. Multiplication of matrices

Consider two matrices $A_{m\times n}$ and $B_{p\times q}$, then the multiplication AB is defined if and only if

a) $m = n = p = q$, both matrices are square and of the same order. The resultant matrix will have the same order too.
b) $n = p$, the resultant matrix will be of $(m \times q)$ order.

Then the product AB is given by

$$ent_{ij}(AB) = \sum_{k=1}^{n} a_{ik}b_{kj};\ i = 1,2 \dots m, j = 1,2 \dots q \tag{1.5}$$

It means that an element of the product AB is obtained by summing a row's product by the corresponding position elements of a column. For example,

$$A_{2\times3}B_{3\times2} = \begin{bmatrix} 2 & 1 & 6 \\ 3 & 4 & 2 \end{bmatrix}\begin{bmatrix} 1 & 2 \\ 3 & 4 \\ 5 & 6 \end{bmatrix} = \begin{bmatrix} [2 \quad 1 \quad 6]\begin{bmatrix} 1 \\ 3 \\ 5 \end{bmatrix} & [2 \quad 1 \quad 6]\begin{bmatrix} 2 \\ 4 \\ 6 \end{bmatrix} \\ [3 \quad 4 \quad 2]\begin{bmatrix} 1 \\ 3 \\ 5 \end{bmatrix} & [3 \quad 4 \quad 2]\begin{bmatrix} 2 \\ 4 \\ 6 \end{bmatrix} \end{bmatrix} = \begin{bmatrix} \sum_{i=1}^{3} a_{1i}b_{i1} & \sum_{i=1}^{3} a_{1i}b_{i2} \\ \sum_{i=1}^{3} a_{2i}b_{i1} & \sum_{i=1}^{3} a_{2i}b_{i2} \end{bmatrix}$$

$$A_{2\times3}B_{3\times2} = \begin{bmatrix} 35 & 44 \\ 25 & 34 \end{bmatrix}$$

Probably, an essential feature of matrix multiplication is that, in general, the products AB and BA may not be the same or equal. For instance, in the previous example, the size of the product BA will be three instead of two. Besides, it may be the case that $AB \neq BA$ even if they have the same size. However, there are cases where $AB = BA$, and it is a difficult task to find out whether they are the same or not without carrying out the calculation. In the terminology dictionary, it is said that A and B commute if $AB = BA$.

A second feature of the multiplication is the representing of a linear algebraic system of equations $AX = K$, where A is a coefficient matrix with m rows (number of equations) and n columns (number of variables). The column vectors X and K are the variable vectors and the input vector, respectively. To illustrate, consider the following system.

$$5x_1 - x_2 + 2x_3 = 10$$
$$3x_1 + 2x_2 - 2x_3 = 8$$
$$-3x_2 + 4x_3 = -4$$

Using the multiplication property, we can write

$$\begin{bmatrix} 5 & -1 & 2 \\ 3 & 2 & -2 \\ 0 & -3 & 4 \end{bmatrix}\begin{bmatrix} x_1 \\ x_2 \\ x_3 \end{bmatrix} = \begin{bmatrix} 10 \\ 8 \\ -4 \end{bmatrix}$$

Similarly, we can write a matrix equation for a system of first-order linear differential equations. It is an essential property because it will be possible to apply unique matrix-based theories for solving these equations. Also, the associative and the distributive laws are valid for matrix multiplication while the cancelation law is not necessary to hold, i.e.,

1. If $A, B,$ and C are any matrices such that the products AB and BC are defined, then the associated law states the equality relation $A(BC) = (AB)C$.
2. If $A, B,$ and C are any matrices such that the sum $A + B$ and the product AB are defined, then the distributive law states $(A + B)C = AC + BC$.
3. The product of two nonzero matrices may be a zero matrix, so the cancelation law may not hold, i.e., $AB = AC, A \neq O$ does not, in general, imply that $B = C$.
4. The direction of performing the multiplication should not be changed arbitrarily. For instance, it is not permissible in general to write $(A + B)C = AC + CB$.
5. The proof of these properties is straightforward.

Definition 1.3. A diagonal matrix and tridiagonal matrix

A diagonal matrix Dg is a square matrix with all off-diagonal entries equal to zero, and at least one of the entries of the on-diagonal entries is not equal to zero. Thus,

$$\text{ent}_{ij}\,(\text{Dg}) = \begin{cases} \text{value}, i = j \\ 0, i \neq j \end{cases} \tag{1.6}$$

A general k- dimension tridiagonal matrix A_n is a square matrix whose entries are defined as

$$a_{ij} = \begin{cases} a_i, i = j, \\ b_j, i = j + k, \\ c_i, j = i + k, \\ 0, \text{otherwise} \end{cases} \tag{1.7}$$

A special tridiagonal is the Toeplitz tridiagonal matrix in which $a_1 = a_2 = \cdots = a_n = a$, $b_1 = b_2 = \cdots = b_n = b$, $c_1 = c_2 = \cdots = c_n = c$, and $k = 1$. Tridiagonal matrices have many significant features for computing the determinants, eigenvalues, eigenvectors, etc. They appear in many practical problems in signal processing and applications like an extensive sparse linear system.

Definition 1.4. Unity or identity matrix I

A unity matrix I is an especial case of the diagonal matrix, and it is defined as

$$\text{ent}_{ij}(I) = \begin{cases} 1, i = j \\ 0, i \neq j \end{cases} \tag{1.8a}$$

The following multiplication operations are defined.

$$I_n \times I_n \ldots \times I_n = I_n \tag{1.8b}$$

$$I_n \times A_n = A_n \tag{1.8c}$$

Definition 1.5. Sparse matrices

It is the matrix that most of its elements are zero. Sparsity corresponds to systems, which are loosely coupled.

Definition 1.6. Lower and upper triangular matrices

A lower triangular matrix is that whose entries above the diagonal are all zeros, while the upper triangular matrix has all entries below the diagonal equal to zero, i.e.,

$$\text{ent}_{ij}\,(L) = \begin{cases} \text{value}, i \geq j \\ 0, i < j \end{cases} \tag{1.9}$$

$$\text{ent}_{ij}\,(U) = \begin{cases} \text{value}, i \leq j \\ 0, i > j \end{cases} \tag{1.10}$$

Definition 1.7. Submatrix, a minor of a matrix, partitioned matrix, and augmented matrix

Consider a matrix, *A* then, the submatrix is a matrix obtained from A by deleting individual rows or/and columns of A. A minor matrix is obtained by deleting an ith row and j^{th} column of A, and it is called the minor of the (i, j) position of A. The minor is denoted by $\mathrm{Min}_{ij}(A)$. A partitioned matrix is viewed as it consists of many smaller order matrices. It is convenient that for a large order matrix, partitioning is performed for easy manipulation. Partitioning style is made in correspondence to the computation or description desires. When modeling large systems, a specific partitioning style may reflect the interconnections between the system's subsystems. Sometimes it is required to work simultaneously the same operations for more than one matrix; therefore, we create what is called the augmented matrix. Augmentation can be arranged horizontally or vertically, providing an equal number of rows or columns, respectively. For example, given the following matrices $A_{m\times n}$, $B_{m\times p}$, $C_{m\times n}$, and $D_{q\times n}$, then we can augment as follows.

$$[A \quad \vdots \quad B], \text{and} \begin{bmatrix} C \\ \cdots \\ D \end{bmatrix}$$

Definition 1.8. A transpose of a matrix

Given a matrix A of order *m* by *n*, then the transpose of A denoted A^T, is the *n* by *m* matrix defined by

$$\mathrm{ent}_{ij}(A^T) = \mathrm{ent}_{ji}(A) = a_{ji} \tag{1.11}$$

In other words, a transpose of a matrix is to interchange rows and columns. The transpose operation has the following properties.

1. For every matrix A, it is valid that $(A^T)^T = A$.
2. Given two conformable for addition matrices *A* and *B*, then $(A + B)^T = A^T + B^T$.
3. For any scalar constant *c* and any matrix A, it holds$(cA)^T = cA^T$.
4. Given two conformable for multiplication matrices A and B, then $(AB)^T = B^T A^T$.

Definition 1.9. Conjugate matrix and tranjugate matrix

If $A_{m\times n}$ is a complex matrix (at least one entry is a complex number), then the conjugate of $A_{m\times n}$ is an $(m \times n)$ complex matrix $\bar{A}$ defined by

$$\mathrm{ent}_{ij}(\bar{A}) = \overline{ent_{ij}(A)} = \overline{a_{ij}} \tag{1.12}$$

The tranjugate or Hermitian conjugate of $A_{m\times n}$ is an $(m \times n)$ complex matrix defined by

$$A^* = \bar{A}^T = \overline{A^T} \tag{1.13}$$

In a similar manner with transpose matrices, the tranjugate matrices have the following properties.

1. For every complex matrix, it is valid $(A^*)^* = A$.
2. Given two conformable for addition matrices *A* and *B*, then $(A + B)^* = A^* + B^*$.
3. For any scalar constant *c* and any matric *A*, it holds $(cA)^* = cA^*$.
4. Given two conformable for multiplication matrices *A* and *B*, then $(AB)^* = B^*A^*$.

Definition 1.10. Symmetric and skew-symmetric matrices

For a real square matrix A, it is called symmetric if $A^T = A$, and it is called skew-symmetric if it holds $A^T = -A$.

Definition 1.11. Hermitian and skew Hermitian matrices

For a complex square matrix A, it is called Hermitian if $A^* = A$, and it is called skew Hermitian if it holds $A^* = -A$.

Before we continue, let us introduce elementary examples to illustrate some of the previous definitions.

Example 1.1.

Consider the following three matrices

$$A = \begin{bmatrix} 1 & 0 & -3 & 5 \\ 2 & -1 & 4 & 11 \\ 2 & 7 & 10 & 3 \\ 5 & 1 & 0 & 9 \end{bmatrix}, B = \begin{bmatrix} 1+j & -5 & 6+2j \\ 4 & 1 & 3 \\ 2-j & 7 & -4 \end{bmatrix}, C = \begin{bmatrix} 2 & 1 & 3 \\ 1 & 7 & 9 \\ 3 & 9 & 8 \end{bmatrix}$$

- The matrix A is a square matrix with four rows and four columns, while the matrix C is a square symmetric matrix of third-order. The matrix A is a real matrix because all its entries are real numbers, and matrix B is a square complex matrix of order equals 3.
- A diagonal matrix from A is

$$\mathrm{Dg}(A) = \begin{bmatrix} 1 & 0 & 0 & 0 \\ 0 & -1 & 0 & 0 \\ 0 & 0 & 10 & 0 \\ 0 & 0 & 0 & 9 \end{bmatrix} = \mathrm{Dg}([1 \quad -1 \quad 10 \quad 9])$$

- The minor of matrix A at position (2, 3) is obtained by deleting the second row and the third column of matrix A.

$$\text{Min}_{23}(A) = \begin{bmatrix} 1 & 0 & 5 \\ 2 & 7 & 3 \\ 5 & 1 & 9 \end{bmatrix}$$

- The Lower and upper triangular matrices obtained from matrix A are, respectively

$$L(A) = \begin{bmatrix} 1 & 0 & 0 & 0 \\ 2 & -1 & 0 & 0 \\ 2 & 7 & 10 & 0 \\ 5 & 1 & 0 & 9 \end{bmatrix}, U(A) = \begin{bmatrix} 1 & 0 & -3 & 5 \\ 0 & -1 & 4 & 11 \\ 0 & 0 & 10 & 3 \\ 0 & 0 & 0 & 9 \end{bmatrix}$$

- The matrix A can be partitioned into four submatrices as follows.

$$A = \begin{bmatrix} A_{11} & A_{12} \\ A_{21} & A_{22} \end{bmatrix}, \qquad \text{where}$$

$$A_{11} = \begin{bmatrix} 1 & 0 \\ 2 & -1 \end{bmatrix}, A_{12} = \begin{bmatrix} -3 & 5 \\ 4 & 11 \end{bmatrix}, A_{21} = \begin{bmatrix} 2 & 7 \\ 5 & 1 \end{bmatrix}, A_{22} = \begin{bmatrix} 10 & 3 \\ 0 & 9 \end{bmatrix}$$

Alternatively, it can be partitioned as

$$A = \begin{bmatrix} A_{11} \\ A_{21} \end{bmatrix}, \qquad \text{where}$$

$$A_{11} = \begin{bmatrix} 1 & 0 & -3 & 5 \\ 2 & -1 & 4 & 11 \end{bmatrix}, A_{21} = \begin{bmatrix} 2 & 7 & 10 & 3 \\ 5 & 1 & 0 & 9 \end{bmatrix}$$

- The transpose of matrix A is

$$A^T = \begin{bmatrix} 1 & 2 & 2 & 5 \\ 0 & -1 & 7 & 1 \\ -3 & 4 & 10 & 0 \\ 5 & 11 & 3 & 9 \end{bmatrix}$$

- The matrix B has a tranjugate given by

$$\bar{B} = \begin{bmatrix} 1-j & -5 & 6-2j \\ 4 & 1 & 3 \\ 2+j & 7 & 4 \end{bmatrix}$$

- The matrix B has a Hermitian matrix given by

$$B^* = \begin{bmatrix} 1-j & 4 & 2+j \\ -5 & 1 & 7 \\ 6-2j & 3 & 4 \end{bmatrix}$$

- The two matrices A and B are not conformable for algebraic addition and multiplication. The two matrices B and C are conformable for algebraic addition and multiplication.
- The product

$$\begin{bmatrix} 2 & 3 & 1 \\ 0 & 4 & 5 \end{bmatrix} C$$

is a conformable multiplication because the number of columns in the first matrix is equal to the number of rows in matrix C.

Any matrix has several essential scalar-valued functions as well as several vector (or matrix) functions. In other words, a matrix can be described uniquely by several scalar-valued and vectors or matrix-valued functions. We will define some of these functions; in the next sections, more of these functions will be considered.

Definition 1.12. A trace of a matrix

A trace of a square matrix $A_{n \times n}$ is the sum of all its entries lying on the diagonal of A, i.e.

$$\operatorname{tr}(A) = \sum_{i=1}^{n} \operatorname{ent}_{ii}(A) = \sum_{i=1}^{n} a_{ii} \tag{1.14}$$

For instance, in example 1.1, we have $\operatorname{tr}(C) = 2 + 7 + 8 = 17$. The trace is a linear function; therefore, the properties of the trace can be deduced.

Definition 1.13. Determinant of a matrix

For a square matrix A, the determinant is one of the most considerable scalar-valued functions on matrices. The value of the determinant of a matrix of order $n = 2$ is given by

$$\det(A_{2\times2}) = a_{11}a_{22} - a_{12}a_{21} \tag{1.15}$$

For a matrix of order $n \geq 3$, the determinant is given by

$$\det(A) = \sum_{k=1}^{n} a_{1k} \operatorname{cof}_{1k}(A) \tag{1.16}$$

where, in general,

$$\operatorname{cof}_{ij}(A) = (-1)^{i+j} \det\left(Min_{ij}(A)\right) \tag{1.17}$$

It is the cofactor of a_{ij}, the cofactor of the (i,j) position of A. For instance, for $n = 3$, we have

$$\det(A_3) = a_{11}\,\text{cof}_{11}(A) + a_{12}\,\text{cof}_{12}(A) + a_{13}\,\text{cof}_{13}(A)$$

Then,

$$\det(A_3) = a_{11}\det\left(\begin{bmatrix} a_{22} & a_{23} \\ a_{32} & a_{33} \end{bmatrix}\right) - a_{12}\det\left(\begin{bmatrix} a_{21} & a_{23} \\ a_{31} & a_{33} \end{bmatrix}\right) + a_{13}\det\left(\begin{bmatrix} a_{21} & a_{22} \\ a_{31} & a_{32} \end{bmatrix}\right)$$

For instance, in example 1.1, we have

$$\det(C) = 2(7 \times 8 - 9 \times 9) - 1(1 \times 8 - 9 \times 3) + 3(1 \times 9 - 7 \times 3) = -67$$

Clearly, as the order increases, the computation of the determinant becomes more and more time-consuming. Therefore, to save time, the calculation by formula (1.16) must be replaced by another technique. Some of the determinant properties offer a fast computation of the determinant value, as will be explained. Furthermore, in the next section, we will give an efficient way to compute the determinant value.

Properties of a determinant are:

1. For any square matrix A and a real constant c, the following hold

$$\det(c\,\text{Row}_i(A)) = c\det(A)$$

$$\det(c\,\text{Col}_i(A)) = c\det(A)$$

$$\det(cA) = c^n\det(A)$$

2. For any square matrix A, it holds $\det(A^T) = \det(A)$.
3. For any triangular (lower or upper) matrix, the determinant is equal to the entries' product on the diagonal.
4. For a matrix with a zero row or column (all row or column elements are equal to zero), the determinant value is equal to zero.
5. The determinant value is equal to zero if a matrix has two (or more) identical rows or columns.
6. The determinant value equals zero for a matrix with two (or more) linearly dependent rows or linearly dependent columns.
7. For any square matrix A, interchanging an odd number of rows or columns changes the sign of its determinant; an even number of interchanging does not alter the value sign.
8. For any square matrix A, multiplying a specific row (column) by a constant and adding the result to another row (column) will not change the determinant's value. For example,

$$A = \begin{bmatrix} 1 & -4 & 2 \\ 5 & 2 & 0 \\ 3 & 1 & 1 \end{bmatrix} \xrightarrow{\overline{2R_1 + R_2 = R_2}} \begin{bmatrix} 1 & -4 & 2 \\ 7 & -6 & 4 \\ 3 & 1 & 1 \end{bmatrix} = B$$

We can easily find that

$$\det(A) = \det(B) = 20$$

9. If the product AB is defined, then

$$\det(AB) = \det(A)\det(B) \tag{1.18}$$

10. For any square matrix A_n, the following two equations are valid for computing the determinant:

$$\det(A) = \sum_{k=1}^{n} a_{ik}\mathrm{cof}_{ik}(A)\,;\ \text{for each } i = 1,2 \ldots n \tag{1.19a}$$

$$\det(A) = \sum_{i=1}^{n} a_{ik}\mathrm{cof}_{ik}(A)\,;\ \text{for each } k = 1,2 \ldots n \tag{1.19b}$$

Properties number 3, 7, and 8 (named elementary raw (column) reduced operations) can be used to constitute a fast way of computing the determinant without using the formula (1.16), as we will illustrate in the next section. As a final note, sometimes, for denoting determinant, it is preferred to use the two vertical lines | | instead of writing $\det(A)$; for vector analysis, we will use the latter.

Definition 1.14. A singular matrix
A square matrix is said to be singular if its determinant is equal to zero.

Definition 1.15. A matrix inverse
Consider two square nonsingular matrices A_n and B_n, if their product is defined and gives a unit matrix of order *n*, then the matrix B is the inverse of the matrix A (sometimes called multiplicative inverse); it is denoted by A^{-1}. The ratio of two matrices say M/N is a meaningless term, while MN^{-1} is a correct meaningful term. The inverse has the following properties:

1. If A^{-1} exists, then it holds $(A^{-1})^{-1} = A$.
2. For a constant $c \neq 0$, and that A^{-1} exists, then it holds $(cA)^{-1} = (1/c)A^{-1}$.
3. If both A^{-1} and B^{-1} exist, then it holds $(AB)^{-1} = B^{-1}A^{-1}$.
4. If A is a symmetric nonsingular matrix, then A^{-1} is symmetric too.

In the next section, the computation of the matrix inverse will be introduced.

Definition 1.16. An adjoint matrix

This is another matrix-valued function on a matrix. For *n*-order matrix A, the adjoint of A, written $\mathrm{adj}(A)$, is defined by

$$\mathrm{ent}_{ij}\big(\mathrm{adj}(A)\big) = \mathrm{cof}_{ji}(A) = (-1)^{i+j}\det(\mathrm{Min}_{ji}(A)) \tag{1.20}$$

Expression (1.20) explains that it is to replace each entry of the matrix by its position cofactor and then to transpose the resulting matrix. For instance, the matrix

$$A = \begin{bmatrix} 2 & 1 & -1 \\ 3 & 5 & 2 \\ 1 & 7 & 10 \end{bmatrix}; \ \det(A) = 28$$

The adjoint of A is given as

$$\mathrm{adj}(A) = \begin{bmatrix} \det(\begin{bmatrix} 5 & 2 \\ 7 & 10 \end{bmatrix}) & -\det(\begin{bmatrix} 3 & 2 \\ 1 & 10 \end{bmatrix}) & \det(\begin{bmatrix} 3 & 5 \\ 1 & 7 \end{bmatrix}) \\ -\det(\begin{bmatrix} 1 & -1 \\ 7 & 10 \end{bmatrix}) & \det(\begin{bmatrix} 2 & -1 \\ 1 & 10 \end{bmatrix}) & -\det(\begin{bmatrix} 2 & 1 \\ 1 & 7 \end{bmatrix}) \\ \det(\begin{bmatrix} 1 & -1 \\ 5 & 2 \end{bmatrix}) & -\det(\begin{bmatrix} 2 & -1 \\ 3 & 2 \end{bmatrix}) & \det(\begin{bmatrix} 2 & 1 \\ 3 & 5 \end{bmatrix}) \end{bmatrix}^T = \begin{bmatrix} 36 & -28 & 16 \\ -17 & 22 & -13 \\ 7 & -7 & 7 \end{bmatrix}^T$$

$$\mathrm{adj}(A) = \begin{bmatrix} 36 & -17 & 7 \\ -28 & 21 & -7 \\ 16 & -13 & 7 \end{bmatrix}$$

For a square matrix, the next theorem relates the scalar-valued function to matrix functions.

Theorem 1.1.

A fundamental theorem that relates the adjoint to the inverse of a matrix states that

$$\mathrm{adj}(A) = \det(A)A^{-1} \tag{1.21a}$$

Relation (1.21a) offers a closed-form method for finding the inverse of a matrix. Next, if we multiply this relation from right by the matrix itself, we obtain

$$\mathrm{adj}(A)A = \det(A)A^{-1}A = \det(A)I \tag{1.21b}$$

In our example, relation (1.21b) shows the following results, which proves the validity of applying theorem 1.1.

$$\operatorname{adj}(A)A = \begin{bmatrix} 36 & -17 & 7 \\ -28 & 21 & -7 \\ 16 & -13 & 7 \end{bmatrix}\begin{bmatrix} 2 & 1 & -1 \\ 3 & 5 & 2 \\ 1 & 7 & 10 \end{bmatrix} = \begin{bmatrix} 28 & 0 & 0 \\ 0 & 28 & 0 \\ 0 & 0 & 28 \end{bmatrix} = 28\, I_3$$

Definition 1.17. Equivalency of matrices

Two matrices A and B are equivalent (written $A \sim B$) if matrix B can be obtained from matrix A by a finite sequence of elementary row or/and column reduced operations (later, it will be studied in detail). For example,

$$A = \begin{bmatrix} 1 & 3 & -5 \\ -1 & 7 & 7 \\ 2 & 1 & 6 \end{bmatrix} \xrightarrow[R_1 + R_2]{} \begin{bmatrix} 1 & 3 & -5 \\ 0 & 10 & 2 \\ 2 & 1 & 6 \end{bmatrix} \xrightarrow[-2R_1 + R_3]{} \begin{bmatrix} 1 & 3 & -5 \\ 0 & 10 & 2 \\ 0 & -5 & 16 \end{bmatrix}$$

$$\xrightarrow[0.5R_2 + R_3]{} \begin{bmatrix} 1 & 3 & -5 \\ 0 & 10 & 2 \\ 0 & 0 & 17 \end{bmatrix} \sim B$$

The target of the above sequence of the elementary operation (multiplying a row and summing to another row) is to obtain an upper triangular matrix B equivalent to A. Further, it can be proved that two matrices A and B are equivalent if there are two nonsingular matrices P and Q such that the following equality holds $B = PAQ$. This equivalency provides a technique to reduce the computation burden in several applications.

The next example demonstrates a daily application of matrices in our life economics.

Example 1.2.

A contractor wants to buy five construction items for a five-day work, as shown in Table 1.1. For each item, the cost of three construction material suppliers is as shown in Table 1.2. How can the contractor get these items with minimum payment?

Table 1.1. Quantities of items per day

	Item 1	**Item 2**	**Item 3**	**Item 4**	**Item 5**
Monday	10	14	11	17	20
Tuesday	13	16	14	15	22
Wednesday	12	16	19	15	21
Thursday	10	12	16	12	20
Friday	8	10	7	12	15

Table 1.2. Cost (in unit price) of items per supplier

	Supplier 1	**Supplier 2**	**Supplier 3**
Item 1	56.00	80.00	44.00
Item 2	67.00	55.00	70.00
Item 3	61.00	56.00	78.00
Item 4	59.00	66.00	63.00
Item 5	48.00	42.00	39.00

First, these tables are represented by demand and cost matrices.

$$D = \begin{bmatrix} 10 & 14 & 11 & 17 & 20 \\ 13 & 16 & 14 & 15 & 22 \\ 12 & 16 & 19 & 15 & 21 \\ 10 & 12 & 16 & 12 & 20 \\ 8 & 10 & 10 & 12 & 15 \end{bmatrix}, P = \begin{bmatrix} 56 & 80 & 44 \\ 67 & 55 & 70 \\ 61 & 56 & 78 \\ 59 & 66 & 63 \\ 48 & 42 & 39 \end{bmatrix}$$

To solve the problem, we have to make a price matrix R as

$$R = D_{5\times5}P_{5\times3} = \begin{bmatrix} 4132 & 4148 & 4129 \\ 4595 & 4618 & 4587 \\ 4796 & 4776 & 4894 \\ 4008 & 3988 & 4064 \\ 3156 & 3172 & 3173 \end{bmatrix}$$

Then, we perform

$$\min\{r_{i1}, r_{i2}, r_{i3}\}\,;\, i = 1,2..5$$

Therefore, the contractor has to buy from the suppliers by the minimum payment per day, as shown in Table 1.3. Thus, the contractor should buy from the third supplier on Monday and Tuesday, and the second supplier only on Wednesday and from the first supplier on Thursday and Friday. The total payment is 20,656.00 units price for the five days, while the total payment from the first, second, and third suppliers is 20,687.00, 20,702.00, and 20,847.00 units' price respectively. In practice, similarly, we can solve a one-month problem with several suppliers. Therefore, keep touch with market prices can save a lot.

Table 1.3. Minimum Cost per day and the corresponding supplier

Days	Payment	Supplier
Monday	4129	3
Tuesday	4587	3
Wednesday	4776	2
Thursday	4008	1
Friday	3156	1

Definition 1.18. Similarity of matrices

Two matrices A, and B are said to be similar (written $A \approx B$) if there exists a nonsingular matrix P so that $B = P^{-1}AP$. Accordingly, it is easy to conclude that similar matrices are also equivalent. When two matrices are similar, then the matrix P is not unique. Similar matrices must have equal determinants and traces. The use of the mentioned definitions will be more discussed.

There are other particular types of beneficial matrices such as the companion, Vandermonde, echelon, etc. matrices, and some other scalar or vector-valued functions on matrices, which we will consider in other sections of this chapter.

1.4. Computation by Elementary Operations

We have already explained the arithmetic operations on matrices, and we mentioned another type of operation through the determinant properties. This second type is a set called elementary reduced operations (simply, elementary operations). When these operations are applied on rows (columns), they are elementary row (column) reduced operations. An example of an elementary row reduced operation was used in definition 1.13 and definition 1.17. We will limit the discussion and use it only to elementary row reduced operations to shorten the text. Students can realize that everything is similar to elementary column reduced operations.

There are three elementary row operations, which can reach an equivalent matrix with specific properties. These are:

1. Interchanging rows, all entries of a particular row are interchanged in places with entries of another row. It is denoted $R_i \rightleftarrows R_j$. The effect upon the considered matrix's determinant value is the change of the determinant sign every time the row interchanging is performed. Therefore, an even number of row-interchanging does not change the original sign of the considered matrix.
2. Multiplying a row by a constant, all entries of a row are multiplied by a constant value. It is denoted kR_i. The determinant value of the considered matrix is scaled by $(1/k)$.
3. Multiplying a row by a constant k and adding the results to another row. It is denoted $kR_i + R_j$. As was said previously, this operation does not affect the matrix's determinant value, i.e., the equivalent matrix has the same determinant as the considered one.

Therefore, the row operations' use represents an efficient way of computing the determinant, especially for matrices of order greater than three. The next example illustrates this efficient way.

Example 1.3.

Evaluate the determinant of the fourth-order matrix

$$A = \begin{bmatrix} 3 & 2 & -2 & 1 \\ 5 & 0 & -2 & 2 \\ 1 & 4 & 6 & 2 \\ 2 & 3 & 1 & 6 \end{bmatrix}$$

$$\begin{bmatrix} 3 & 2 & -2 & 1 \\ 5 & 0 & -2 & 2 \\ 1 & 4 & 6 & 2 \\ 2 & 3 & 1 & 6 \end{bmatrix} \xrightarrow{R_1 \rightleftarrows R_3} \begin{bmatrix} 1 & 4 & 6 & 2 \\ 5 & 0 & -2 & 2 \\ 3 & 2 & -2 & 1 \\ 2 & 3 & 1 & 6 \end{bmatrix} \xrightarrow[\substack{-3R_1 + R_3 \\ -2R_1 + R_4}]{-5R_1 + R_2} \begin{bmatrix} 1 & 4 & 6 & 2 \\ 0 & -20 & -32 & -8 \\ 0 & -10 & -20 & -5 \\ 0 & -5 & -11 & 2 \end{bmatrix}$$

Using property 10 of the determinant, we can consider the cofactor of the $(1,1)$ position to continue the computation of the determinant of the given matrix, i.e.,

$$\begin{bmatrix} -20 & -32 & -8 \\ -10 & -20 & -5 \\ -5 & -11 & 2 \end{bmatrix} \xrightarrow{-\frac{1}{4}R_1} \begin{bmatrix} 5 & 8 & 2 \\ -10 & -20 & -5 \\ -5 & -11 & 2 \end{bmatrix} \xrightarrow{R_1 + R_3} \begin{bmatrix} 5 & 8 & 2 \\ -10 & -20 & -5 \\ 0 & -3 & 4 \end{bmatrix}$$

$$\xrightarrow{2R_1 + R_2} \begin{bmatrix} 5 & 8 & 2 \\ 0 & -4 & -1 \\ 0 & -3 & 4 \end{bmatrix} \xrightarrow{\frac{1}{5}R_1} \begin{bmatrix} 1 & 8/5 & 2/5 \\ 0 & -4 & -1 \\ 0 & -3 & 4 \end{bmatrix} \sim B$$

Again using property 10 to obtain

$$\det(B) = \det\left(\begin{bmatrix} -4 & -1 \\ -3 & 4 \end{bmatrix}\right) = -16 - 3 = -19$$

Therefore, the determinant of matrix A is obtained by considering the effects of the used elementary row operations multiplied by the determinant of B, i.e.,

$$\det(A) = (-1)(-4)(5)(-19) = -380$$

Note that there are many other possible sequences of using the elementary row operations to get the same result. For instance, a sequence leads to an equivalent upper triangular matrix, and hence the determinant is obtained by the product of the elements on the diagonal; see definition 1.17. The best sequence is that of a minimum number of arithmetic operations.

Example 1.4.

Reduce the matrix A to obtain a unity matrix and obtain the determinant of A.

$$A = \begin{bmatrix} 2 & 6 & 8 \\ 3 & 19 & 15 \\ 4 & 14 & 17 \end{bmatrix} \xrightarrow{\frac{1}{2}R_1} \begin{bmatrix} 1 & 3 & 4 \\ 3 & 19 & 15 \\ 4 & 14 & 17 \end{bmatrix} \xrightarrow[-4R_1 + R_3]{-3R_1 + R_2} \begin{bmatrix} 1 & 3 & 4 \\ 0 & 10 & 3 \\ 0 & 2 & 1 \end{bmatrix} \xrightarrow{-\frac{1}{5}R_2 + R_3} \begin{bmatrix} 1 & 3 & 4 \\ 0 & 10 & 3 \\ 0 & 0 & 2/5 \end{bmatrix}$$

$$\xrightarrow[\frac{5}{2}R_3]{\frac{1}{10}R_2} \begin{bmatrix} 1 & 3 & 4 \\ 0 & 1 & 3/10 \\ 0 & 0 & 1 \end{bmatrix} \xrightarrow[-4R_3 + R_1]{-\frac{10}{3}R_3 + R_2} \begin{bmatrix} 1 & 3 & 0 \\ 0 & 1 & 0 \\ 0 & 0 & 1 \end{bmatrix} \xrightarrow{-3R_2 + R_1} \begin{bmatrix} 1 & 0 & 0 \\ 0 & 1 & 0 \\ 0 & 0 & 1 \end{bmatrix} = I$$

Since only the three-row operations affect the value of the determinant and $\det(I) = 1$, then

$$\det(A) = 2 \times 10 \times \frac{2}{5} \times 1 = 8$$

It is also possible to stop when we have an upper triangular matrix.

It can be concluded that every nonsingular square matrix can be reduced to an equivalent unit matrix, i.e., by applying a finite number of elementary operations

$$[A_n] \underset{R}{\tilde{}} [I_n] \tag{1.22}$$

For the adjoint formula, the elementary operations can be used in a comparable fewer arithmetic operation for computing the matrix inverse. The following sequence of row operations summarizes the procedure of computing a matrix inverse by elementary operations. The column operation can also be used or even a combination of both operations. However, for hand computation, only row operations are recommended.

$$[A \;\vdots\; I] \underset{R}{\tilde{}} [I \;\vdots\; P] \Rightarrow A^{-1} = P \tag{1.23}$$

The left-hand side augmented matrix has to be converted to an equivalent right-hand side augmented matrix in which the unity matrix becomes the preceding one. In contrast, the latter matrix represents the desired inverse matrix. In other words, the elementary operations are implemented in such a way to reduce the given matrix to a unity matrix, and the same operations on the latter unity matrix will provide the desired inverse. It is worth noting that when all the matrix entries are integers, the entries of its inverse are all ratios of integers to the determinant's integer value.

Similarly, other significant formulas can be stated. Given three matrices $A, B,$ and C, the use of elementary operations can provide the following.

$$\text{if } [A \;\vdots\; I] \underset{R}{\tilde{}} [B \;\vdots\; C] \text{ then } CA = B \tag{1.24a}$$

$$\text{if } [A \;\vdots\; B] \underset{R}{\tilde{}} [I \;\vdots\; C] \text{ then } C = A^{-1}B \tag{1.24b}$$

The results of expressions (1.24) represent a unique technique to solve linear systems, as will be discussed later.

Example 1.5.

Determine the inverse of the matrix

$$A = \begin{bmatrix} 2 & -1 & 4 \\ 1 & -2 & 5 \\ 3 & 3 & -2 \end{bmatrix}$$

We set the augmented matrix

$$\begin{bmatrix} 2 & -1 & 4 & \vdots & 1 & 0 & 0 \\ 1 & -2 & 5 & \vdots & 0 & 1 & 0 \\ 3 & 3 & -2 & \vdots & 0 & 0 & 1 \end{bmatrix} \xrightarrow[R_1 \rightleftarrows R_2]{} \begin{bmatrix} 1 & -2 & 5 & \vdots & 0 & 1 & 0 \\ 2 & -1 & 5 & \vdots & 1 & 0 & 0 \\ 3 & 3 & -2 & \vdots & 0 & 0 & 1 \end{bmatrix} \xrightarrow[-3R_1+R_3]{-2R_1+R_2}$$

$$\begin{bmatrix} 1 & -2 & 5 & \vdots & 0 & 1 & 0 \\ 0 & 3 & -6 & \vdots & 1 & -2 & 0 \\ 0 & 9 & -17 & \vdots & 0 & -3 & 1 \end{bmatrix} \xrightarrow[-3R_2+R_3]{} \begin{bmatrix} 1 & -2 & 5 & \vdots & 0 & 1 & 0 \\ 0 & 3 & -6 & \vdots & 1 & -2 & 0 \\ 0 & 0 & 1 & \vdots & -3 & 3 & 1 \end{bmatrix} \xrightarrow{\frac{1}{3}R_2}$$

$$\begin{bmatrix} 1 & -2 & 5 & \vdots & 0 & 1 & 0 \\ 0 & 1 & -2 & \vdots & 1/3 & -2/3 & 0 \\ 0 & 0 & 1 & \vdots & -3 & 3 & 1 \end{bmatrix} \xrightarrow[-5R_3+R_1]{-2R_3+R_2} \begin{bmatrix} 1 & -2 & 0 & \vdots & 15 & 14 & -5 \\ 0 & 1 & 0 & \vdots & -17/3 & 16/3 & 2 \\ 0 & 0 & 1 & \vdots & -3 & 3 & 1 \end{bmatrix}$$

$$\xrightarrow[2R_2+R_1]{} \begin{bmatrix} 1 & 0 & 0 & \vdots & 11/3 & -10/3 & -1 \\ 0 & 1 & 0 & \vdots & -17/3 & 16/3 & 2 \\ 0 & 0 & 1 & \vdots & -3 & 3 & 1 \end{bmatrix} \underset{R}{\sim} [I \;\vdots\; P]$$

According to the formula (1.23), we obtain

$$A^{-1} = \begin{bmatrix} 11/3 & -10/3 & -1 \\ -17/3 & 16/3 & 2 \\ -3 & 3 & 1 \end{bmatrix}$$

Furthermore, we can compute $\det(A) = -3$, and hence also, according to formula (1.21a); consequently, we have

$$\operatorname{adj}(A) = \begin{bmatrix} -11 & 10 & 3 \\ 17 & -16 & -6 \\ 9 & -9 & -3 \end{bmatrix}$$

For calculating the determinant, any arbitrary right sequence of the three types of elementary row operations will lead to obtaining the correct result. However, for finding the matrix inverse, it is recommended to use a particular right sequence, which includes, as possible, only the first and the third types of row operations, leaving the second type of operation to the last. This procedure will avoid complicated calculations and incorrect results. Therefore, unless the matrix is rigged (as in example 1.5), finding the right sequence may not be straightforward. The next example illustrates how the sequence of operations is carefully selected.

Example 1.6.

Determine the inverse of the matrix given in example 1.4.

First, the reader is encouraged to work the augmented matrix $[A \;\vdots\; I]$ using the row operations of example 1.4. Here, we will use another set of row operations as follows:

$$\begin{bmatrix} 2 & 6 & 8 & & 1 & 0 & 0 \\ 3 & 19 & 15 & \vdots & 0 & 1 & 0 \\ 4 & 14 & 17 & & 0 & 0 & 1 \end{bmatrix} \xrightarrow[-R_2+R_3]{} \begin{bmatrix} 2 & 6 & 8 & & 1 & 0 & 0 \\ 3 & 19 & 15 & \vdots & 0 & 1 & 0 \\ 1 & -5 & 2 & & 0 & -1 & 1 \end{bmatrix} \xrightarrow[R_1 \rightleftarrows R_3]{}$$

$$\begin{bmatrix}1 & -5 & 2 & \vdots & 0 & -1 & 1\\ 3 & 19 & 15 & \vdots & 0 & 1 & 0\\ 2 & 6 & 8 & & 1 & 0 & 0\end{bmatrix}\xrightarrow[-2R_1+R_3]{-3R_1+R_2}\begin{bmatrix}1 & -5 & 2 & & 0 & -1 & 1\\ 0 & 34 & 9 & \vdots & 0 & 4 & -3\\ 0 & 16 & 4 & & 1 & 2 & -2\end{bmatrix}\xrightarrow{-2R_3+R_2}$$

$$\begin{bmatrix}1 & -5 & 2 & & 0 & -1 & 1\\ 0 & 2 & 1 & \vdots & -2 & 0 & 1\\ 0 & 16 & 4 & & 1 & 2 & -2\end{bmatrix}\xrightarrow{-8R_2+R_3}\begin{bmatrix}1 & -5 & 2 & & 0 & -1 & 1\\ 0 & 2 & 1 & \vdots & -2 & 0 & 1\\ 0 & 0 & -4 & & 17 & 2 & -10\end{bmatrix}\begin{matrix}\xrightarrow{4R_1}\\ 2R_2\\ -R_3\end{matrix}$$

$$\begin{bmatrix}4 & -20 & 8 & & 0 & -4 & 4\\ 0 & 4 & 2 & \vdots & -4 & 0 & 2\\ 0 & 0 & 4 & & -17 & -2 & 10\end{bmatrix}\begin{matrix}\xrightarrow{(1/4)R_1}\\ (1/4)R_2\\ (1/4)R_3\end{matrix}\begin{bmatrix}1 & -5 & 2 & & 0 & -1 & 1\\ 0 & 1 & 1/2 & \vdots & -1 & 0 & 1/2\\ 0 & 0 & 1 & & -17/4 & -1/2 & 10/4\end{bmatrix}$$

$$\xrightarrow[-2R_3+R_1]{-(1/2)R_3+R_2}\begin{bmatrix}1 & -5 & 0 & & 17/2 & 0 & -4\\ 0 & 1 & 0 & \vdots & 9/8 & 1/4 & -3/4\\ 0 & 0 & 1 & & -17/4 & -1/2 & 10/4\end{bmatrix}\xrightarrow{5R_2+R_1}$$

$$\begin{bmatrix}1 & 0 & 0 & & 113/8 & 5/4 & -31/4\\ 0 & 1 & 0 & \vdots & 9/8 & 1/4 & -3/4\\ 0 & 0 & 1 & & -17/4 & -1/2 & 10/4\end{bmatrix}$$

Accordingly, we have

$$A^{-1}=\frac{1}{8}\begin{bmatrix}113 & 10 & -62\\ 9 & 2 & -6\\ -34 & -4 & 20\end{bmatrix}$$

One can check the result by testing the adjoint formula (1.21a). The reader can recognize the ambiguity in selecting the right set of elementary row operations.

Example 1.7.

Use elementary row operations to find a nonsingular matrix C such that $C = A^{-1}B$, where

$$A=\begin{bmatrix}2 & 1 & 0\\ 1 & 1 & 2\\ 1 & 1 & 1\end{bmatrix}, B=\begin{bmatrix}1 & 2 & 3\\ 4 & 2 & -1\\ 1 & 1 & 2\end{bmatrix}$$

It is to work the formula (1.24b); therefore, we set the augmented matrix and let the matrix A to change to a unity matrix.

$$\begin{bmatrix}2 & 1 & 0 & \vdots & 1 & 2 & 3\\ 1 & 1 & 2 & \vdots & 4 & 2 & -1\\ 1 & 1 & 1 & \vdots & 1 & 1 & 2\end{bmatrix}\xrightarrow[-(1/2)R_1+R_3]{-(1/2)R_1+R_2}\begin{bmatrix}2 & 1 & 0 & \vdots & 1 & 2 & 3\\ 0 & 1/2 & 2 & \vdots & 7/2 & 1 & -5/2\\ 0 & 1/2 & 1 & \vdots & 1/2 & 0 & 1/2\end{bmatrix}\xrightarrow[-R_2+R_3]{-2R_2+R_1}$$

$$\begin{bmatrix} 2 & 0 & -4 & \vdots & -6 & 0 & 8 \\ 0 & 1/2 & 2 & \vdots & 7/2 & 1 & -5/2 \\ 0 & 0 & -1 & \vdots & -3 & -1 & 3 \end{bmatrix} \xrightarrow[2R_3+R_2]{-4R_3+R_1} \begin{bmatrix} 2 & 0 & 0 & \vdots & 6 & 4 & -4 \\ 0 & 1/2 & 0 & \vdots & -5/2 & -1 & 7/2 \\ 0 & 0 & -1 & \vdots & -3 & -1 & 3 \end{bmatrix}$$

$$\xrightarrow{\substack{(1/2)R_1 \\ 2R_2 \\ -R_3}} \begin{bmatrix} 1 & 0 & 0 & \vdots & 3 & 2 & -2 \\ 0 & 1 & 0 & \vdots & -5 & -2 & 7 \\ 0 & 0 & 1 & \vdots & 3 & 1 & -3 \end{bmatrix}$$

Therefore, we obtain

$$C = \begin{bmatrix} 3 & 2 & -2 \\ -5 & -2 & 7 \\ 3 & 1 & -3 \end{bmatrix}$$

This example explains a method to solve an algebraic matrix system $AC = B$, where C is the unknown matrix. In our example, all matrices are squares; however, matrix C can have a dimension $(3 \times r)$; $r \geq 1$. The solution of the algebraic system of linear equations represents the case $r = 1$. This method is known as the Gauss-Jordan elimination method. We will consider this subject thoroughly in the next section.

Example 1.8.

In this example, the basic idea of sending a coding message will be explored utilizing the matrix concepts. Random numbers in a specific range (a, b); $a, b \neq 0$ are assigned to the alphabetic letters and some additional punctuation. The letters of the message are coded by numbers, which are arranged sequentially in a square matrix. The zero is assigned to complete the square matrix when the message letters and spaces do not fill the whole square matrix. Before sending the message, the letters' matrix is manipulated in a prescribed way. On the receiving side, an inverse sequence of the performed manipulation is performed to retrieve the message.

For example, in the rang $(1, 30)$ values, the coding is described as follows.

A	B	C	D	E	F	G	H	I	J	K	L	M	N	O	P	Q	R	S	T	U	V	X	Y	W	Z	
6	20	30	19	17	27	28	18	7	25	22	9	14	3	15	26	21	8	2	10	1	5	12	11	13	16	4

The message is "ISRA UNIV" which has eight letters and one space. This massage completes filling square 3rd-order matrix entries. A third-order matrix can hold words and spaces of less than nine characters, such as "RED BOOK," and the zero value can be used for the last unused entries. Then, in our example, the massage matrix becomes

$$M = \begin{bmatrix} 7 & 2 & 8 \\ 6 & 4 & 1 \\ 3 & 7 & 5 \end{bmatrix}$$

A matrix C and a constant k are used to formulate a ciphering process, such that the sending coded matrix S becomes

$$S = C(M + kN_{3\times 3})$$

where $N_{3\times 3}$ is a full one's matrix, to add the constant k to each element of the matrix M before multiplying by the matrix C. The matrix C and the constant k are

$$C = \begin{bmatrix} 1 & 0 & 1 \\ 2 & 3 & 9 \\ 4 & 1 & 6 \end{bmatrix}, k = 2$$

Then,

$$S = \begin{bmatrix} 14 & 13 & 17 \\ 87 & 107 & 92 \\ 74 & 76 & 85 \end{bmatrix}$$

On the receiving side, the message is retrieved by

$$M = C^{-1}S - kN_{3\times 3} = \begin{bmatrix} -9 & -1 & 3 \\ -24 & -2 & 7 \\ 10 & 1 & -3 \end{bmatrix}\begin{bmatrix} 14 & 13 & 17 \\ 87 & 107 & 92 \\ 74 & 76 & 85 \end{bmatrix} - 2[\begin{bmatrix} 1 & 1 & 1 \\ 1 & 1 & 1 \\ 1 & 1 & 1 \end{bmatrix} = \begin{bmatrix} 7 & 2 & 8 \\ 6 & 4 & 1 \\ 3 & 7 & 5 \end{bmatrix}$$

Of course, complicated mathematical manipulation gives stronger ciphering. As the message length increases, the matrix dimension is increased; nevertheless, with the aid of nowadays-micro technology, a larger message length can be efficiently handled.

1.5. The Solution of Linear Algebraic Systems

The matrix concept is a more efficient notation to represent a system of algebraic linear equations than the classical written style. The number of variables and the number of equations classifies a specific class of systems. When the two numbers are equal, then a regular system is defined. It is also possible to consider, in this content, systems that have several variables larger than several equations. The solution of these two classes will be explored in this section utilizing the matrix theory. Systems named overdetermined (several equations larger than the number of variables) will not be considered. Nevertheless, an equivalent solution can be obtained in some exceptional practical cases.

Perhaps, in practice, the subject of solving a linear system has the largest number of applications. It includes not only engineering disciplines but also several other scientific fields. We have already seen how one of the matrix multiplication features can represent a system of algebraic linear equations. Generally, we consider the following system.

$$a_{11}x_1 + a_{12}x_2 + \cdots + a_{1n}x_n = k_1$$
$$a_{21}x_1 + a_{22}x_2 + \cdots + a_{2n}x_n = k_2$$
$$\vdots$$
$$a_{m1}x_1 + a_{m2}x_2 + \cdots + a_{mn}x_n = k_m \tag{1.25}$$

In matrix notations, system (1.25) is written as

$$A_{m\times n}X_{n\times 1} = K_{m\times 1}$$

where:

$A_{m\times n}$ is the coefficient matrix of the system given as

$$\begin{bmatrix} a_{11} & a_{12} & \cdots & a_{1n} \\ a_{21} & a_{22} & \cdots & a_{2n} \\ \vdots & \vdots & \vdots & \vdots \\ a_{m1} & a_{m2} & \cdots & a_{mn} \end{bmatrix}$$

$X_{n\times 1}$ is the variable vector given as $[x_1 \quad x_2 \quad \cdots \quad x_n]^T$
$K_{m\times 1}$ is the input or force vector given as $[k_1 \quad k_2 \quad \ldots \quad k_m]^T$
m is the number of equations
n is the number of variables and $n \geq m$.

The task is to solve system (1.25) for a given coefficient matrix and input vector; i.e., we have to find the variable vector. A more elaborated problem investigation includes the answer to three questions. The first is whether a solution exists, the second one is whether the solution is unique or not, and finally, how to find the solution(s). For regular systems ($m = n$), the answer to the first question is to determine whether the coefficient matrix is singular. A unique solution exists if and only if the coefficient matrix is nonsingular ($\det(A) \neq 0$); furthermore, the solution is unique. The answer is found by

$$X = A^{-1}K \tag{1.26}$$

Alternatively, in viewing expression (1.24b), we have a computational tool to find the solution X by carrying out the elementary row operations.

$$[A \;\vdots\; K] \underset{R}{\sim} [I \;\vdots\; X]$$

Let us consider, at this stage, an introductory example.

Example 1.9.
Solve the system

$$2x_1 + x_2 = 5$$
$$x_1 + x_2 + 2x_3 = 11$$
$$x_1 + x_2 + x_3 = 6$$

First, we write the system in matrix notation $AX = K$, and applying the multiplication feature yields

$$\begin{bmatrix} 2 & 1 & 0 \\ 1 & 1 & 2 \\ 1 & 1 & 1 \end{bmatrix} \begin{bmatrix} x_1 \\ x_2 \\ x_3 \end{bmatrix} = \begin{bmatrix} 5 \\ 11 \\ 6 \end{bmatrix}$$

As it can be noticed, the coefficient matrix is the same as that considered in example 1.7. We already know that A^{-1} exists, and hence the same set of elementary row operations will be utilized for the augmented matrix $[A \;\vdots\; K]$. Since the given system is regular, then a unique solution will be obtained. Thus,

$$\begin{bmatrix} 2 & 1 & 0 & \vdots & 5 \\ 1 & 1 & 2 & \vdots & 11 \\ 1 & 1 & 1 & \vdots & 6 \end{bmatrix} \xrightarrow[-(1/2)R_1 + R_3]{-(1/2)R_1 + R_2} \begin{bmatrix} 2 & 1 & 0 & \vdots & 5 \\ 0 & 1/2 & 2 & \vdots & 17/2 \\ 0 & 1/2 & 1 & \vdots & 7/2 \end{bmatrix} \xrightarrow[-R_2 + R_3]{-2R_2 + R_1}$$

$$\begin{bmatrix} 2 & 0 & -4 & \vdots & -12 \\ 0 & 1/2 & 2 & \vdots & 17/2 \\ 0 & 0 & -1 & \vdots & -5 \end{bmatrix} \xrightarrow[2R_3 + R_2]{-4R_3 + R_1} \begin{bmatrix} 2 & 0 & 0 & \vdots & 8 \\ 0 & 1/2 & 0 & \vdots & -3/2 \\ 0 & 0 & -1 & \vdots & -5 \end{bmatrix}$$

$$\xrightarrow[\substack{2R_2 \\ -R_3}]{(1/2)R_1} \begin{bmatrix} 1 & 0 & 0 & \vdots & 4 \\ 0 & 1 & 0 & \vdots & -3 \\ 0 & 0 & 1 & \vdots & 5 \end{bmatrix} = [I \;\vdots\; X]$$

Thus, the variable vector is $X = [4 \quad -3 \quad 5]^T$; i.e., the solution is $x_1 = 4, x_2 = -3, x_3 = 5$. Now let us write the equivalent system when the coefficient matrix becomes an upper triangular matrix, i.e.,

$$2x_1 + 0x_2 - 4x_3 = -12$$
$$0x_1 + \frac{1}{2}x_2 + 2x_3 = \frac{17}{2}$$
$$0x_1 + 0x_2 - x_3 = -5$$

For the third equation, we have $x_3 = 5$. By substituting in the second equation for x_3, we find

$$\frac{1}{2}x_2 + 2(5) = \frac{17}{2} \Rightarrow x_2 = -3$$

Again by substituting in the first equation for x_3 and x_2, we find

$$2x_1 + 0(-3) - 4(5) = -12 \Rightarrow x_1 = 4$$

The previously used technique is called the back-substitution or Gauss elimination method.

The technique used in example 1.9 can be generally described by the following theorem.

Theorem 1.2.

The systems $AX = K$ and $BX = H$ have precisely the same solution if and only if

$$[A \;\vdots\; K] \underset{R}{\sim} [B \;\vdots\; H]$$

That is, they have augmented matrices, which are row equivalent. For this equivalency, the most straightforward row reduced matrix is named as a row-reduction matrix (simply echelon) E, i.e., to have $[B \;\vdots\; H] = E$. Solving by reducing to echelon matrix is called the Gauss-Jordan elimination method. The following characteristics describe the echelon matrix.

The first nonzero element of each row is 1, and the column in which it appears is a column of the unity matrix (or above them all zeros).

If the leading ones in the nonzero rows occur in positions $(1, c_1), (2, c_2) \ldots (r, c_r)$, then $c_1 < c_2 \ldots < c_r$.

The zero rows of E, if any, come last.

For instance, the following is a form of a matrix E

$$E = \begin{bmatrix} 1 & 5 & 0 & 0 & 2 & 2 \\ 0 & 0 & 1 & 0 & 5 & 1 \\ 0 & 0 & 0 & 1 & 8 & 4 \\ 0 & 0 & 0 & 0 & 0 & 0 \end{bmatrix}$$

It is an echelon because the three leading ones appear as in unity matrices of positions (1, 1), (2, 3), and (3, 4), and the last row contains only zeros. Any nonsingular square matrix has its row-reduced echelon matrix as a unity matrix of the same order.

Definition 1.19. Rank of a matrix

Matrix rank is another scalar-valued function on matrices, denote as $\mathrm{rank}(A)$. A matrix's rank is defined as the largest square submatrix's order with a nonzero determinant. In other words, it is the maximum number of linearly independent vectors (rows or columns) in a matrix. Alternatively and for easy computation, the rank value of the matrix $A_{m \times n}$ is equal to the number of the nonzero rows in the row-reduced echelon matrix of the matrix A. For non-square matrices, it holds that

$$r = \text{rank}(A_{m \times n}) \leq \min\{m, n\}$$

For instance, the rank of the previous E matrix is $r = \text{rank}(E_{4\times 5}) \leq \min\{4,5\}$, or $r = 3$; it is the number of the nonzero rows, hence a rank deficient. On the other hand, full rank (row or column) terminology is used for matrices that have ranks equal to the minimum value. The followings are some properties of the rank:

1. If $A + B$ is defined, then the $\text{rank}(A + B)$ depends on the entry values of both matrices, and hence there are different possibilities.
2. If AB is defined, then $\text{rank}(AB) \leq \text{rank}(A)$, and $\text{rank}(AB) \leq \text{rank}(B)$.
3. For a nonsingular matrix of order *n*, the rank is equal to *n*. Inversely if the rank of a matrix is equal to the matrix order, then the inverse exists.
4. For two row-equivalent matrices, the rank will be the same.
5. For a non-square matrix A, it holds that $\text{rank}(A) = \text{rank}(A^T)$.
6. If two matrices P and Q are invertible and PAQ is defined, then $\text{rank}(PAQ) = \text{rank}(A)$.

Definition 1.20. Consistency of linear system

A linear system $AX = K$ is called consistent if it has at least one nonzero solution. Therefore, linear systems of nonsingular coefficient matrices are consistent.

Theorem 1.3.

Let the coefficient matrix be $A_{m \times n}$ where $n > m$, and it has a rank *r*, then the non-regular system $AX = K$ is consistent if and only if

$$r = \text{rank}(A) = \text{rank}([A \ \vdots \ K]) \tag{1.27}$$

In this case, there are $(n - r)$ degrees of freedom (independent arbitrary constants) in the most general solution, and hence there is an infinite number of solutions.

To make use of theorem 1.3 for answering the mentioned three questions, we apply the elementary row operation on the augmented matrix $[A \ \vdots \ K]$ to reach a row reduced echelon augmented matrix, and hence the ranks of both A and $[A \ \vdots \ K]$ can be readily read.

Example 1.10.
Solve the system

$$\begin{bmatrix} 2 & 0 & 1 & -1 \\ 1 & 3 & 2 & 4 \\ 1 & 1 & 1 & 1 \end{bmatrix} \begin{bmatrix} x_1 \\ x_2 \\ x_3 \\ x_4 \end{bmatrix} = \begin{bmatrix} 2 \\ 3 \\ 5 \end{bmatrix}$$

Set the augmented matrix and apply elementary row operations as follows:

$$\begin{bmatrix} 2 & 0 & 1 & -1 & \vdots & 2 \\ 1 & 3 & 2 & 4 & \vdots & 3 \\ 1 & 1 & 1 & 1 & & 5 \end{bmatrix} \xrightarrow{R_1 \rightleftarrows R_3} \begin{bmatrix} 1 & 1 & 1 & 1 & & 5 \\ 1 & 3 & 2 & 4 & \vdots & 3 \\ 2 & 0 & 1 & -1 & & 2 \end{bmatrix} \xrightarrow[-2R_1 + R_3]{-R_1 + R_2}$$

$$\begin{bmatrix} 1 & 1 & 1 & 1 & & 5 \\ 0 & 2 & 1 & 3 & \vdots & -2 \\ 0 & -2 & -1 & -3 & & -8 \end{bmatrix} \xrightarrow{R_2 + R_3} \begin{bmatrix} 1 & 1 & 1 & 1 & & 5 \\ 0 & 2 & 1 & 3 & \vdots & -2 \\ 0 & 0 & 0 & 0 & & -10 \end{bmatrix} \xrightarrow{-(1/2)R_2 + R_1}$$

$$\begin{bmatrix} 1 & 0 & 1/2 & -1/2 & & 6 \\ 0 & 2 & 1 & 3 & \vdots & -2 \\ 0 & 0 & 0 & 0 & & -10 \end{bmatrix} \xrightarrow[-(1/10)R_3]{(1/2)R_2}$$

$$\begin{bmatrix} 1 & 0 & 1/2 & -1/2 & & 6 \\ 0 & 1 & 1/2 & 3/2 & \vdots & -1 \\ 0 & 0 & 0 & 0 & & 1 \end{bmatrix} \xrightarrow[-6R_3 + R_1]{-R_3 + R_2} \begin{bmatrix} 1 & 0 & 1/2 & -1/2 & & 0 \\ 0 & 1 & 1/2 & 3/2 & \vdots & 0 \\ 0 & 0 & 0 & 0 & & 1 \end{bmatrix} \approx E$$

The ranks of A and $[A \ \vdots \ K]$ are 2 and 3, respectively. Hence the system is not consistent. In this example, although one can argue that the ranks can be obtained only after applying three times the elementary row operations; however, in general, to complete the solution, it is necessary to reach the row reduced echelon matrix.

Example 1.11.
Solve the system of three equations and five variables described by

$$\begin{bmatrix} 2 & -4 & -2 & 6 & 8 \\ 3 & -6 & 1 & 17 & 0 \\ 1 & -2 & 0 & 5 & 1 \end{bmatrix} \begin{bmatrix} x_1 \\ x_2 \\ x_3 \\ x_4 \\ x_5 \end{bmatrix} = \begin{bmatrix} -20 \\ -10 \\ -5 \end{bmatrix}$$

Set the augmented matrix and apply elementary row operations as follows:

$$\begin{bmatrix} 2 & -4 & -2 & 6 & 8 & & -20 \\ 3 & -6 & 1 & 17 & 0 & \vdots & -10 \\ 1 & -2 & 0 & 5 & 1 & & -5 \end{bmatrix} \xrightarrow{R_1 \rightleftarrows R_3} \begin{bmatrix} 1 & -2 & 0 & 5 & 1 & & -5 \\ 3 & -6 & 1 & 17 & 0 & \vdots & -10 \\ 2 & -4 & -2 & 6 & 8 & & -20 \end{bmatrix} \xrightarrow[-2R_1 + R_3]{-3R_1 + R_2}$$

$$\begin{bmatrix} 1 & -2 & 0 & 5 & 1 & \vdots & -5 \\ 0 & 0 & 1 & 2 & -3 & \vdots & 5 \\ 0 & 0 & -2 & -4 & 6 & & -10 \end{bmatrix} \xrightarrow{2R_2 + R_3} \begin{bmatrix} 1 & -2 & 0 & 5 & 1 & & -5 \\ 0 & 0 & 1 & 2 & -3 & \vdots & 5 \\ 0 & 0 & 0 & 0 & 0 & & 0 \end{bmatrix}$$

The final augmented matrix is a row-reduced echelon. Thus, we find

$$r(A) = r([A \;\vdots\; K]) = 2$$

Therefore, the system is consistent. The degree of freedom is $(5 - 2) = 3$; then two variables are determined in terms of the three arbitrary variables, which make the solution not unique. The answers to the first two questions are accomplished. The equivalent system is

$$x_1 - 2x_2 + 5x_4 + x_5 = -5$$
$$x_3 + 2x_4 - 3x_5 = 5$$

The variables x_4 and x_5 appear in both equations; hence, they must be in the set of the arbitrary variables, and consequently, the variable x_3 is one of the two determined variables. The second determining variable may be either x_1 or x_2 letting the other arbitrarily. Let the selected arbitrary variables be $x_2 = c_1, x_4 = c_2, x_5 = c_3$. The answer to the third question about the most general solution becomes

$$X = \begin{bmatrix} x_1 \\ x_2 \\ x_3 \\ x_4 \\ x_5 \end{bmatrix} = \begin{bmatrix} 2c_1 - 5c_2 - c_3 - 5 \\ c_1 \\ -2c_2 + 3c_3 + 5 \\ c_2 \\ c_3 \end{bmatrix} = c_1 \begin{bmatrix} 2 \\ 1 \\ 0 \\ 0 \\ 0 \end{bmatrix} + c_2 \begin{bmatrix} -5 \\ 0 \\ -2 \\ 1 \\ 0 \end{bmatrix} + c_3 \begin{bmatrix} -1 \\ 0 \\ 3 \\ 0 \\ 1 \end{bmatrix} + 5 \begin{bmatrix} -1 \\ 0 \\ 1 \\ 0 \\ 0 \end{bmatrix}$$

$$X = c_1X_1 + c_2X_2 + c_3X_3 + X_p$$

where X_1, X_2, and X_3 are linearly independent solutions (none can be obtained as a linear combination of the others) of the homogenous part $AX = 0$, and X_p is a particular solution that corresponds to the input vector K. A similar concept will be explained in chapter two when we discuss a nonhomogeneous differential equation's solution. It is remarkable to note that these linearly independent solutions are

$$[2 \quad -4 \quad 0 \quad 6 \quad 8]X_1 = 0,$$

$$[3 \quad -6 \quad 1 \quad 17 \quad 0]X_2 = 0,$$

$$[1 \quad -2 \quad 0 \quad 5 \quad 1]X_3 = 0$$

Finally, the reader may check the solution correctness by selecting any of three equations, for example; in the first equation, we substitute for finding

$$2(2c_1 - 5c_2 - c_3 - 5) - 4c_1 - 2(-2c_2 + 3c_3 + 5) + 6c_2 + 8c_3 = -20$$

Furthermore, the reader may want to determine and check the other possible solution by interchanging the variables x_1 and x_2.

The three previous examples summarize the three possible cases of solving the linear system $AX = K$. For a coefficient matrix $A_{m\times n}$, these cases are:

1. When $m = n$ and the ranks of A and $[A \;\vdots\; K]$ are equal to *n*, then the system is consistent, the solution is unique, and it is given by $X = A^{-1}K$.
2. When $n \geq m$, and $r = \text{rank}(A) = \text{rank}([A \;\vdots\; K])$ is less than *n*, then the system is consistent, and there is an infinite number of solutions, where the most general answer is in terms of $(n - r)$ arbitrary independent constant, i.e.,

$$X = c_1X_1 + c_2X_2 + \cdots + c_{n-r}X_{n-r} + c_pX_p \tag{1.28}$$

3. Otherwise, if the ranks are not equal, then the system is inconsistent, and no solutions exist.

Example 1.12.

Consider the passive electric circuit shown in Figure 1.2, where the DC voltage source causes currents to pass through each resistor. For a given source voltage and resistor values, it is desired to find these currents and also the power delivered to the load resistance R_L.

In the electric circuit theory, Kirchhoff's law states that the sum of the voltages around any loop is equal to zero (the sum of the currents is zero at a node). On the other hand, Ampere's law states the following three equations (some additional circuit theory concepts are not mentioned).

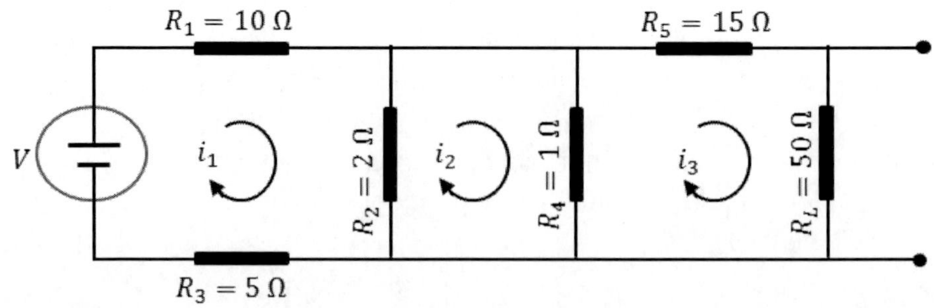

Figure 1.2. A DC electric circuit (Example 1.12).

$$V - i_1R_1 - (i_1 - i_2)R_2 - i_1R_3 = 0$$
$$-(i_2 - i_3)R_4 - (i_2 - i_1)R_2 = 0$$
$$-i_3R_5 - i_3R_L - (i_3 - i_2)R_4 = 0$$

Let the unknown vector be $X = [i_1 \quad i_2 \quad i_3]^T$, a force (input) vector $K = [V \quad 0 \quad 0]^T$, and a coefficient matrix as

$$A = \begin{bmatrix} R_1 + R_2 + R_3 & -R_2 & 0 \\ -R_2 & R_4 + R_2 & -R_4 \\ 0 & -R_4 & R_4 + R_5 + R_L \end{bmatrix}$$

The system $AX = K$ is

$$\begin{bmatrix} R_1 + R_2 + R_3 & -R_2 & 0 \\ -R_2 & R_4 + R_2 & -R_4 \\ 0 & -R_4 & R_4 + R_5 + R_L \end{bmatrix}\begin{bmatrix} i_1 \\ i_2 \\ i_3 \end{bmatrix} = \begin{bmatrix} V \\ 0 \\ 0 \end{bmatrix}$$

For 100 volt source and substituting the given values of resisters, we get

$$\begin{bmatrix} 17 & -2 & 0 \\ -2 & 3 & -1 \\ 0 & -1 & 66 \end{bmatrix}\begin{bmatrix} i_1 \\ i_2 \\ i_3 \end{bmatrix} = \begin{bmatrix} 100 \\ 0 \\ 0 \end{bmatrix}$$

When the augmented matrix is row reduced to its echelon matrix, we get

$$\begin{bmatrix} 17 & -2 & 0 & : & 100 \\ -2 & 3 & -1 & : & 0 \\ 0 & -1 & 66 & : & 0 \end{bmatrix} \xrightarrow[Row\ reduced]{} \begin{bmatrix} 1 & 0 & 0 & : & 6.3857 \\ 0 & 1 & 0 & : & 4.2788 \\ 0 & 0 & 1 & : & 0.0648 \end{bmatrix}$$

Therefore, the three currents are $i_1 = 6.3857\ \text{A}, i_2 = 4.2788\ \text{A}, i_3 = 0.0648\ \text{A}$. The delivered power to the load is

$$P = i_3^2 R_L = 0.21\ \text{Watt}$$

To complete the discussion of solving a linear system, we have to consider the solution of the homogenous case ($K = 0$). It is evident that such systems have trivial solution $X = 0$; nevertheless, these systems can have a nontrivial solution (nonzero solutions) as stated by the next theorem.

Theorem 1.4.

The system $AX = 0$ has nonzero solutions (not unique) if and only if

$$r = \text{rank}(A) < n \tag{1.29}$$

Thus, there are $(n-r)$ degrees of freedom in the general solution. Consequently, if $m < n$, then the homogenous linear system has a nonzero solution. The available solution will be in the form

$$X = c_1 X_1 + c_2 X_2 + \cdots + c_{n-r} X_{n-r} \tag{1.30}$$

The most used technique for obtaining the general solution is to obtain the row-reduced echelon of the augmented matrix.

One essential application of the linear homogenous system is to test the linear dependency of matrices or vectors. A set of *n* vectors of the same size p (or matrices) are linearly independent if

$$\sum_{i=1}^{n} c_i X_i = O_{n\times 1} \text{ only for } c_1 = c_2 = \cdots = c_n = 0 \tag{1.31}$$

The summation (1.31) forms a homogeneous system of the variables $c_1, c_2 \ldots c_n$. When the vector size $p < n$, then the vectors are linearly independent according to theorem 1.4. On the other hand, if $p = n$, then the vectors are linearly independent if and only if the determinant of the coefficient matrix of the homogeneous system is not equal to zero or the rank of the coefficient matrix is equal to the length of the vector.

Example 1.13.

Matrices play a vital role in a chemical reaction. The balancing of a chemical equation can be performed by a matrix method, i.e., to determine the stoichiometric coefficients. The matrix method is powerful compared to the usual trial-and-error procedure, especially with many different species. In this example, the gunpowder reaction will illustrate the matrix method of balancing a reaction equation. Without going into many details, we write the equation as

$$\mathrm{KNO_3 + S + C \rightarrow K_2CO_3 + K_2SO_4 + CO_2 + N_2}$$

Together with the stoichiometric coefficients, the reaction equation becomes

$$a\,\mathrm{KNO_3} + b\,\mathrm{S} + c\,\mathrm{C} \rightarrow d\,\mathrm{K_2CO_3} + e\,\mathrm{K_2SO_4} + f\,\mathrm{CO_2} + g\,\mathrm{N_2}$$

For each material, we write the following equations:

$$\begin{aligned}
&\mathrm{K}: 1\,a + 0\,b + 0\,c = 2\,d + 2\,e + 0\,f + 0\,g \\
&\mathrm{S}: 0\,a + 1\,b + 0\,c = 0\,d + 1\,e + 0\,f + 0\,g \\
&\mathrm{C}: 0\,a + 0\,b + 1\,c = 1\,d + 0\,e + 1\,f + 0\,g \\
&\mathrm{O}: 3\,a + 0\,b + 0\,c = 3\,d + 4\,e + 2\,f + 0\,g \\
&\mathrm{N}: 1\,a + 0\,b + 0\,c = 0\,d + 0\,e + 0\,f + 2\,g
\end{aligned}$$

The above equations can be rewritten as a homogenous linear system of equations.

$$\begin{bmatrix} 1 & 0 & 0 & -2 & -2 & 0 & 0 \\ 0 & 1 & 0 & 0 & -1 & 0 & 0 \\ 0 & 0 & 1 & -1 & 0 & -1 & 0 \\ 3 & 0 & 0 & -3 & -4 & -2 & 0 \\ 1 & 0 & 0 & 0 & 0 & 0 & -2 \end{bmatrix} \begin{bmatrix} a \\ b \\ c \\ d \\ e \\ f \\ g \end{bmatrix} = 0_{5\times 1}$$

The stoichiometric coefficient matrix can be row-reduced to its echelon matrix to obtain the following equivalent homogenous system

$$\begin{bmatrix} 1 & 0 & 0 & 0 & 0 & 0 & -2 \\ 0 & 1 & 0 & 0 & 0 & 2 & -3 \\ 0 & 0 & 1 & 0 & 0 & -3 & 2 \\ 0 & 0 & 0 & 1 & 0 & -2 & 2 \\ 0 & 0 & 0 & 0 & 1 & 2 & -3 \end{bmatrix} \begin{bmatrix} a \\ b \\ c \\ d \\ e \\ f \\ g \end{bmatrix} = 0_{5\times 1}$$

None of the stoichiometric coefficients is equal to zero; at the same time, two coefficients are arbitrary. The system reduces to

$$a = 2g$$

$$b = e = -2f + 3g$$

$$c = 3f - 2g$$

$$d = 2f - 2g$$

Hence, the smallest positive integer solution is obtained by selecting $f = 6$ and $g = 5$. Thus, the reset coefficients are $a = 10, b = 3, c = 8, d = 2, e = 3$. The balanced reaction equation is

$$10\ KNO_3 + 3\ S + 8\ C \rightarrow 2\ K_2CO_3 + 3\ K_2SO_4 + 6\ CO_2 + 5\ N_2$$

Another application is to determine whether two matrices are similar or not. The similarity test is reduced to check a homogenous system of n^2 equations. The system must have a nonzero solution to conclude that the similarity occurs, and the solution determines the matrix P stated in definition 1.18.

Example 1.14.

Cheek whether the following vectors are linearly independent or linearly dependent.

$$X_1 = \begin{bmatrix} 2 \\ 1 \\ 3 \end{bmatrix}, X_2 = \begin{bmatrix} -1 \\ 2 \\ -4 \end{bmatrix}, X_3 = \begin{bmatrix} 1 \\ -3 \\ 5 \end{bmatrix}$$

We have to set the system

$$\sum_{i=1}^{3} c_i X_i = O_{3\times1}$$

$$c_1 \begin{bmatrix} 2 \\ 1 \\ 3 \end{bmatrix} + c_2 \begin{bmatrix} -1 \\ 2 \\ -4 \end{bmatrix} + c_3 \begin{bmatrix} 1 \\ -3 \\ 5 \end{bmatrix} = \begin{bmatrix} 2c_1 - c_2 + c_3 \\ c_1 + 2c_2 - 3c_3 \\ 3c_1 - 4c_2 + 5c_3 \end{bmatrix} = \begin{bmatrix} 0 \\ 0 \\ 0 \end{bmatrix}$$

The coefficient matrix of the homogeneous equation is

$$A = \begin{bmatrix} 2 & -1 & 1 \\ 1 & 2 & -3 \\ 3 & -4 & 5 \end{bmatrix}$$

Applying the elementary row operations on A yields

$$\begin{bmatrix} 2 & -1 & 1 \\ 1 & 2 & -3 \\ 3 & -4 & 5 \end{bmatrix} \xrightarrow{R_1 \rightleftarrows R_2} \begin{bmatrix} 1 & 2 & -3 \\ 2 & -1 & 1 \\ 3 & -4 & 5 \end{bmatrix} \xrightarrow[-3R_1 + R_3]{-2R_1 + R_2} \begin{bmatrix} 1 & 2 & -3 \\ 0 & -5 & 7 \\ 0 & -10 & 14 \end{bmatrix} \xrightarrow{-2R_2 + R_3} \begin{bmatrix} 1 & 2 & -3 \\ 0 & -5 & 7 \\ 0 & 0 & 0 \end{bmatrix}$$

The coefficient matrix's determinant is zero (the rank equals 2, which is less than the vector length 3). Therefore, the vectors are linearly dependent. How these vectors are depending on each other requires finding the general solution of the system $AC = O_{3\times1}$, where $C = [c_1 \quad c_2 \quad c_3]^T$. Using the back substitution technique and setting $c_3 = k$, where k is an arbitrary constant, then the general solution vector can be easily found to be

$$C = \begin{bmatrix} c_1 \\ c_2 \\ c_3 \end{bmatrix} = \begin{bmatrix} \left(-\frac{14}{5}\right)k + 3k \\ \left(\frac{7}{5}\right)k \\ k \end{bmatrix}$$

Therefore, equation (1.31) is valid not only for $c_1 = c_2 = \cdots = c_n = 0$. However, it is valid for an infinite number of sets of these constants.

As an obvious extension beyond solving a linear system, it is to consider a matrix system. It is a system with a variable matrix X and an input matrix K of dimensions $(m \times p)$. For instance, with $p = 2$, a matrix system

$$\begin{bmatrix} a_{11} & a_{12} & a_{13} \\ a_{21} & a_{22} & a_{23} \\ a_{31} & a_{32} & a_{33} \end{bmatrix} \begin{bmatrix} x_{11} & x_{12} \\ x_{21} & x_{22} \\ x_{31} & x_{32} \end{bmatrix} = \begin{bmatrix} k_{11} & k_{12} \\ k_{21} & k_{22} \\ k_{31} & k_{32} \end{bmatrix} \tag{1.32a}$$

Everything we have learned for the linear algebraic equations solution is applied for a matrix equation also. For instance, the solution can be obtained by elementary row operations, i.e.,

$$\begin{bmatrix} a_{11} & a_{12} & a_{13} & & k_{11} & k_{12} \\ a_{21} & a_{22} & a_{23} & \vdots & k_{21} & k_{22} \\ a_{31} & a_{32} & a_{33} & & k_{31} & k_{32} \end{bmatrix} \xrightarrow{\text{Row Operations}} \begin{bmatrix} 1 & 0 & 0 & & x_{11} & x_{12} \\ 0 & 1 & 0 & \vdots & x_{21} & x_{22} \\ 0 & 0 & 1 & & x_{31} & x_{32} \end{bmatrix}$$

Another rare version of solving linear systems, which, for example, can be found in linear electric circuits, is

$$AX = BZ \tag{1.32b}$$

where A is $(n \times n)$ coefficient matrix, X is $(n \times 1)$ the output vector of variables (for example, voltages at nodes), B is a matrix of dimension $(n \times m)$, and Z is another vector of variables (such as sources of current or/and voltages) of dimensions $(m \times 1)$. The solution has the form

$$X = A^{-1}BZ \tag{1.32c}$$

Example 1.15.

A truss analysis problem is whether a particular truss is stable or not when an arbitrary external force configuration is applied to the truss joints. The stability problem arises when the truss basic building blocks are not only of triangular shape. Consider the well-known truss shown in Figure 1.3. The joints are numbered from one to six, and the member forces are labeled from *A* to *H* and two supported joints. The left-right arrows are used to indicate the direction of member forces towards the considered joints. It is required to determine the set of static equilibrium equations, assuming compressive forces (the member force is inward the joint) and arbitrary loads at each of the six joints. Then to determine whether the truss is stable or not (move or not).

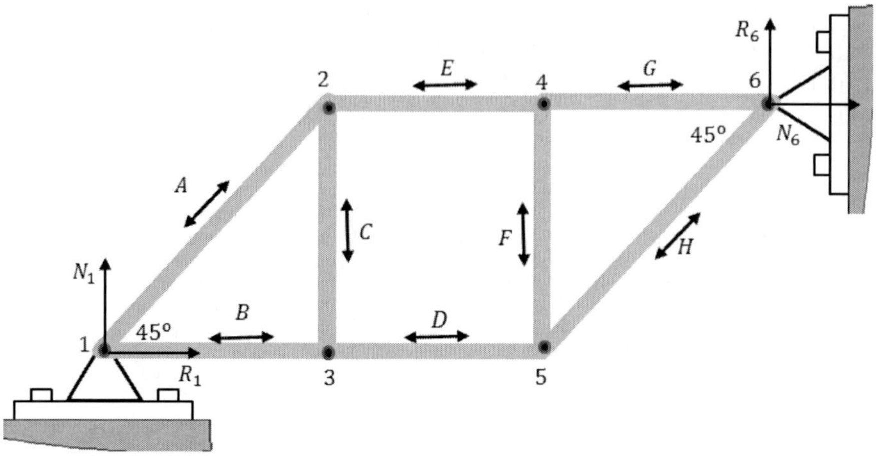

Figure 1.3. A specific fixed truss (Example 1.15).

Two forces (a vertical and horizontal) equilibrium equations are valid at each joint, one along the horizontal *x*-axis and the other along the vertical *y*-axis. Then, 12 equations of 12 unknowns are obtained. These are:

$$F_{1x} = R_1 - A\cos 45 - B$$

$$F_{1y} = N_1 - A\sin 45$$

$$F_{2x} = A\cos 45 - E$$

$$F_{2y} = A\sin 45 + C$$

$$F_{3x} = B - D$$

$$F_{3y} = -C$$

$$F_{4x} = E - G$$

$$F_{4y} = F$$

$$F_{5x} = D - H\cos 45$$

$$F_{5y} = -F - H\sin 45$$

$$F_{6x} = N_6 + G + H\cos 45$$

$$F_{6y} = R_6 + H\sin 45$$

In matrix form, we can write the system of equations in the form $AX = K$.

$$\begin{bmatrix} 1 & 0 & -c & -1 & 0 & 0 & 0 & 0 & 0 & 0 & 0 & 0 \\ 0 & 1 & -s & 0 & 0 & 0 & 0 & 0 & 0 & 0 & 0 & 0 \\ 0 & 0 & c & 0 & 0 & 0 & -1 & 0 & 0 & 0 & 0 & 0 \\ 0 & 0 & s & 0 & 1 & 0 & 0 & 0 & 0 & 0 & 0 & 0 \\ 0 & 0 & 0 & 1 & 0 & -1 & 0 & 0 & 0 & 0 & 0 & 0 \\ 0 & 0 & 0 & 0 & -1 & 0 & 0 & 0 & 0 & 0 & 0 & 0 \\ 0 & 0 & 0 & 0 & 0 & 0 & 1 & 0 & -1 & 0 & 0 & 0 \\ 0 & 0 & 0 & 0 & 0 & 0 & 0 & 1 & 0 & 0 & 0 & 0 \\ 0 & 0 & 0 & 0 & 0 & 1 & 0 & 0 & 0 & -c & 0 & 0 \\ 0 & 0 & 0 & 0 & 0 & 0 & 0 & -1 & 0 & -s & 0 & 0 \\ 0 & 0 & 0 & 0 & 0 & 0 & 0 & 0 & 1 & c & 0 & 1 \\ 0 & 0 & 0 & 0 & 0 & 0 & 0 & 0 & 0 & s & 1 & 0 \end{bmatrix} \begin{bmatrix} R_1 \\ N_1 \\ A \\ B \\ C \\ D \\ E \\ F \\ G \\ H \\ R_6 \\ N_6 \end{bmatrix} = \begin{bmatrix} F_{1x} \\ F_{1y} \\ F_{2x} \\ F_{2y} \\ F_{3x} \\ F_{3y} \\ F_{4x} \\ F_{4y} \\ F_{5x} \\ F_{5y} \\ F_{6x} \\ F_{6y} \end{bmatrix}$$

where $c = \cos 45,$ and $s = \sin 45$. The stability of the truss is secured if and only if A^{-1} exists. The inverse existence is checked by finding one of the following:

- Reduce *A* to an upper triangular matrix and read the diagonal elements to be nonzero values; thus, the determinant is not equal to zero.
- Reduce to echelon matrix and read the rank to be equal to 12.

For the given values of $F_{ix},$ and F_{iy}; $i = 1, 2, 3, 4, 5, 6$, calculating the inverse itself is needed to find the member forces for a given load. Thus, we can calculate

$$[A \;\vdots\; K] \xrightarrow{\text{row reduction}} [I_{12\times12} \;\vdots\; X]$$

For, $F_{ix} = F_{iy} = 1;\ i = 1, 2, 3, 4, 5, 6$, We obtain

$$X = [3 \quad 3 \quad 2.828 \quad 0 \quad -1 \quad -1 \quad 1 \quad 1 \quad 0 \quad -2.828 \quad 3 \quad 3]^T$$

Therefore, in this example, we conclude that the reaction forces at the supported joints are all equal, and the internal forces of beams *B* and *G* are equal to zero. The internal forces of beams *C*, *D*, and *H* are in the opposite direction to the assumed directions in the equations, while beams *A*, *E*, and *F* are in the assumed direction.

1.6. Basics of Linear Algebra

First, we have to say that the topic of linear algebra is more extended than we will give here because linear algebra deals with a general treatment of sets of polynomials, vectors, matrices, and the solution space of differential equations. Furthermore, linear algebra studies the similarities and differences between these sets and the linear transformation theory. However, we have already discussed an essential part of the linear algebra topic in previous sections, and we will give the most fundamental concepts, definitions, and theories in this section.

We shall begin with the basic definition that describes some fundamental concepts.

Definition 1.21. Addition and scalar multiplication

A central property in linear algebra is that for a set, if a linear combination of some members creates a new member in the set, then the set is closed under addition and scalar multiplication. To illustrate, let us consider a set containing $X_1, X_2, X_3 \dots X_p$ vectors, and two arbitrary constants *a*, and *b*. Then, for instance, the vector $aX_1 + bX_2$ is a vector, which belongs to the set. There will always be two constants, which create a zero vector.

Definition 1.22. Vector spaces

A real vector space $\phi = \{\alpha, \beta, \gamma \dots\}$ represents a set of objects called vectors (i.e., a matrix as an element of the set is also called a vector, etc.). The properties of addition and scalar multiplication are valid. The addition and multiplication properties that we have seen for matrices in section 1.2 are valid for the real vector space objects. Consequently, we denote a set of matrix objects by the set $\mathcal{R}_{m\times n}$ and a set of polynomials object of finite degree with real coefficients by the set $\mathcal{P}$. We denote a set containing all continuous functions having n^{th} derivatives in the interval $[a, b]$ by $\mathcal{C}^n(a, b)$.

Definition 1.23. Subspaces

We use $\varphi \subseteq \phi$ to denote that φ is a subset, which has the same properties of the set ϕ, i.e., the subspace is also closed under arbitrary linear combination. For example, the null space is a subspace, while the set of solutions of the nonhomogeneous system is not a subspace.

Definition 1.24. Spanning

Let $\alpha_1, \alpha_2 \dots \alpha_p$ be arbitrary vectors in real vector space ϕ. The smallest subspace of the space ϕ containing all given vectors is called the subspace spanned by the vector α_i or the subspace generated by the α_i. We denote such subspace as $\mathrm{Span}\{\alpha_1, \alpha_2 \dots \alpha_p\}$.

Definition 1.25. Null spaces

The null space is that set of all solutions of the homogeneous system $AX = 0$, and it is denoted by

$$\mathrm{NS}(A) = \{X \; in \; \mathcal{R}_{n\times 1} \,|AX = 0\} \qquad (1.33)$$

Consequently, the solution of the nonhomogeneous system is not a null space.

Theorem 1.5.

The objects or elements of a subspace that is spanned by $\alpha_1, \alpha_2 \dots \alpha_p$ are precisely the linear combination of the generating vectors $\alpha_1, \alpha_2 \dots \alpha_p$, i.e.,

$$\text{Span}\{\alpha_1, \alpha_2 \dots \alpha_p\} = \{\textstyle\sum_{i=1}^{p} c_i \alpha_i \,|c_i \text{ in } \mathcal{R}\} \tag{1.34}$$

It is convenient to test whether a specific vector is in $\text{Span}\{\alpha_1, \alpha_2 \dots \alpha_p\}$ or not. To do this, we have to determine whether a related linear algebraic system has a solution or not. We have only to test for the solution's existence and not find a solution, i.e., to test the condition (1.29).

Example 1.16.

Given the following three vectors in $\mathcal{R}_{3\times1}$

$$X_1 = \begin{bmatrix} 2 \\ -3 \\ 1 \end{bmatrix}, X_2 = \begin{bmatrix} 1 \\ 5 \\ 1 \end{bmatrix}, X_3 = \begin{bmatrix} -1 \\ -1 \\ 4 \end{bmatrix}$$

Is the vector $x = [5 \quad 10 \quad 15]^T$ in $\text{Span}\{X_1, X_2, X_3\}$?

It is to decide whether or not there exist three scalars $c_1, c_2, \text{and } c_3$ such that

$$x = c_1 X_1 + c_2 X_2 + c_3 X_3$$

Or,

$$\begin{bmatrix} 5 \\ 10 \\ 15 \end{bmatrix} = c_1 \begin{bmatrix} 2 \\ -3 \\ 1 \end{bmatrix} + c_2 \begin{bmatrix} 1 \\ 5 \\ 1 \end{bmatrix} + c_3 \begin{bmatrix} -1 \\ -1 \\ 4 \end{bmatrix}$$

Equivalently, we can write the related linear system as

$$\begin{bmatrix} 2 & 1 & -1 \\ -3 & 5 & -1 \\ 1 & 1 & 4 \end{bmatrix} \begin{bmatrix} c_1 \\ c_2 \\ c_3 \end{bmatrix} = \begin{bmatrix} 5 \\ 10 \\ 15 \end{bmatrix}$$

Let us apply the elementary row reduced operations on the coefficient matrix,

$$\begin{bmatrix} 2 & 1 & -1 \\ -3 & 5 & -1 \\ 1 & 1 & 4 \end{bmatrix} \xrightarrow{R_1 \rightleftarrows R_3} \begin{bmatrix} 1 & 1 & 4 \\ -3 & 5 & -1 \\ 2 & 1 & -1 \end{bmatrix} \xrightarrow[-2R_1 + R_3]{3R_1 + R_2} \begin{bmatrix} 1 & 1 & 4 \\ 0 & 8 & 11 \\ 0 & -1 & -9 \end{bmatrix} \xrightarrow{\frac{1}{8}R_2 + R_3} \begin{bmatrix} 1 & 1 & 4 \\ 0 & 8 & 11 \\ 0 & 0 & -\frac{61}{8} \end{bmatrix}$$

It is clear that the reached matrix is nonsingular (has a determinant equal to 61), and hence the rank equals the order of the system. Therefore, the system is consistent and has a unique solution. Thus, the answer is yes, and in turn, any arbitrary real vector $y = [a \quad b \quad c]$ is in $\text{Span}\{X_1, X_2, X_3\}$. Furthermore, the complete solution can be determined to express x in terms of the linear combination of $X_1, X_2, \text{and } X_3$, i.e.,

$$x = 1.8852X_1 + 3.6066X_2 + 2.377X_3$$

Obviously, for an arbitrary vector, the linear combination constants will differ from the above values.

Definition 1.26. A basis

A set of linearly independent vectors (in their general meaning) that span ϕ is called a basis for ϕ. A famous basis is a standard in the order $e_1: (1,0,0)$, $e_2: (0,1,0)$, and $e_3: (0,0,1)$.

Theorem 1.6.

A linear combination of its basis uniquely expresses every vector in space ϕ is. Conversely, if every vector in ϕ is uniquely expressed as a linear combination of a particular set of vectors, these vectors form a basis of ϕ.

The first part of theorem 1.6 was demonstrated in example 1.16. The test of a set of vectors to be linearly independent has already been explained in theorem 1.4 (also, see example 1.10). Therefore, similarly, the test of a set of vectors based on a specific space can be carried out. However, the procedure in example 1.14 can be carried out for a space of polynomials, as in the following example.

Example 1.17.

Consider the polynomial space of finite degree $\mathcal{P}$. Are the following polynomials linearly independent?

$$p_1(x) = x^2 + 5x + 6$$

$$p_2(x) = x^2 + 3x + 2$$

$$p_3(x) = x^2 + 3x - 4$$

We set the following equation

$$c_1(x^2 + 5x + 6) + c_2(x^2 + 3x + 2) + c_3(x^2 + 3x - 4) = 0$$

We arrange according to the power of x

$$(c_1 + c_2 + c_3)x^2 + (5c_1 + 3c_2 + 3c_2)x + (6c_1 + 2c_2 - 4c_2) = 0x^2 + 0x + 0$$

Equating the coefficients of the equal power gives the homogeneous linear system.

$$\begin{bmatrix} 1 & 1 & 1 \\ 5 & 3 & 3 \\ 6 & 2 & -4 \end{bmatrix} \begin{bmatrix} c_1 \\ c_2 \\ c_3 \end{bmatrix} = \begin{bmatrix} 0 \\ 0 \\ 0 \end{bmatrix}$$

It is easy to show that the determinant of the coefficient matrix is not equal to zero by applying elementary row-reduced operations, and hence the only possible solution is

$$c_1 = c_2 = c_3 = 0$$

Therefore, the given polynomials are linearly independent.

Example 1.18.

Consider the vectors r, s, and v from $C[0, \pi]$, where

$$r(t) = \sin t\,, s(t) = \cos t\,, v(t) = e^{t/\pi}$$

We want to show whether these functions are linearly independent vectors. Let $x_1 r + x_2 s + x_3 v = 0$, then for any $t_0, 0 \leq t_0 \leq \pi$, we have

$$x_1 r(\,t_0) + x_2 s(\,t_0) + x_3 v(\,t_0) = 0$$

We need to calculate three values in the considered interval: $t_0 = 0, t_0 = \frac{1}{2}\pi$, and $t_0 = \pi$

$$x_1 \sin(0) + x_2 \cos(0) + x_3 e^0 = 0$$

$$x_1 \sin(\frac{1}{2}\pi) + x_2 \cos(\frac{1}{2}\pi) + x_3 e^{\frac{1}{2}} = 0$$

$$x_1 \sin(\pi) + x_2 \cos(\pi) + x_3 e^1 = 0$$

The matrix form of these equations is

$$\begin{bmatrix} 0 & 1 & 1 \\ 1 & 0 & e^{0.5} \\ 0 & -1 & e^1 \end{bmatrix} \begin{bmatrix} x_1 \\ x_2 \\ x_3 \end{bmatrix} = \begin{bmatrix} 0 \\ 0 \\ 0 \end{bmatrix}$$

It can be determined that the coefficient matrix is nonsingular, and hence the only solution of the homogenous system is the trivial solution $X = 0$. Thus, the three functions r, s, and v are linearly independent.

An essential topic from linear algebra is a linear transformation. The reader may remember the linear transformation in reducing the complexity of algebraic equations or transforming the axes, etc., linearly. We now introduce the linear transformation in a more general framework, particularly the linear transformation between vector spaces.

In a vector space, an essential relationship between vectors is the orthogonality and orthonormality properties, stated in the following definitions.

Definition 1.27. Orthogonal spaces

A set of vectors $\{V_1, V_2, \ldots V_n\}$ from a vector space $\mathcal{R}_{n\times 1}$ is said to form an orthogonal set of vectors if it holds that

$$V_i \,.\, V_j = V_i^T \,.\, V_j = 0, \forall\, i \neq j \tag{1.35}$$

Moreover, the orthogonal set will be orthonormal if it holds $V_i^T \,.\, V_i = 1$. The most famous set of orthogonal and orthonormal vectors is the euclidean three vectors, usually named $e_1, e_2,$ and e_3 in space $\mathcal{R}_{3\times 1}$.

Definition 1.28. Orthogonal matrices

A real matrix R is an orthogonal matrix if and only if $R^T R = I$. Furthermore, two matrices A and B are said to be orthogonal if there exists an orthogonal matrix R such that

$$B = R^{-1} A R = R^T A R \tag{1.36}$$

To check a matrix M to be orthogonal, it is enough to perform by direct computation the multiplication $M^T M$ to see whether it is equal to a unity matrix or not. For instance, a well-known orthogonal matrix is

$$M = \begin{bmatrix} \cos\theta & -\sin\theta \\ \sin\theta & \cos\theta \end{bmatrix}$$

$$M^T M = \begin{bmatrix} \cos\theta & \sin\theta \\ -\sin\theta & \cos\theta \end{bmatrix} \begin{bmatrix} \cos\theta & -\sin\theta \\ \sin\theta & \cos\theta \end{bmatrix} = \begin{bmatrix} \cos^2\theta + \sin^2\theta & 0 \\ 0 & \cos^2\theta + \sin^2\theta \end{bmatrix} = \begin{bmatrix} 1 & 0 \\ 0 & 1 \end{bmatrix}$$

A basic property is that the product of two orthogonal matrices gives an orthogonal matrix as well.

An efficient formula for determining an orthogonal matrix for a given unit vector X in $\mathcal{R}_{n\times 1}$ is described by letting the unit vector partitioning as

$$X = \begin{bmatrix} x_1 \\ x_2 \\ \vdots \\ x_n \end{bmatrix} = \begin{bmatrix} x_1 \\ \cdots \\ Q \end{bmatrix}; \; x_1 \neq 1$$

Then, an orthogonal matrix R is given by the partitioned matrix

$$R = \begin{bmatrix} x_1 & \vdots & Q^T \\ \cdots & \cdots & \cdots \\ Q & \vdots & I - \dfrac{1}{1 - x_1} Q Q^T \end{bmatrix} \tag{1.37}$$

For example, for a unit vector $X = (1/3)[2 \quad 2 \quad 1]^T$, by direct applying of formula (1.37), we find

$$R = \begin{bmatrix} \frac{2}{3} & \frac{2}{3} & \frac{1}{3} \\ \frac{2}{3} & -\frac{1}{3} & -\frac{2}{3} \\ \frac{1}{3} & -\frac{2}{3} & \frac{2}{3} \end{bmatrix}$$

It can be easily checked that $R^T R = R\, R^T = I$.

1.7. Linear Transformations

Perhaps the linear transformation applications are the most frequent than other types of transformations. For instance, the two and three-dimension rotations by an angle or the reflection to one axis are examples. One feature of a linear transformation is the one-to-one correspondence between the matrices and linear transformations based on the vector-matrix multiplication. Throughout the next discussion, we will be concerned with this topic. However, several other topics concerning linear transformations can be found within the specialist books of linear algebra.

Definition 1.29. Linear transformations (or linear mapping)

Consider two spaces $\mathcal{v}$ and $\mathcal{w}$, two arbitrary vectors α and β in $\mathcal{v}$, and a constant c. Then we call a linear transform $\mathbf{T}$ from $\mathcal{v}$ to $\mathcal{w}$ (or in symbol $\mathcal{v} \to \mathcal{w}$) if the following hold:

$$\mathbf{T}(\alpha + \beta) = \mathbf{T}(\alpha) + \mathbf{T}(\beta) \tag{1.38a}$$

$$\mathbf{T}(c\alpha) = c\mathbf{T}(\alpha) \tag{1.38b}$$

Correspondingly,

$$\mathbf{T}(0) = 0 \tag{1.38c}$$

$$\mathbf{T}(-x) = -\mathbf{T}(x) \tag{1.38d}$$

$$\mathbf{T}(c_1\alpha + c_2\beta) = c_1\mathbf{T}(\alpha) + c_2\mathbf{T}(\beta) \tag{1.38e}$$

$$\mathbf{T}\left[\sum_{i=1}^{n} a_i x_i\right] = \sum_{i=1}^{n} a_i \mathbf{T}(x_i) \tag{1.38f}$$

In other words, $\mathbf{T}$ is a function that carries elements from the vector space $\mathcal{v}$ (called the domain) to the vector space $\mathcal{w}$ (called the codomain), which satisfies equations (1.38). Any defined transformation (by a matrix or a vector) can be checked for linearity by working equations (1.38) to be held for arbitrary two vectors. Meaningful to say is that applying a linear transformation to a vector may change the vector's direction in the new basis.

Example 1.19.

Is the transformation $\mathbf{T}$ linear?

$$\mathbf{T} = \begin{bmatrix} 1 \\ x_1 + 2x_2 \\ 3x_1 + x_2 - 4x_3 \end{bmatrix}$$

Let us take two arbitrary third-order vectors $\alpha = [a_1 \quad a_2 \quad a_3]^T$ and $\beta = [b_1 \quad b_2 \quad b_3]^T$, then we determine

$$(\alpha + \beta) = \begin{bmatrix} a_1 + b_1 \\ a_2 + b_2 \\ a_3 + b_3 \end{bmatrix}$$

$$\mathbf{T}(\alpha + \beta) = \begin{bmatrix} 1 \\ a_1 + b_1 + 2(a_2 + b_2) \\ 3(a_1 + b_1) + a_2 + b_2 - 4(a_3 + b_3) \end{bmatrix}$$

$$\mathbf{T}(\alpha) = \begin{bmatrix} 1 \\ a_1 + 2a_2 \\ 3a_1 + a_2 - 4a_3 \end{bmatrix}, \mathbf{T}(\beta) = \begin{bmatrix} 1 \\ b_1 + 2b_2 \\ 3b_1 + b_2 - 4b_3 \end{bmatrix}$$

$$\mathbf{T}(\alpha) + \mathbf{T}(\beta) = \begin{bmatrix} 2 \\ a_1 + 2a_2 + b_1 + 2b_2 \\ 3a_1 + a_2 - 4a_3 + 3b_1 + b_2 - 4b_3 \end{bmatrix}$$

Since $\mathbf{T}(\alpha + \beta) \neq \mathbf{T}(\alpha) + \mathbf{T}(\beta)$, then the transformation is not linear because the first elements are not equal. It is not necessary to check equality in equation (1.38b). In general, any constant value in the given vector will prevent the linearity of the transformation.

Example 1.20.

Consider the linear transformation $\mathbf{T}: \mathcal{R}_{3\times 1} \to \mathcal{R}_{3\times 1}$ given as

$$\mathbf{T}\begin{bmatrix} x_1 \\ x_2 \\ x_3 \end{bmatrix} = \begin{bmatrix} x_1 \\ 3x_1 + x_2 \\ x_1 + x_2 + 3x_3 \end{bmatrix}$$

Moreover, a vector $\alpha = [1 \quad -1 \quad 1]^T$, determine $\mathbf{T}(\mathbf{T}(\alpha))$.

We proceed sequentially,

$$\mathbf{T}(\alpha) = \begin{bmatrix} 1 \\ 3-1 \\ 1-1+3 \end{bmatrix} = \begin{bmatrix} 1 \\ 2 \\ 3 \end{bmatrix}$$

$$\mathbf{T}\big(\mathbf{T}(\alpha)\big) = \begin{bmatrix} 1 \\ 3+2 \\ 1+2+9 \end{bmatrix} = \begin{bmatrix} 1 \\ 5 \\ 12 \end{bmatrix}$$

The example explains the possibilities of determining different expressions of linear transformations such as $3\mathbf{T}(\alpha) + 2\mathbf{T}(\mathbf{T}(\alpha))$, etc. However, such expressions exclude these containing constant terms only.

Example 1.21.

Consider the transformation $\mathbf{T}$ from a third-order polynomial to a matrix of order two.

$$\mathbf{T}(\mathrm{a}x^3 + bx^2 + cx + d) = \begin{bmatrix} d+c & d-2b \\ a & c-a \end{bmatrix}$$

Show that $\mathbf{T}$ is a linear transform.

Let us take two arbitrary polynomials

$$p_1(x) = a_1x^3 + \ b_1x^2 + c_1x + d_1,\ p_2(x) = a_2x^3 + \ b_2x^2 + c_2x + d_2$$

If the $\mathbf{T}$ is a linear transformation, then we have to verify the conditions in equations (1.38a and 1.38b). Otherwise, it is not linear.

$$\mathbf{T}(p_1 + p_2) = \mathbf{T}((a_1x^3 + \ b_1x^2 + c_1x + d_1) + (a_2x^3 + \ b_2x^2 + c_2x + d_2))$$

$$\mathbf{T}(p_1 + p_2) = \mathbf{T}\big((a_1 + a_2)x^3 + (b_1 + b_2)x^2 + (c_1 + c_2)\mathrm{x} + (d_1 + d_2)\big)$$

$$= \begin{bmatrix} (d_1 + d_2) + (c_1 + c_2) & (d_1 + d_2) - 2(b_1 + b_2) \\ (a_1 + a_2) & (c_1 + c_2) - (a_1 + a_2) \end{bmatrix}$$

$$= \begin{bmatrix} (d_1 + c_1) + (d_2 + c_2) & (d_1 - 2b_1) + (d_2 - 2b_2) \\ (a_1 + a_2) & (c_1 - a_1) + (c_2 - a_2) \end{bmatrix}$$

$$= \begin{bmatrix} (d_1 + c_1) & (d_1 - 2b_1) \\ a_1 & (c_1 - a_1) \end{bmatrix} + \begin{bmatrix} (d_2 + c_2) & (d_2 - 2b_2) \\ a_2 & (c_2 - a_2) \end{bmatrix}$$

The result is

$$\begin{aligned}\mathbf{T}(p_1+p_2) = {} & \mathbf{T}(a_1x^3+\ b_1x^2+c_1x+d_1)+\mathbf{T}(a_2x^3+\ b_2x^2+c_2x+d_2)\\ & =\mathbf{T}(p_1)+\mathbf{T}(p_2)\end{aligned}$$

Therefore, the condition in equation (1.38a) is satisfied. Now, we proceed to the next condition given in equation (1.38b).

$$\mathbf{T}(qp(x))=\mathbf{T}\big(q(\mathrm{a}x^3+bx^2+cx+d)\big)=\mathbf{T}((q\mathrm{a})x^3+(qb)x^2+(qc)x+qd)$$

$$=\begin{bmatrix}(qd)+(qc) & (qd)-2(qb)\\ (qa) & (qc)-(qa)\end{bmatrix}=\begin{bmatrix}q(d+c) & q(d-2b)\\ qa & q(c-a)\end{bmatrix}$$

$$=q\begin{bmatrix}(d+c) & (d-2b)\\ a & (c-a)\end{bmatrix}=q\mathbf{T}(\mathrm{a}x^3+bx^2+cx+d)=q\mathbf{T}(p(x))$$

By definition (1.29), $\mathbf{T}$ is a linear transformation.

Example 1.22.
Let M be the matrix transformation $\mathcal{R}_{3\times 1}\to\mathcal{R}_{3\times 1}$ with the following matrix

$$M=\begin{bmatrix}1 & 0 & 2\\ -1 & 3 & 1\\ 5 & 2 & 1\end{bmatrix}$$

For $v=[2\quad -1\quad 3]^T$, compute $\mathbf{M}(v)$.
Although the solution can be obtained by the product Mv using the multiplication properties, we compute by the meaning of the linear transformation concept, i.e.,

$$\mathbf{M}(v)=2\begin{bmatrix}1\\ -1\\ 5\end{bmatrix}+(-1)\begin{bmatrix}0\\ 3\\ 2\end{bmatrix}+3\begin{bmatrix}2\\ 1\\ 1\end{bmatrix}=\begin{bmatrix}8\\ -2\\ 11\end{bmatrix}$$

A geometric interpretation is if we assume that the vector v is originated at the origin in (x,y,z) coordinate-system and ending at the point $(2,-1,3)$ with a length of $\sqrt{14}$ units, then it rotates such to have its end at the point $(8,-2,11)$, and its length is extended to $\sqrt{189}$ units. Similarly, as we said in example 1.14, we can perform linear transformation when the transform matrix becomes a function of M, such as $(M^2+5M)v$, etc. Such expressions are nothing but another matrix that is operating on the given vector.

Example 1.23.

Several matrix operations, such as translations, rotations, scaling, etc., are used in computer graphics and related applications such as video game graphics. In this illustrative example, the rotation operation will be considered. In the Cartesian coordinate system, a point, which is represented by column vector P, is rotated around the *x*-axis, y-axis, or *z*-axis with an angle θ by the following three matrices:

$$R_x(\theta) = \begin{bmatrix} 1 & 0 & 0 \\ 0 & \cos\theta & -\sin\theta \\ 0 & \sin\theta & \cos\theta \end{bmatrix}, R_y(\theta) = \begin{bmatrix} \cos\theta & 0 & \sin\theta \\ 0 & 1 & 0 \\ -\sin\theta & 0 & \cos\theta \end{bmatrix}, R_z(\theta)$$

$$= \begin{bmatrix} \cos\theta & -\sin\theta & 0 \\ \sin\theta & \cos\theta & 0 \\ 0 & 0 & 1 \end{bmatrix}$$

Let a point be represented by

$$P = \begin{bmatrix} x_p \\ y_p \\ z_p \end{bmatrix}$$

For instance, considering the rotation about the *x*-axis, the point will be transformed to the coordinates given by

$$\begin{bmatrix} 1 & 0 & 0 \\ 0 & \cos\theta & -\sin\theta \\ 0 & \sin\theta & \cos\theta \end{bmatrix} \begin{bmatrix} x_p \\ y_p \\ z_p \end{bmatrix} = \begin{bmatrix} x_p \\ y_p\cos\theta - z_p\sin\theta \\ y_p\sin\theta + z_p\cos\theta \end{bmatrix}$$

Consequently, a set of points representing a body in space can all be transformed about one axis or another to obtain a body rotation about that axis. Given a tetrahedron, whose vertices are the four points are:

$$O: (0,0,0), A: (1,0,0), B: (0,1,0), \text{and } C: (0,0,1).$$

It is required to rotate the tetrahedron about the *y*-axis with an angle equal to $\pi/2$ (a positive angle means an anticlockwise rotation).

Using the transform matrix

$$R_y(\pi/2) = \begin{bmatrix} \cos\pi/2 & 0 & \sin\pi/2 \\ 0 & 1 & 0 \\ -\sin\pi/2 & 0 & \cos\pi/2 \end{bmatrix} = \begin{bmatrix} 0 & 0 & 1 \\ 0 & 1 & 0 \\ -1 & 0 & 0 \end{bmatrix}$$

Upon using the transformation $R_y(\pi/2)$, the four vertices are transformed into new coordinates as follows.

$$O: (0,0,0): \begin{bmatrix} 0 & 0 & 1 \\ 0 & 1 & 0 \\ -1 & 0 & 0 \end{bmatrix} \begin{bmatrix} 0 \\ 0 \\ 0 \end{bmatrix} = \begin{bmatrix} 0 \\ 0 \\ 0 \end{bmatrix} \rightarrow (0,0,0)$$

$$A: (1,0,0): \begin{bmatrix} 0 & 0 & 1 \\ 0 & 1 & 0 \\ -1 & 0 & 0 \end{bmatrix} \begin{bmatrix} 1 \\ 0 \\ 0 \end{bmatrix} = \begin{bmatrix} 0 \\ 0 \\ -1 \end{bmatrix} \rightarrow (0,0,-1)$$

$$B: (0,1,0): \begin{bmatrix} 0 & 0 & 1 \\ 0 & 1 & 0 \\ -1 & 0 & 0 \end{bmatrix} \begin{bmatrix} 0 \\ 1 \\ 0 \end{bmatrix} = \begin{bmatrix} 0 \\ 1 \\ 0 \end{bmatrix} \rightarrow (0,1,0)$$

$$C: (0,0,1): \begin{bmatrix} 0 & 0 & 1 \\ 0 & 1 & 0 \\ -1 & 0 & 0 \end{bmatrix} \begin{bmatrix} 0 \\ 0 \\ 1 \end{bmatrix} = \begin{bmatrix} 1 \\ 0 \\ 0 \end{bmatrix} \rightarrow (1,0,0)$$

In Figure 1.4, the above tetrahedron is the one given in the example. The below one is rotated about the *y*-axis by an angle equal $\pi/2$ and moved down by 0.6. Therefore, the four vertices will have the following Cartesian coordinates; note that the -0.6 down movement is made to depict the rotation of the tetrahedron in the figure, $O: (0,0,-0.6), A: (0,0,-1.6), B: (0,1,-0.6), C: (1,0,-0.6)$. Figure 1.14 is obtained using MATLAB toolboxes (the MATLAB function ***line***). A sequence of transformations around the axes with different angles will result in different tetrahedron orientations in space.

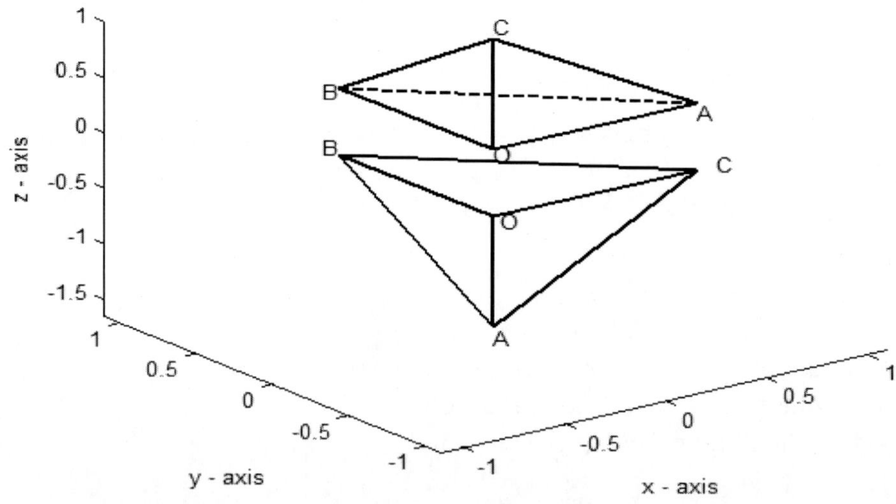

Figure 1.4. Rotating a tetrahedron in space (Example 1.23).

1.8. EIGENVALUE AND EIGENVECTOR PROBLEMS

The prefix "Eigen" is a German word, which means characteristic or likewise. It represents an essential characteristic of matrices and correspondingly of systems of linear algebraic and differential equations. In a few books and references, it is also named characteristic value and characteristic vector. These terminologies are initially used in mechanics to study the principal axes of a rigid body's rotational motion. Eigenvalues and eigenvectors have a wide range of applications in different disciplines. Here, we will focus on the evaluation of these characteristics and will interpret some physical meaning. First, we will give the following definition.

Definition 1.30. Eigenvalue and Eigenvector

A scalar λ is called an eigenvalue of the square matrix A if there exists a nonzero solution of the system $AX = \lambda X$. Any $X \neq 0$ such that $AX = \lambda X$ is called eigenvector associated with the eigenvalue λ.

Definition 1.30 suggests a direct method of computing the eigenvalues and eigenvectors of a matrix. We can write

$$\lambda X - AX = (\lambda I - A)X = 0, \text{or } (A - \lambda I)X = 0 \tag{1.39}$$

The coefficient matrix of the system (1.39) is $[\lambda I - A]$, and we recall theorem 1.4 to say that the coefficient matrix should be a singular one to obtain a nonzero solution, thus

$$\det([\lambda I - A]) = 0 \tag{1.40}$$

For a particular *n*-order matrix *A*, the result of carrying out equation (1.40) gives a single polynomial algebraic equation (named characteristic polynomial and characteristic equation) of order *n* in terms of the parameter λ.

$$c(\lambda) = \lambda^n + a_{n-1}\lambda^{n-1} + \cdots + a_1\lambda + a_0 = 0 \tag{1.41}$$

Therefore, the polynomial equation's roots (1.41) are precisely *n* roots (*n* eigenvalues). For each of these eigenvalues, the homogeneous system $[\lambda_i I - A]X_i = 0$ can be solved to obtain the nonzero corresponding eigenvector X_i. The eigenvalues and their associated eigenvectors can be real or complex numbers. In this book, we consider the cases of real eigenvalues. When the matrix order is large, the computation of the eigenvalues and vectors involves a large amount of calculation. Nevertheless, several numerical techniques override this difficulty. Furthermore, essential theoretical results were derived to facilitate the computation procedure. The above discussion implies the following theorem.

Theorem 1.7.

For a particular square matrix of order *n*, there are precisely *n* eigenvalues and *n* associated linearly independent eigenvectors.

Before we continue in more detail, we demonstrate the computation procedure through an introductory example.

Example 1.24.

Consider the matrix

$$A = \begin{bmatrix} 2 & 0 & 0 \\ 2 & 2 & 1 \\ 1 & 1 & 2 \end{bmatrix}$$

First, we determine the characteristic equation

$$\det\left(\left[\lambda\begin{bmatrix} 1 & 0 & 0 \\ 0 & 1 & 0 \\ 0 & 0 & 0 \end{bmatrix} - \begin{bmatrix} 2 & 0 & 0 \\ 2 & 2 & 1 \\ 1 & 1 & 2 \end{bmatrix}\right]\right) = \det(\begin{bmatrix} \lambda-2 & 0 & 0 \\ -2 & \lambda-2 & -1 \\ -1 & -1 & \lambda-2 \end{bmatrix})$$
$$= (\lambda-2)((\lambda-2)^2-1)$$

$$= \lambda^3 - 6\lambda^2 + 11\lambda - 6 = (\lambda-3)(\lambda-2)(\lambda-1) = 0$$

Therefore, the three eigenvalues are $\lambda_1 = 3$, $\lambda_2 = 2$, and $\lambda_3 = 1$. Usually, they are arranged in descending order. Therefore, this matrix has distinct eigenvalues. The next step is to determine the corresponding eigenvectors, i.e., to solve the homogeneous system three times. For $\lambda_1 = 3$, we will determine X_1 by solving the homogeneous system $[3I - A]X_1 = 0$.

$$\begin{bmatrix} 1 & 0 & 0 \\ -2 & 1 & -1 \\ -1 & -1 & 1 \end{bmatrix}\begin{bmatrix} x_1 \\ x_2 \\ x_3 \end{bmatrix} = \begin{bmatrix} 0 \\ 0 \\ 0 \end{bmatrix}$$

$$\begin{bmatrix} 1 & 0 & 0 \\ -2 & 1 & -1 \\ -1 & -1 & 1 \end{bmatrix} \tilde{R} \begin{bmatrix} 1 & 0 & 0 \\ 0 & 1 & -1 \\ 0 & 0 & 0 \end{bmatrix}$$

Thus, $x_3 = \text{arbitrary} = c, x_2 = c, x_1 = 0$; if we set, for example, $c = 1$, then

$$X_1 = [0 \quad 1 \quad 1]^T$$

Similarly, for $\lambda_2 = 2$, we solve the homogeneous system

$$[2I - A]X_2 = 0$$

$$\begin{bmatrix} 0 & 0 & 0 \\ -2 & 0 & -1 \\ -1 & -1 & 0 \end{bmatrix} \begin{bmatrix} x_1 \\ x_2 \\ x_3 \end{bmatrix} = \begin{bmatrix} 0 \\ 0 \\ 0 \end{bmatrix}$$

Directly, from the coefficient matrix, we obtain

$$x_3 = c, x_2 = \frac{1}{2}c, \text{and } x_3 = \frac{1}{2}, \text{ if we set } c = 2,$$

Thus, we get the second eigenvector

$$X_2 = [-1 \quad 1 \quad 2]^T$$

Finally, for $\lambda_3 = 1$, we solve $[1I - A]X_3 = 0$

$$\begin{bmatrix} -1 & 0 & 0 \\ -2 & -1 & -1 \\ -1 & -1 & -1 \end{bmatrix} \begin{bmatrix} x_1 \\ x_2 \\ x_3 \end{bmatrix} = \begin{bmatrix} 0 \\ 0 \\ 0 \end{bmatrix}$$

The solution is

$$X_3 = [0 \quad -1 \quad 1]^T$$

Let us check the linear dependency of the three eigenvectors, as we have worked in example 1.16. We set a matrix *P* with $\mathrm{Col}_i(P) = X_i$.

$$P = \begin{bmatrix} 0 & -1 & 0 \\ 1 & 1 & -1 \\ 1 & 2 & 1 \end{bmatrix}$$

The determinant of the matrix *P* is equal to 2. The matrix is nonsingular and has a rank equal to 3. Therefore, based on theorem 1.4, the eigenvectors are linearly independent. This result is cast in the next theorem.

Based on the linear matrix transformation on a vector, say $A(v)$, if a specific constant only scales the resultant vector without changing the direction, such a vector is an eigenvector. The scalar constant is the eigenvalue. For instance, let us apply the procedure in example 1.22 when the matrix is that given in example 1.24, and the vector is $v = [-1 \quad 1 \quad 2]^T$, thus

$$A(v) = \begin{bmatrix} 2 & 0 & 0 \\ 2 & 2 & 1 \\ 1 & 1 & 2 \end{bmatrix} \begin{bmatrix} -1 \\ 1 \\ 2 \end{bmatrix} = \begin{bmatrix} -2 \\ 2 \\ 4 \end{bmatrix} = 2 \begin{bmatrix} -1 \\ 1 \\ 2 \end{bmatrix}$$

Accordingly, *v* is an eigenvector associated with $\lambda = 2$, as found in example 1.24.

Since finding eigenvalues and eigenvectors passes through a determinant solution, some useful computational notes can be stated for a matrix *A* of order *n* as follows.

1. The characteristic polynomial always has the form.

$$c(\lambda) = \lambda^n + \cdots + (-1)^n \det(A) \tag{1.42a}$$

Besides, it is proved elsewhere that for A_3, the characteristic polynomial is given by

$$c(\lambda) = \lambda^3 - \mathrm{tr}(A)\lambda^2 - \left(-\mathrm{tr}^2(A) + \mathrm{tr}(A^2)\right)\lambda - \det(A) \tag{1.42b}$$

2. The matrix *A* is singular if and only if $\lambda = 0$ is one of its eigenvalues. It can be seen clearly from equation (1.42) since when $\det(A) = 0$, then λ can be factorized from the characteristic polynomial and set to zero.
3. The eigenvectors are not unique; in example 1.24, one can select other arbitrary constants.
4. For upper or lower triangular matrices, their eigenvalues are the elements on the diagonal.

Theorem 1.8.

Eigenvectors associated with distinct eigenvalues are always linearly independent.

The word "always" means that the required linear independence is guaranteed. In this context, for the case of multiple eigenvalues, the procedure of finding the eigenvectors should contain a rule (or a procedure) for making them linearly independent. When the matrix order is small, the linear independence can be obtained by appreciated selection of the arbitrary constants; otherwise, a safety technique should be incorporated for a large order. The more serious situation is when the matrix has many sets of multiple eigenvalues. The interested reader can refer to special references to learn about this problem. The next example illustrates a simple case of multiple eigenvalues.

Example 1.25.

Consider the matrix

$$A = \begin{bmatrix} 2 & 1 & 1 \\ 2 & 3 & 2 \\ 1 & 1 & 2 \end{bmatrix}$$

$$\det\left(\left[\lambda\begin{bmatrix} 1 & 0 & 0 \\ 0 & 1 & 0 \\ 0 & 0 & 1 \end{bmatrix} - \begin{bmatrix} 2 & 1 & 1 \\ 2 & 3 & 2 \\ 1 & 1 & 2 \end{bmatrix}\right]\right) = \lambda^3 - 7\lambda^2 + 11\lambda - 5 = (\lambda - 5)(\lambda - 1)^2 = 0$$

The three eigenvalues are $\lambda_1 = 5$, $\lambda_2 = \lambda_3 = 1$. Thus, they are not distinct. Therefore, based on theorem 1.8, linear independence is not guaranteed. For $\lambda_1 = 5$, we will determine X_1 by solving the homogeneous system $[5I - A]X_1 = 0$, that is

$$\begin{bmatrix} 3 & -1 & -1 \\ -2 & 2 & -2 \\ -1 & -1 & 3 \end{bmatrix} \begin{bmatrix} x_1 \\ x_2 \\ x_3 \end{bmatrix} = \begin{bmatrix} 0 \\ 0 \\ 0 \end{bmatrix}$$

$$\begin{bmatrix} 3 & -1 & -1 \\ -2 & 2 & -2 \\ -1 & -1 & 3 \end{bmatrix} \widetilde{\underset{\text{operations}}{\text{Row}}} \begin{bmatrix} 1 & 1 & -3 \\ 0 & 1 & -2 \\ 0 & 0 & 0 \end{bmatrix} \Rightarrow X_1 = \begin{bmatrix} c \\ 2c \\ c \end{bmatrix}, c = 1 \Rightarrow X_1 = \begin{bmatrix} 1 \\ 2 \\ 1 \end{bmatrix}$$

For $\lambda_2 = \lambda_3 = 1$, we have the system

$$\begin{bmatrix} -1 & -1 & -1 \\ -2 & -2 & -2 \\ -1 & -1 & -1 \end{bmatrix} \begin{bmatrix} x_1 \\ x_2 \\ x_3 \end{bmatrix} = \begin{bmatrix} 0 \\ 0 \\ 0 \end{bmatrix}$$

Then the equivalent system is

$$\begin{bmatrix} 1 & 1 & 1 \\ 0 & 0 & 0 \\ 0 & 0 & 0 \end{bmatrix} \begin{bmatrix} x_1 \\ x_2 \\ x_3 \end{bmatrix} = \begin{bmatrix} 0 \\ 0 \\ 0 \end{bmatrix}$$

Now we have two degrees of freedom; hence we have to select two arbitrary constants for two variables, say, $x_3 = p$, and $x_2 = q$, and hence $x_1 = -p - q$. Therefore, to have two linearly independent eigenvectors, a careful twice choosing of the set (*p*, *q*) has to be made. If we select the two sets (0, 1) and (1, 0), then the two eigenvectors become

$$X_2 = \begin{bmatrix} -1 \\ 1 \\ 0 \end{bmatrix}, X_3 = \begin{bmatrix} -1 \\ 0 \\ 1 \end{bmatrix}$$

As in example 1.24, let us check the linear dependency of the three eigenvectors. We set the matrix *P*.

$$P = \begin{bmatrix} 1 & -1 & -1 \\ 2 & 1 & 0 \\ 1 & 0 & 1 \end{bmatrix} \Rightarrow \det(P) \neq 0$$

Therefore, the three obtained eigenvectors are linearly independent. This simple procedure of choosing arbitrary constants may not be such an easy way when there are larger degrees of freedom.

Theorem 1.9.

Two matrices with the same set of distinct eigenvalues are similar.

A significant result is that if *P* is the matrix whose columns are the matrix *A*'s eigenvectors, then the similar matrix *B* is a diagonal matrix whose elements are the eigenvalues of matrix *A*. However, it should be learned that not for every matrix there exists a similar matrix. For instance, in example 1.25, we can find by direct computation.

$$\frac{1}{2}\begin{bmatrix} 3 & 1 & 1 \\ -2 & 0 & 0 \\ 1 & -1 & 1 \end{bmatrix}\begin{bmatrix} 2 & 1 & 1 \\ 2 & 3 & 2 \\ 1 & 1 & 2 \end{bmatrix}\begin{bmatrix} 1 & -1 & -1 \\ 2 & 1 & 0 \\ 1 & 0 & 1 \end{bmatrix} = \begin{bmatrix} 3 & 0 & 0 \\ 0 & 2 & 0 \\ 0 & 0 & 1 \end{bmatrix} = \mathrm{Dg}[\lambda_1 \quad \lambda_2 \quad \lambda_3]$$

The result of equation (1.41) represents a full check on the computation of the Eigen problem. This property is called in literature eigenvalue decomposition, and a great advantage is established for solving a system of first-order differential equations. The application of similarity is the primary tool to uncouple these systems. In many applications, the computation of the Eigen problem is less cumbersome for several special matrices. One particular matrix is the companion matrix, which is given by the form

$$C = \begin{bmatrix} 0 & 1 & 0 & \dots & 0 \\ 0 & 0 & 1 & \dots & 0 \\ \vdots & \vdots & \vdots & \vdots & \vdots \\ 0 & 0 & 0 & \dots & 1 \\ -a_n & -a_{n-1} & -a_{n-2} & \dots & -a_1 \end{bmatrix} \tag{1.43}$$

The characteristic polynomial can be obtained directly as

$$c(\lambda) = \lambda^n + a_1\lambda^{n-1} + a_2\lambda^{n-2} + \dots + a_{n-1}\lambda + a_n \tag{1.44}$$

Moreover, the most fantastic result is the computation of eigenvectors. For every eigenvalue λ_i, the associated eigenvector is given immediately by the vector

$$X_i = [1 \quad \lambda_i \quad \lambda_i^2 \quad \dots \quad \lambda_i^{n-1}]^T \tag{1.45}$$

The matrix $V = [X_1 \quad X_2 \quad \dots \quad X_n]$ is called a Vandermonde matrix, which is a nonsingular (eigenvectors are linearly independent) matrix satisfying

$$V^{-1}C\,V = \mathrm{Dg}[\lambda_1 \quad \lambda_2 \quad \dots \quad \lambda_n] \tag{1.46}$$

For *n*-multiple eigenvalue, i.e., $c(\lambda) = (\lambda - \alpha)^n = 0$ then we have a generalized Vandermonde matrix, which has the form

$$G = \left[V(\alpha) \quad V'(\alpha) \quad \frac{1}{2!}V''(\alpha) \quad \dots \quad \frac{1}{(n-1)}V^{(n-1)}(\alpha) \right]^T \tag{1.47}$$

where $V(\alpha) = [1 \quad \alpha \quad \alpha^2 \quad \dots \quad \alpha^{n-1}]^T$, and G is a lower triangular matrix of 1 on the diagonal. It can be shown that $J = G^{-1}CG$, where the matrix J is called the Jordan canonical matrix, and it is given by

$$J = \begin{bmatrix} \alpha & 1 & 0 & \cdots & 0 \\ 0 & \alpha & 1 & \dots & 0 \\ 0 & 0 & \alpha & & 0 \\ \vdots & \vdots & \vdots & \ddots & 0 \\ 0 & 0 & \dots & & \alpha \end{bmatrix}$$

Example 1.26.

1. Consider the companion matrix

$$C = \begin{bmatrix} 0 & 1 & 0 \\ 0 & 0 & 1 \\ -24 & -26 & -9 \end{bmatrix}$$

The characteristic equation is

$$\lambda^3 + 9\lambda^2 + 26\lambda + 24 = 0 \Rightarrow \lambda_1 = -4, \lambda_2 = -3, \lambda_3 = -2$$

The Vandermonde matrix is

$$V = \begin{bmatrix} 1 & 1 & 1 \\ -2 & -3 & -4 \\ 4 & 9 & 16 \end{bmatrix}; V^{-1} = \frac{1}{2}\begin{bmatrix} 6 & -5 & 1 \\ -16 & 12 & -2 \\ 12 & -7 & 1 \end{bmatrix}$$

$$\frac{1}{2}\begin{bmatrix} 12 & 7 & 1 \\ -16 & -12 & -2 \\ 6 & 5 & 1 \end{bmatrix}\begin{bmatrix} 0 & 1 & 0 \\ 0 & 0 & 1 \\ -24 & -26 & -9 \end{bmatrix}\begin{bmatrix} 1 & 1 & 1 \\ 4 & 3 & 2 \\ 16 & 9 & 4 \end{bmatrix} = \begin{bmatrix} -2 & 0 & 0 \\ 0 & -3 & 0 \\ 0 & 0 & -4 \end{bmatrix}$$

2. Consider the companion matrix

$$C = \begin{bmatrix} 0 & 1 & 0 \\ 0 & 0 & 1 \\ -27 & -27 & -9 \end{bmatrix}$$

The characteristic equation is

$$\lambda^3 + 9\lambda^2 + 27\lambda + 27 = (\lambda + 3)^3 = 0$$

The three eigenvalues are $\lambda_1 = \lambda_2 = \lambda_3 = -3$, and the generalized Vandermonde matrix and its inverse are

$$G = \begin{bmatrix} 1 & 0 & 0 \\ -3 & 1 & 0 \\ 9 & -6 & 1 \end{bmatrix}; \; G^{-1} = \begin{bmatrix} 1 & 0 & 0 \\ 3 & 1 & 0 \\ 9 & 6 & 1 \end{bmatrix}$$

The Jordan canonical matrix is computed from

$$\begin{bmatrix} 1 & 0 & 0 \\ 3 & 1 & 0 \\ 9 & 6 & 1 \end{bmatrix} \begin{bmatrix} 0 & 1 & 0 \\ 0 & 0 & 1 \\ -27 & -27 & -9 \end{bmatrix} \begin{bmatrix} 1 & 0 & 0 \\ -3 & 1 & 0 \\ 9 & -6 & 1 \end{bmatrix} = \begin{bmatrix} -3 & 1 & 0 \\ 0 & -3 & 1 \\ 0 & 0 & -3 \end{bmatrix}$$

The ease of computation is not limited to a companion matrix, but also many of the difficulties that occur in a general eigenvalue problem disappear with real symmetric matrices. The most useful feature of a real symmetric matrix is that it has always been real eigenvalues. Furthermore, eigenvectors associated with distinct eigenvalues of a real symmetric matrix are orthogonal. These two facts have an impact on the solution of many practical problems. Later in this chapter, we will elaborate deeply on the matter of real symmetric matrices.

Another simplification in computing the eigenvalues is found for the tridiagonal 1-dimension Toeplitz matrix (see definition 1.3). The eigenvalues can be calculated exactly from the following formula.

$$\lambda_k = a - 2\sqrt{bc}\cos\left(\frac{\pi k}{n+1}\right), k = 1,2 \dots n \tag{1.48}$$

In literature, other analytic formulas are found for calculating the 2-dimension Toeplitz matrix.

Example 1.27.

An aircraft longitudinal motion's stability can be determined efficiently from the knowledge of the eigenvalues of the mode coefficient matrix. Then the long-period and short-period modes of motion can be analyzed for dynamic stability. Let us consider the coefficient matrix of a passenger transport aircraft as

$$A = \begin{bmatrix} -0.033 & 0.0001 & 0 & -9.81 \\ 0.1680 & -0.387 & 260 & 0 \\ 55 \times 10^{-4} & -0.0064 & -0.551 & 0 \\ 0 & 0 & 1 & 0 \end{bmatrix}$$

The fourth-order characteristic polynomial is factorized usually into two multiplied quadratic polynomials that belong to these modes. Hence, the characteristic equation becomes

$$(\lambda^2 + 0.9111\,\lambda + 1.848)(\lambda^2 + 0.0599\,\lambda + 0.0056) = 0$$

Thus,

$$\lambda_1, \lambda_2 = -0.4556 \pm j1.2808$$
$$\lambda_3, \lambda_4 = -0.0299 \pm j0.0658$$

Therefore, the long-period mode, characterized by a very small damping ratio and low frequency (0.0658 rad/second), is stable, and the short-period high-frequency mode is stable too.

1.9. Cayley-Hamilton Theorem and Minimum Polynomial of a Matrix

We begin by asking ourselves the following two questions: Are functions of a matrix such as $\sin(A)$, or e^A, etc. exist? How are these functions computed for a particular matrix? The answer to the first question is yes. For the second question, it is sufficient now to say that the result is not to apply the function on each element of the matrix separately, i.e., for example,

$$\sin\left(\begin{bmatrix} a & b \\ c & d \end{bmatrix}\right) \neq \begin{bmatrix} \sin a & \sin b \\ \sin c & \sin d \end{bmatrix}$$

However, the integer power operation of a square matrix means multiplicating the matrix by itself several times. Based on this, we can define the value of a polynomial $p(x)$ at the square matrix A of order *n* as

$$p(A) = c_k A^k + c_{k-1} A^{k-1} + \cdots + c_1 A + c_0 I \tag{1.49}$$

Where the integer *k* is not necessarily equal to *n*. The sum and product operations on two polynomials in *x* are also valid on two polynomials at a matrix A, i.e.,

$$s(x) = p(x) + w(s) \;\equiv\; s(A) = p(A) + w(A) \tag{1.50a}$$

$$m(x) = p(x)w(x) \;\equiv\; m(A) = p(A)w(A) \tag{1.50b}$$

A unique polynomial of the form (1.49) is the characteristic polynomial on its matrix. For this, two mathematicians William Hamilton and Arthur Cayley gave the following theorem.

Theorem 1.10.
Given a square matrix *A* of order *n*, whose characteristic polynomial is

$$c(\lambda) = \det([\lambda I - A]) = \lambda^n + a_1\lambda^{n-1} + a_2\lambda^{n-2} + \cdots + a_{n-1}\lambda + a_n$$

Then

$$c(A) = A^n + a_1A^{n-1} + a_2A^{n-2} + \cdots + a_{n-1}A + a_nI = O_{n\times n} \tag{1.51}$$

The matrix equation (1.51) has excellent advantages in computing many expressions of functions of a square matrix. There are different approaches to prove this theorem, which can be found in mathematics references.

Example 1.28.
From equation (1.51), it holds

$$A^n = -(a_1A^{n-1} + a_2A^{n-2} + \cdots + a_{n-1}A + a_nI)$$

Thus, multiplying both sides by *A*

$$\begin{aligned} A^{n+1} &= -A(a_1A^{n-1} + a_2A^{n-2} + \cdots + a_{n-1}A + a_nI) \\ &= -(a_1A^n + a_2A^{n-1} + \cdots + a_{n-1}A^2 + a_nA) \\ A^{n+1} &= -a_1\big(-(a_1A^{n-1} + a_2A^{n-2} + \cdots + a_{n-1}A + a_nI)\big) \\ &\quad -(a_2A^{n-1} + \cdots + a_{n-1}A^2 + a_nA) \\ A^{n+1} &= (a_1^2 - a_2)A^{n-1} + \cdots + (-a_1a_{n-1} - a_n)A - a_1a_nI \end{aligned}$$

Therefore, in general, we can obtain

$$A^{n+k} = f_{n-1}(a_1, a_2, \dots)A^{n-1} + f_{n-2}(a_1, a_2, \dots)A^{n-2} + \cdots + f(a_1, a_2, \dots)A + g(a_1, a_2, \dots)I$$

For instance, let us consider

$$A = \begin{bmatrix} 2 & 1 \\ -2 & 3 \end{bmatrix}$$

It is required to compute A^3. Then the characteristic polynomial is

$$c(\lambda) = \lambda^2 - 5\lambda + 8$$

Applying the Cayley-Hamilton theorem yields

$$A^2 - 5A + 8I = \begin{bmatrix} 0 & 0 \\ 0 & 0 \end{bmatrix} \Rightarrow A^2 = 5A - 8I$$

$$A^3 = 5A^2 - 8A = 5(5A - 8I) - 8A = 17A - 40I = 17\begin{bmatrix} 2 & 1 \\ -2 & 3 \end{bmatrix} - 40\begin{bmatrix} 1 & 0 \\ 0 & 1 \end{bmatrix}$$
$$= \begin{bmatrix} -6 & 17 \\ -34 & 11 \end{bmatrix}$$

Higher power or linear combinations of different powers can be similarly computed as in the above example. However, for a square matrix A_n, whose eigenvalues $\lambda_1, \lambda_2 \ldots \lambda_n$ are distinct, the result stated by Sylvester's Identity can be used to compute a polynomial of matrices $p(A)$ directly (as given in equation (1.49)) of order $k > n$. Sylvester's Identity formula is

$$p(A) = \sum_{q=1}^{n} \frac{p(\lambda_q)}{\prod_{\substack{i=1 \\ i \neq q}}^{n} (\lambda_q - \lambda_i)} \prod_{\substack{i=1 \\ i \neq q}}^{n} (A - \lambda_i I) \tag{1.52}$$

Furthermore, it can be proved that the known series of the elementary functions (for instance, Maclaurin series) are also valid for matrices, for example,

$$e^A = I + \frac{1}{1!}A + \frac{1}{2!}A^2 + \cdots + \frac{1}{m!}A^m + \cdots$$

Thus, using the Cayley-Hamilton theorem, a closed-form expression can be quickly determined. We can write for the matrix exponential or any other elementary function a similar expression as (1.52), i.e., for instance,

$$e^A = e_{n-1}A^{n-1} + e_{n-2}A^{n-2} + \cdots + e_1 A + e_0 I$$

$$\sin(A) = s_{n-1}A^{n-1} + s_{n-2}A^{n-2} + \cdots + s_1 A + s_0$$

Only the coefficients are changed from one elementary function to another. For computing these functions, the next example is chosen.

Example 1.29.
Compute the matrix exponential of

$$A = \begin{bmatrix} -10 & 1 & 7 \\ -9 & 4 & 5 \\ -17 & 1 & 12 \end{bmatrix}$$

First, we find the characteristic polynomial.

$$c(\lambda) = \lambda^3 - 6\lambda^2 + 11\lambda - 6$$

$$c(\lambda) = 0 \Rightarrow \lambda_1 = 3, \lambda_2 = 2, \lambda_3 = 1$$

Based on the above discussion, we have

$$e^A = e_2A^2 + e_1A + e_0I$$

To determine the three coefficients, let us define a polynomial $r(x) = e_2x^2 + e_1x + e_0$. The technique is to evaluate $e^\lambda = r(\lambda)$ for each eigenvalue that is

$$e^3 = 9e_2 + 3e_1 + e_0$$
$$e^2 = 4e_2 + 2e_1 + e_0$$
$$e^1 = e_2 + e_1 + e_0$$

Solving for the coefficients,

$$[e_2 \quad e_1 \quad e_0]^T = [4.0129 \quad -7.3678 \quad 6.0732]^T$$

Substituting these coefficients, we compute

$$e^A = \begin{bmatrix} -32.6100 & -3.3549 & 24.6705 \\ -58.0897 & 24.7568 & 31.3803 \\ -47.3021 & -11.3807 & 38.0466 \end{bmatrix}$$

For the case of multiple eigenvalues, the computation procedure of the previous example has to be a little modified. For *m* multiplicity and (*n*–*m*) single eigenvalues case, the number of the coefficient algebraic equations is equal only to (*n*–*m*+1), which is sufficient to determine the unique set of coefficients. The number of equations is completed to *n* by applying

$$\frac{d^k f(x)}{dx^k}\Big|_{x=\lambda} = \frac{d^k r(x)}{dx^k}\Big|_{x=\lambda}, k = 1,2 \dots m-1 \tag{1.53}$$

Example 1.30.
Compute $\sin(A)$, where

$$A = \begin{bmatrix} 2 & 1 & 1 \\ 1 & 2 & 1 \\ 1 & 1 & 2 \end{bmatrix}$$

For this real symmetric matrix, the characteristic polynomial is

$$c(\lambda) = \lambda^3 - 6\lambda^2 + 9\lambda - 4$$

$$c(\lambda) = 0 \implies \lambda_1 = 4, \lambda_2 = 1, \lambda_3 = 1$$

Based on the Cayley-Hamilton theory, we have

$$\sin(A) = s_2 A^2 + s_1 A + s_0 I$$

Thus,

$$r(x) = s_2 x^2 + s_1 x + s_0$$

The multiplicity is 2, so the first two coefficient equations from $f(\lambda) = \sin(\lambda) = r(\lambda)$ are

$$\sin(4) = 16 s_2 + 4 s_1 + s_0$$

$$\sin(1) = s_2 + s_1 + s_0$$

The third equation is obtained by applying equation (1.53).

$$\cos(1) = 2 s_2 + s_1$$

Solving for the coefficients yields

$$\begin{bmatrix} s_2 \\ s_1 \\ s_0 \end{bmatrix} = \begin{bmatrix} -0.3577 \\ 1.2557 \\ -0.0565 \end{bmatrix}$$

Substituting these coefficients, we compute

$$\sin(A) = \begin{bmatrix} 0.3087 & -0.5328 & -0.5328 \\ -0.5328 & 0.3087 & -0.5328 \\ -0.5328 & -0.5328 & 0.3087 \end{bmatrix}$$

Another application is to compute a matrix's inverse (if it is nonsingular, i.e., the eigenvalues set does not contain a zero eigenvalue). To do that, we follow the below procedure.

$$I = -\frac{1}{a_n}(A^n + a_1 A^{n-1} + a_2 A^{n-2} + \cdots + a_{n-1} A)$$

Multiplying both sides by A^{-1} from right

$$IA^{-1} = -\frac{1}{a_n}(A^n + a_1 A^{n-1} + a_2 A^{n-2} + \cdots + a_{n-1} A)A^{-1}$$

$$A^{-1} = -\frac{1}{a_n}(A^{n-1} + a_1 A^{n-2} + a_2 A^{n-3} + \cdots + a_{n-1} I) \qquad (1.54a)$$

Finally, for the calculation of eigenvalues of A_n, two significant relations can be proved through the application of the Cayley-Hamilton theorem. They are:

$$\lambda_i(A^m) = \lambda_i^m(A), i = 1,2 \dots n \qquad (1.54b)$$

$$\lambda_i(A^{-1}) = \frac{1}{\lambda_i(A)}, i = 1,2 \dots n \qquad (1.54c)$$

Example 1.31.

Consider the matrix of example 1.24

$$A = \begin{bmatrix} 2 & 0 & 0 \\ 2 & 2 & 1 \\ 1 & 1 & 2 \end{bmatrix}$$

The characteristic polynomial is

$$c(\lambda) = \lambda^3 - 6\lambda^2 + 11\lambda - 6$$

Note that, based on theorem 1.9, the matrix in this example is similar to that in example 1.29. Both have the same determinant, trace, and eigenvalues but different inverse. Applying equation (1.54a) yields

$$A^{-1} = \frac{1}{6}(A^2 - 6A + 11I) = \frac{1}{6}\left(\begin{bmatrix} 4 & 0 & 0 \\ 9 & 5 & 4 \\ 6 & 4 & 5 \end{bmatrix} - 6\begin{bmatrix} 2 & 0 & 0 \\ 2 & 2 & 1 \\ 1 & 1 & 2 \end{bmatrix} + 11\begin{bmatrix} 1 & 0 & 0 \\ 0 & 1 & 0 \\ 0 & 0 & 1 \end{bmatrix}\right)$$

$$= \begin{bmatrix} 1/2 & 0 & 0 \\ -1/2 & 2/3 & -1/3 \\ 0 & -1/3 & 2/3 \end{bmatrix}$$

Suppose it is also desired to compute the matrix polynomial

$$p(A) = A^5 - 2A^3 - 3A$$

Since the eigenvalues are distinct $\lambda_1 = 3$, $\lambda_2 = 2$, and $\lambda_3 = 1$, we can use Sylvester's Identity directly for computing the matrix polynomial $p(A)$. We have

$$p(\lambda_1) = p(3) = 180, p(\lambda_2) = p(2) = 10, p(\lambda_3) = p(1) = -4$$

$$p(A) = \frac{p(3)}{(3-1)(3-2)}\begin{bmatrix} 0 & 0 & 0 \\ 2 & 0 & 1 \\ 1 & 1 & 0 \end{bmatrix}\begin{bmatrix} 1 & 0 & 0 \\ 2 & 1 & 1 \\ 1 & 1 & 1 \end{bmatrix}$$

$$+\frac{p(2)}{(2-3)(2-1)}\begin{bmatrix} -1 & 0 & 0 \\ 2 & -1 & 1 \\ 1 & 1 & -1 \end{bmatrix}\begin{bmatrix} 1 & 0 & 0 \\ 2 & 1 & 1 \\ 1 & 1 & 1 \end{bmatrix}$$

$$+\frac{p(1)}{(1-3)(1-2)}\begin{bmatrix} -1 & 0 & 0 \\ 2 & -1 & 1 \\ 1 & 1 & -1 \end{bmatrix}\begin{bmatrix} 0 & 0 & 0 \\ 2 & 0 & 1 \\ 1 & 1 & 0 \end{bmatrix} = \begin{bmatrix} 10 & 0 & 0 \\ 262 & 88 & 92 \\ 248 & 92 & 88 \end{bmatrix}$$

We have already learned that for a square matrix *A* of order *n*, its characteristic polynomial is equal to a zero matrix of order *n*. However, there may be other polynomials of *A* that have a zero matrix value. The following definition considers this property.

Definition 1.31. Minimal polynomials

The minimal polynomial of a matrix *A* of order *n* is a unique monic polynomial of the lowest degree (power of the variable *x*) whose value at *A* is a zero matrix. For a matrix, there may be more than one minimal polynomial but of different degrees. Usually, the minimal polynomial of *A* is denoted by $m_A(x)$.

Let the characteristic polynomial be factorized as

$$c(x) = (x - p_1)^r(x - p_2)^s \dots (x - p_n)^t$$

where $r + s + \dots + t = n$. Then, in general, there are many possible lowest degree polynomials. For instance, a polynomial $m_k(x) = (x - p_1)^r(x - p_2)^s$ has a degree $(r + s) < n$, and the question is whether this polynomial is minimal or not. That is, based on the Cayley-Hamilton theorem, if

$$m_k(A) = (A - p_1I)^r(A - p_2I)^s = O_{n\times n}$$

then, this is a minimal polynomial, and there is no other one.

The subscript *k* is used to indicate the *k*th lowest degree polynomial in the set of all possible polynomials. When the powers $r = s = \dots = t = 1$, and $p_1 \neq p_2 \neq \dots \neq p_n$, then we say that the characteristic polynomial has distinct linear factors.

In general, a minimal polynomial determination is a difficult task. There are several features, which make things less complicated. For example, two similar systems have the same minimal polynomial. For a diagonal matrix, or a matrix similar to a diagonal matrix such as the symmetric matrix, the minimal polynomial is a product of distinct linear factors. Furthermore, if $c(x)$ has distinct linear factors, then the minimal polynomial is $c(x)$ itself. The following example illustrates a workable procedure for finding the minimal polynomial.

Example 1.32.
Let the characteristic polynomial of a specific fourth-order matrix be given by

$$c(x) = \det([xI - A]) = (x-2)^3(x-3)$$

There are three possibilities for the minimal polynomial

$$m_1(x) = (x-2)(x-3)$$
$$m_2(x) = (x-2)^2(x-3)$$
$$m_3(x) = c(x)$$

To find the minimal polynomial, we follow the following procedure:

First, we compute $m_1(A) = (A-2I)(A-3I)$, if $m_1(A) = 0$, then it is the desired minimal polynomial. Otherwise, we compute $m_2(A) = (A-2I)\, m_1(x)$, and if $m_2(A) = 0$, then it is the minimal polynomial. If not, then the minimal polynomial is the characteristic polynomial itself. Of course, according to the given matrix, the results are altered.

Clearly, for matrices of high orders, the procedure of finding the minimal polynomial becomes much involved. However, since the minimal polynomial is unique, checking zero at a matrix plays a central role in minimizing the computation efforts. That is to include this condition in any used algorithm.

Example 1.33
Find the minimal polynomial of the matrix

$$A = \begin{bmatrix} 2 & 0 & 0 & 1 \\ 2 & -2 & 1 & 0 \\ -2 & 4 & 1 & 0 \\ 0 & 2 & 2 & -2 \end{bmatrix}$$

The characteristic polynomial is

$$c(x) = x^4 + x^3 - 10x^2 - 4x + 24 = (x-2)^2(x+2)(x+3)$$

The possible minimal polynomials are

$$m_1(x) = (x-2)(x+2)(x+3)$$
$$m_2(x) = (x-2)^2(x+2)(x+3) = m_1(x)(x-2) = c(x)$$

Computing the corresponding matrix polynomials $m_1(A)$ gives

$$m_1(A) = (A - 2I)(A + 2I)(A + 3I) = \begin{bmatrix} 0 & 10 & 10 & 0 \\ 0 & 4 & 4 & 0 \\ 0 & -4 & -4 & 0 \\ 0 & 0 & 0 & 0 \end{bmatrix} \neq 0$$

Thus, based on the Cayley-Hamilton theorem, the minimal polynomial is the characteristic polynomial itself. As a check, we compute

$$m_1(A)(x - 2I) = \begin{bmatrix} 0 & 10 & 10 & 0 \\ 0 & 4 & 4 & 0 \\ 0 & -4 & -4 & 0 \\ 0 & 0 & 0 & 0 \end{bmatrix} \left(\begin{bmatrix} 2 & 0 & 0 & 1 \\ 2 & -2 & 1 & 0 \\ -2 & 4 & 1 & 0 \\ 0 & 2 & 2 & -2 \end{bmatrix} - \begin{bmatrix} -2 & 0 & 0 & 0 \\ 0 & -2 & 0 & 0 \\ 0 & 0 & -2 & 0 \\ 0 & 0 & 0 & -2 \end{bmatrix} \right)$$

$$m_1(A)(x - 2I) = \begin{bmatrix} 0 & 0 & 0 & 0 \\ 0 & 0 & 0 & 0 \\ 0 & 0 & 0 & 0 \\ 0 & 0 & 0 & 0 \end{bmatrix}$$

Thus, by definition 1.31, our results are correctly computed.

1.10. Symmetric and Orthogonal Matrices

In definition 1.9, we previously defined the symmetric matrices, and due to their existence in many applications, we prefer to consider them in a separate section with related topics. Perhaps, the essential feature of symmetric matrices is stated in the next theorems.

Theorem 1.11.

For a real symmetric matrix, the eigenvalues set contains only real numbers. The symmetric positive definite matrix (a matrix that has a positive determinant and positive determinants of all minor matrices) has positive real eigenvalues.

Theorem 1.12.

Eigenvectors associated with distinct eigenvalues of a symmetric matrix are orthogonal, i.e., according to definition 1.27, all eigenvectors satisfy

$$X_i^T X_j = 0; i \neq j, \forall\, i = 1,2 \dots n, j = 1,2 \dots n$$

Consequently, the matrix whose columns are the eigenvectors (see examples 1.18 and 1.19) is orthogonal. This is a significant result in decoupling systems described by the square symmetric real matrices since the inverse of an orthogonal matrix is the same as the transpose of the matrix itself. Besides, according to formula (1.37), another significant result can be stated for real matrices in the next theorem.

Theorem 1.13.

For a real matrix A_n whose eigenvalue λ_1 is a real number, there exists an orthogonal matrix R such that

$$R^{-1}AR = R^TAR = T = \begin{bmatrix} \lambda_1 & \vdots & T_{12} \\ \cdots & \cdots & \cdots \\ 0 & \vdots & \\ \vdots & \vdots & T_{22} \\ 0 & \vdots & \end{bmatrix} \tag{1.55}$$

where T_{12} is a row of dimensions $(n-1)$ and T_{22} is a square matrix of dimensions $(n-1)$. This result can be easily proved by considering the normalized eigenvector associated with λ_1 as a unit vector for applying formula (1.37). Schur's theorem completes these results.

Theorem 1.14.

For a real matrix A_n whose eigenvalues are all real, there exists an orthogonal matrix R such that

$$R^{-1}AR = R^TAR = T \equiv \text{upper triangular matrix}$$

Since the product of orthogonal matrices results in an orthogonal matrix, the proof of the theorem shows that

$$R = R_1 \begin{bmatrix} 1 & 0 \\ 0 & R_2 \end{bmatrix} \begin{bmatrix} I_2 & 0 \\ 0 & R_3 \end{bmatrix} \cdots \begin{bmatrix} I_{m-2} & 0 \\ 0 & R_{n-1} \end{bmatrix} \tag{1.56}$$

Consequently, Schur's theorem also proves that every real symmetric matrix is orthogonal similar to a diagonal matrix. Finally, the following corollary of Schur's theorem can be established.

Corollary 1.1.

For a square matrix A and an arbitrary polynomial $p(x)$, the eigenvalues of the matrix $p(A)$ are computed directly as $p(\lambda_1), p(\lambda_2), \ldots$ where $\lambda_1, \lambda_2, \ldots$ are the eigenvalues of the matrix A.

In summary, for a real symmetric matrix A, an orthogonal matrix R can be found such that R^TAR is a diagonal matrix. Besides, for a real non-symmetric matrix A, an orthogonal matrix R can be found such that R^TAR is an upper triangular matrix. The following examples illustrate the procedure of finding such orthogonal matrices.

Example 1.34.
Consider the real matrix of example 1.25

$$A = \begin{bmatrix} 2 & 1 & 1 \\ 2 & 3 & 2 \\ 1 & 1 & 2 \end{bmatrix}$$

Find an orthogonal matrix R such that $R^T AR$ is an upper triangular

From the solution of example 1.25, it is found that the three eigenvalues are $\lambda_1 = 5$, $\lambda_2 = \lambda_3 = 1$, and the eigenvector associated with $\lambda_1 = 5$ is $X_1 = [1 \quad 2 \quad 1]^T$; thus, a unit vector is

$$X_1 = \frac{1}{\sqrt{6}} \begin{bmatrix} 1 \\ 2 \\ 1 \end{bmatrix}$$

Using the formula (1.37), an orthogonal matrix R_1 can be calculated as

$$R_1 = \begin{bmatrix} \frac{1}{\sqrt{6}} & \frac{2}{\sqrt{6}} & \frac{1}{\sqrt{6}} \\ \frac{2}{\sqrt{6}} & \frac{2-\sqrt{6}}{6-\sqrt{6}} & -\frac{2}{6-\sqrt{6}} \\ \frac{1}{\sqrt{6}} & -\frac{2}{6-\sqrt{6}} & \frac{5-\sqrt{6}}{6-\sqrt{6}} \end{bmatrix}$$

Next, we compute

$$R_1^T AR_1 = \begin{bmatrix} 5 & \frac{\sqrt{6}-2}{\sqrt{6}-1} & \frac{2}{\sqrt{6}-1} \\ 0 & 1 & 0 \\ 0 & 0 & 1 \end{bmatrix}$$

The last matrix is an upper triangular matrix, which has the eigenvalues of the matrix A lying on the diagonal. It ends the task. However, let us repeat the computation by considering the repeated eigenvalues, say, $\lambda_1 = 1$ with its associated eigenvector $X_1 = [-1 \quad 1 \quad 0]^T$ (as found in example 1.25), then a unit vector is obtained as

$$X_1 = \frac{1}{\sqrt{2}} \begin{bmatrix} -1 \\ 1 \\ 0 \end{bmatrix}$$

Again, using formula (1.37), an orthogonal matrix R_1 can be calculated as

$$R_1 = \begin{bmatrix} -\frac{1}{\sqrt{2}} & \frac{1}{\sqrt{2}} & 0 \\ \frac{1}{\sqrt{2}} & \frac{1}{\sqrt{2}} & 0 \\ 0 & 0 & 1 \end{bmatrix}$$

Now, we compute

$$R_1^T A R_1 = \begin{bmatrix} 1 & 1 & \frac{1}{\sqrt{2}} \\ 0 & 4 & \frac{3}{\sqrt{2}} \\ 0 & \sqrt{2} & 2 \end{bmatrix}$$

Nevertheless, this last matrix is not an upper triangular; therefore, we continue by arranging as it is given in formula (1.55).

$$R_1^T A R_1 = \begin{bmatrix} \lambda_2 & \vdots & & * \\ 0 & \vdots & 4 & \frac{3}{\sqrt{2}} \\ 0 & \vdots & \sqrt{2} & 2 \end{bmatrix} = \begin{bmatrix} \lambda_2 & \vdots & * \\ 0 & \vdots & \\ 0 & \vdots & A_2 \end{bmatrix}$$

And

$$A_2 = \begin{bmatrix} 4 & \frac{3}{\sqrt{2}} \\ \sqrt{2} & 2 \end{bmatrix}$$

If everything is correct, then the eigenvalues of the matrix A_2 are the remaining unused eigenvalues, i.e., $\lambda_1 = 5,\ \lambda_2 = 1$. Since the characteristic polynomial is $c(\lambda) = \lambda^2 - 6 + 5$, then the previous computation is correct. Let us select $\lambda_1 = 5$ to determine the associated eigenvector, i.e., to solve

$$\begin{bmatrix} 1 & -\frac{3}{\sqrt{2}} \\ -\sqrt{2} & 3 \end{bmatrix} \begin{bmatrix} x_1 \\ x_2 \end{bmatrix} = \begin{bmatrix} 0 \\ 0 \end{bmatrix}$$

Therefore, the eigenvector is $X_2 = [3/\sqrt{2} \quad 1]^T$, thus a unit vector is

$$X_2 = \begin{bmatrix} \frac{\sqrt{3}}{\sqrt{7}} \\ -\frac{2}{\sqrt{7}} \end{bmatrix}$$

Using the formula (1.37), an orthogonal matrix R_2 can be computed as

$$R_2 = \begin{bmatrix} \frac{3}{\sqrt{11}} & \frac{\sqrt{2}}{\sqrt{11}} \\ \frac{\sqrt{2}}{\sqrt{11}} & -\frac{3}{\sqrt{11}} \end{bmatrix}, \text{and } R_2^T A_2 R_2 = \begin{bmatrix} 5 & -\frac{1}{\sqrt{2}} \\ 0 & 1 \end{bmatrix}$$

According to the result in equation (1.56), we have

$$R = R_1 \begin{bmatrix} 1 & 0 \\ 0 & R_2 \end{bmatrix} = \begin{bmatrix} -\frac{1}{\sqrt{2}} & \frac{1}{\sqrt{2}} & 0 \\ \frac{1}{\sqrt{2}} & \frac{1}{\sqrt{2}} & 0 \\ 0 & 0 & 1 \end{bmatrix} \begin{bmatrix} 1 & 0 & 0 \\ 0 & \frac{3}{\sqrt{11}} & \frac{\sqrt{2}}{\sqrt{11}} \\ 0 & \frac{\sqrt{2}}{\sqrt{11}} & -\frac{3}{\sqrt{11}} \end{bmatrix} = \begin{bmatrix} -\frac{1}{\sqrt{2}} & \frac{3}{\sqrt{22}} & \frac{1}{\sqrt{11}} \\ \frac{1}{\sqrt{2}} & \frac{3}{\sqrt{22}} & \frac{1}{\sqrt{11}} \\ 0 & \frac{\sqrt{2}}{\sqrt{11}} & -\frac{3}{\sqrt{11}} \end{bmatrix}$$

Finally, we compute to have the upper triangular matrix $R^T AR$

$$R^T AR = \begin{bmatrix} 1 & \frac{4}{\sqrt{11}} & -\frac{1}{\sqrt{22}} \\ 0 & 5 & -\frac{1}{\sqrt{2}} \\ 0 & 0 & 1 \end{bmatrix}$$

It should be pointed out that for the given matrix, two upper triangular matrices are obtained through different computation procedures. However, the first procedure is shorter than the second. The distributions of the eigenvalues on the diagonal follow the computation procedure. Thus, another orthogonal matrix can be found with the distribution of the third eigenvalues. Moreover, computation accepts multiple eigenvalues cases.

For the next example, we repeat the computation for a real symmetric matrix.

Example 1.35.

Given the real symmetric matrix of example 1.30

$$A = \begin{bmatrix} 2 & 1 & 1 \\ 1 & 2 & 1 \\ 1 & 1 & 2 \end{bmatrix}$$

Find an orthogonal matrix R such that $R^T AR$ is a diagonal matrix.

From the solution of example 1.30, it is found that the three eigenvalues are $\lambda_1 = 4$, $\lambda_2 = \lambda_3 = 1$. First, we determine the eigenvector associated with $\lambda_1 = 4$. We write

$$\begin{bmatrix} 2 & -1 & -1 \\ -1 & 2 & -1 \\ -1 & -1 & 2 \end{bmatrix}\begin{bmatrix} x_1 \\ x_2 \\ x_3 \end{bmatrix} = \begin{bmatrix} 0 \\ 0 \\ 0 \end{bmatrix}$$

The equivalent row-reduction system is

$$\begin{bmatrix} 1 & 0 & -1 \\ 0 & 1 & -1 \\ 0 & 0 & 0 \end{bmatrix}\begin{bmatrix} x_1 \\ x_2 \\ x_3 \end{bmatrix} = \begin{bmatrix} 0 \\ 0 \\ 0 \end{bmatrix}$$

A nontrivial solution is $x_1 = x_2 = x_3 = 1$. A unit eigenvector is

$$X_1 = \frac{1}{\sqrt{3}}\begin{bmatrix} 1 \\ 1 \\ 1 \end{bmatrix}$$

Using the formula (1.37), an orthogonal matrix R can be calculated as

$$R = \begin{bmatrix} \frac{1}{\sqrt{3}} & \frac{1}{\sqrt{3}} & \frac{1}{\sqrt{3}} \\ \frac{1}{\sqrt{3}} & \frac{2-\sqrt{3}}{3-\sqrt{3}} & -\frac{1}{3-\sqrt{3}} \\ \frac{1}{\sqrt{3}} & -\frac{1}{3-\sqrt{3}} & \frac{2-\sqrt{3}}{3-\sqrt{3}} \end{bmatrix}$$

Compute the product $R^T AR$

$$R^T AR = \begin{bmatrix} 4 & 0 & 0 \\ 0 & 1 & 0 \\ 0 & 0 & 1 \end{bmatrix} = \mathrm{Dg}([4 \quad 1 \quad 1])$$

Note that the same result can be obtained if we select another value from the eigenvalue set; however, the computation will be longer.

As a final explanatory note, the computation for a transposed matrix is much more rigorous than for the inverse matrix. For the real symmetric matrix, the required diagonal matrix is better obtained by formula (1.23) than to compute the eigenvector column matrix's inverse.

Definition 1.32. Rayleigh ratio

For a matrix A of order *n*, the Rayleigh ratio is the ratio λ_k given by

$$\lambda_k = \frac{X^T AX}{X^T X} \tag{1.57}$$

where X is not an eigenvector. However, for a real symmetric matrix, if X is an eigenvector, then λ_k is the associated eigenvalue. Let us check the Rayleigh ratio for the matrix in example 1.35. Take the eigenvector $X = [1 \quad 1 \quad 1]^T$, then

$$\lambda_k = \frac{[1 \quad 1 \quad 1]\begin{bmatrix} 2 & 1 & 1 \\ 1 & 2 & 1 \\ 1 & 1 & 2 \end{bmatrix}\begin{bmatrix} 1 \\ 1 \\ 1 \end{bmatrix}}{[1 \quad 1 \quad 1]\begin{bmatrix} 1 \\ 1 \\ 1 \end{bmatrix}} = \frac{12}{3} = 4 = \lambda_1$$

As it is explained in definition 1.32.

1.11. Singular Values, Singular Vectors, and Norms of Matrices

As the reader notices, the scalar-valued functions are defined only on square matrices like determinant, trace, and eigenvalues, while others are defined for square and non-square matrices like rank. Another scalar-valued function defined for square as well as non-square matrices is the singular values. Singular values and singular vectors are found in different applications. For instance, they represent a significant topic in robust control theory and image processing. Singular values relate to eigenvalues according to the following definition.

Definition 1.33. Singular values
For a complex rectangular matrix $A_{m \times n}$, the singular values are given by

$$\sigma_i(A) = +\sqrt{\lambda_i(A^* A)};\ i = 1,2 \ldots k \tag{1.58}$$

where A^* is the Hermitian matrix and $k = \min(m, n)$. Usually, the singular values are ordered $\sigma_i \geq \sigma_{i+1}$, and in especial literature, the maximum and minimum singular values are denoted as $\overline{\sigma}(A)$, and $\underline{\sigma}(A)$ respectively.

For real matrices ($A^* = A$), it holds

$$\sum_{i=1}^{k} \sigma_i^2 = \mathrm{tr}\,(A^T A) \tag{1.59}$$

The following are the most essential properties of the singular values:

i. $\sigma_i(A) = \sigma_i(A^T)$
ii. $\overline{\sigma}(A) = 1/\underline{\sigma}(A^{-1})$ and $\underline{\sigma}(A) = 1/\overline{\sigma}(A^{-1})$.
iii. $\overline{\sigma}(cA) \leq |c|\overline{\sigma}(A)$.
iv. $\overline{\sigma}(A+B) \leq \overline{\sigma}(A) + \overline{\sigma}(B)$.
v. $\overline{\sigma}(AB) \leq \overline{\sigma}(A).\overline{\sigma}(B)$.
vi. $\max\{\overline{\sigma}(A), \overline{\sigma}(B)\} \leq \overline{\sigma}(AB) \leq \sqrt{2}\max\{\overline{\sigma}(A), \overline{\sigma}(B)\}$.
vii. $\underline{\sigma}(A) \leq |\lambda_i(A)| \leq \overline{\sigma}(A)$.
viii. The ill-conditioned case is defined by $\mathrm{de}(A) \gg 1$, and a well-conditioned case is defined by $\mathrm{de}(A) \approx 1$.

Definition 1.34. Singular vectors and singular value decomposition (SVD)
For a rectangular matrix $A_{m\times n}$, with a singular value $\sigma(A)$, there is an associated pair of singular vectors, say $u_{m\times 1}$, and $v_{n\times 1}$ such that

$$Av = \sigma u \tag{160a}$$

$$A^T u = \sigma v \tag{160b}$$

With the singular values on a diagonal matrix $D_{m\times n}$ and two-column matrices of the singular vectors u and v as $U_{m\times m}$ and $V_{n\times n}$ matrices, it holds

$$AV = UD \tag{160c}$$

$$A^T U = VD \tag{160d}$$

Since U and V are orthogonal, the singular value decomposition (SVD) becomes.

$$A = UDV^T = \sum_{i=1}^{k} \sigma_i u_i v_i^T \tag{160e}$$

The matrix D is diagonal with singular values on the first diagonal. It has the form

$$D = \begin{bmatrix} \overline{\sigma}(A) & 0 & \dots & 0 & \\ 0 & \ddots & \dots & 0 & \\ \vdots & & \underline{\sigma}(A) & & 0 \\ 0 & \dots & & 0 & \end{bmatrix}$$

The importance of equation (1.60e) is very significant in reducing the number of pixels required to generate an acceptable visible image after illuminating the small values of the singular values corresponding to the image's noise or insignificant details. The reader can refer to the mathematical and discrete signal processing literature to learn more about singular value decomposition and its image processing applications. The next examples illustrate the necessary computation procedure.

Example 1.36.
Given the two complex matrices

$$A = \begin{bmatrix} 3 & 5+6j \\ 2 & 3j \\ 2+3j & 2-j \end{bmatrix}, B = \begin{bmatrix} 1-3j & 3+7j \\ 2 & 3j \\ 4 & 1-j \end{bmatrix}$$

The Hermitian matrix is

$$A^* = \begin{bmatrix} 3 & 2 & 2-3j \\ 5-6j & -3j & 2+j \end{bmatrix}, B^* = \begin{bmatrix} 1+3j & 2 & 4 \\ 3-7j & -3j & 1+j \end{bmatrix}$$

The products A^*A and B^*B are the matrices,

$$A^*A = \begin{bmatrix} 26 & 16+16j \\ 16-16j & 75 \end{bmatrix}, \ B^*B = \begin{bmatrix} 30 & -14+18j \\ -14-18j & 69 \end{bmatrix}$$

The characteristic of the matrix A^*A equation is

$$(\lambda - 26)(\lambda - 75) - (16+16j)(16-16j) = 0$$
$$\lambda^2 - 101\lambda + 1438 = 0 \implies \lambda_1 = 83.8504, \lambda_2 = 17.1496$$

Therefore,

$$\overline{\sigma}(A) = \sqrt{83.8504} = 9.1569, \underline{\sigma}(A) = \sqrt{17.1496} = 4.1412$$

Similarly, for the matrix B^*B, we find that $\overline{\sigma}(A) = 8.9165$, and $\underline{\sigma}(A) = 4.4154$. Quickly, it can be computed to show the validity of the formula (1.58). Also, for some of the mentioned properties, we compute

i. $$(A^*A)^{-1} = \frac{1}{1438}\begin{bmatrix} 75 & -16-16j \\ -16+16j & 26 \end{bmatrix} = M$$

The characteristic equation of the matrix *M* is

$$\lambda^2 - 0.0702\lambda + 0.0007 = 0 \implies \lambda_1 = 0.583, \lambda_2 = 0.0119$$

Therefore,

$$\overline{\sigma}(A) = \sqrt{0.0583} = 0.2414 = \frac{1}{4.1412} = 1/\underline{\sigma}(A^{-1}),$$

$$\underline{\sigma}(A) = \sqrt{0.0119} = 0.10908 = 9.167 = 1/\overline{\sigma}(A^{-1})$$

ii. Assume a constant $c = 1.45$, then the characteristic equation of the matrix 1.45 (A^*A) is

$$\lambda^2 - 106.5\lambda + 3.0234 = 0 \Rightarrow \lambda_1 = 121.5831, \lambda_2 = 24.8669$$

Therefore,

$$\overline{\sigma}(A) = \sqrt{121.5831} = 11.0265 < 1.45\,\overline{\sigma}(A)$$

iii. Let the sum $(A^*A + B^*B) = Q = \begin{bmatrix} 56 & 2+34j \\ 2-34j & 144 \end{bmatrix}$, thus

$$\lambda^2 - 200\lambda + 6904 = 0 \Rightarrow \lambda_1 = 155.6417, \lambda_2 = 44.3583$$

$$\overline{\sigma}(A+B) = \sqrt{155.6417} = 12.4756 < 9.1569 + 8.9156$$

Similarly, we can show other properties. The computation of singular vectors requires a simultaneous solution of equations (1.60a and b), which in turn needs the solution of $(m+n)$ homogenous algebraic equations. To avoid the computation with complex numbers, we will solve this problem in the next example for a real rectangular matrix.

Example 1.37.
Given the matrix

$$A = \begin{bmatrix} 2 & 0 \\ 1 & 3 \\ 2 & 5 \end{bmatrix}$$

Determine the singular values and singular vector pair associated with the maximum singular value.

$$A^TA = \begin{bmatrix} 9 & 13 \\ 13 & 34 \end{bmatrix} \Rightarrow c(\lambda) = \lambda^2 - 43\lambda + 137 = 0, \lambda_1 = 39.5347, \lambda_2 = 3.4653$$

Thus, $\overline{\sigma}(A) = 6.2877$, $\underline{\sigma}(A) = 1.8615$. Let us now find the singular vector pair associated with the value $\overline{\sigma}(A)$.

Let the first singular vector pair (associated with $\overline{\sigma}(A)$) are denoted as

$$u = \begin{bmatrix} \alpha_1 \\ \alpha_2 \\ \alpha_3 \end{bmatrix}, v = \begin{bmatrix} \alpha_4 \\ \alpha_5 \end{bmatrix}$$

Substituting in equations (1.60a and b) yields the following augmented homogenous system.

$$\begin{bmatrix} -6.2877 & 0 & 0 \;\; 2 \;\; 0 & & \\ 0 & -6.2877 & 0 \;\; 1 \;\; 3 & & \\ 0 & 0 & -6.2877 & 2 & 5 \\ 2 & 1 & 2 & -6.2877 & 0 \\ 0 & 3 & 5 & 0 & -6.2877 \end{bmatrix} \begin{bmatrix} \alpha_1 \\ \alpha_2 \\ \alpha_3 \\ \alpha_4 \\ \alpha_5 \end{bmatrix} = \begin{bmatrix} 0 \\ 0 \\ 0 \\ 0 \\ 0 \end{bmatrix}$$

Following the procedure of solving the homogenous system, we can determine

$$[\alpha_1 \quad \alpha_2 \quad \alpha_3 \quad \alpha_4 \quad \alpha_5]^T = [0.1246 \quad 0.5013 \quad 0.8563 \quad 0.3917 \quad 0.9201]^T$$

Thus, the first singular vector pair is

$$u_1 = \begin{bmatrix} 0.1246 \\ 0.5013 \\ 0.8563 \end{bmatrix}, v_1 = \begin{bmatrix} 0.3917 \\ 0.9201 \end{bmatrix}$$

Similarly, the second singular vector pair can be found as

$$u_2 = \begin{bmatrix} -0.9885 \\ 0.1370 \\ 0.0636 \end{bmatrix}, v_2 = \begin{bmatrix} -0.3917 \\ 0.9201 \end{bmatrix}$$

Thus, we can check for the validity of the summation in equation (1.60e) by computing

$$A = 6.2877 \begin{bmatrix} 0.1246 \\ 0.5013 \\ 0.8563 \end{bmatrix} [0.3917 \quad 0.9201] + 1.8615 \begin{bmatrix} -0.9885 \\ 0.1370 \\ 0.0636 \end{bmatrix} [-0.3917 \quad 0.9201]$$

As a final note, the decomposition equation's solution (1.60e) requires a unique technique to determine the two matrices U and V.

Definition 1.35. Norm of a matrix

Let X be a vector space, then a real-valued function denoted by $||.||$ that is defined on X is said to be a norm on X, if it satisfies the following properties:

i. The positivity, i.e., $||X|| \geq 0$.
ii. The positive definiteness $\|X\| = 0$, which holds if and only if $x = 0$.
iii. The homogeneity $\|aX\| = |a|||X||$, for any scalar constant a.
iv. The triangular inequality $\|x + y\| \leq \|x\| + \|y\|, \forall\, x \in X, \text{and } y \in Y$.

Consider a vector of *n* elements $x \in C^n$, and then the vector *p*-norm is given as

$$||X||_p = \left[\sum_{i=1}^{n} |x_i|^p\right]^{1/p} \quad ; 1 \leq p \leq \infty \tag{1.61}$$

In particular, there are three used norms when *p* is equal to 1, or 2, or infinity, which are respectively given by

$$||X||_1 = \sum_{i=1}^{n} |x_i| \tag{1.62a}$$

$$||X||_2 = \sqrt{\sum_{i=1}^{n} |x_i|^2} \tag{1.62b}$$

$$||X||_\infty = \max_{1 \leq i \leq n}\{|x_i|\} \tag{1.62c}$$

The use of a specific norm 1.62a-c depends on the particular application. Similarly, let us consider a matrix A whose elements $a_{ij} \in C^{m\times n}$, then the matrix norm induced by a vector of *p*-norm is defined by

$$\|A\|_p = \sup_{x\neq 0}\left\{\frac{\|AX\|_p}{\|X\|_p}\right\} \tag{1.63}$$

The abbreviation 'sup' stands for the supremum word (supremum is defined as the least upper bound of a set S, defined as a quantity M such that no member of the set exceeds M). For a matrix $A_{m\times n}$, the following basic four types of numerical norms are used:

i. The 1-norm defined by (absolute sum by columns)

$$||A||_1 = \max_{1 \le j \le n} \sum_{i=1}^{m} |a_{ij}| \tag{1.64a}$$

ii. The 2-norm defined by

$$||A||_2 = \max_{i=1,2\ldots n} \{\sqrt{\lambda_i(A^T A)}\} = \overline{\sigma}(A) \tag{1.64b}$$

iii. The ∞-norm defined by (absolute sum by rows)

$$||A||_\infty = \max_{1 \le i \le m} \sum_{j=1}^{n} |a_{ij}| \tag{1.64c}$$

iv. The Frobenius norm defined

$$||A||_{Fr} = \sqrt{\sum_{i=1}^{n} \sum_{j=1}^{m} |a_{ij}|^2} \tag{1.64d}$$

Obviously, for square symmetric matrices, the two norms $||A||_1$ and $||A||_\infty$ are equal. The following two relations summarize an actual result that relates the singular values to the norm given by relation 1.63:

i. The maximum singular value is also given by

$$\overline{\sigma}(A) = \sup_{x \neq 0} \left\{ \frac{||AX||_2}{||X||_2} \right\} = ||A||_2 \tag{1.65a}$$

This relation of computing the $||A||_2$ norm measures how much the mapping included by this matrix can stretch vectors. Therefore, the second norm of a matrix can be computed by evaluating the maximum singular value rather than calculating the supremum.

ii. The minimum singular value is also given by

$$\underline{\sigma}(A) = \inf_{x \neq 0} \left\{ \frac{||AX||_2}{||X||_2} \right\} = ||A^{-1}||_2^{-1} \tag{1.65b}$$

The abbreviation ‘inf’ stands for the supremum word infimum word (it is the greatest lower bound of set S, defined as a quantity m such that no number of the set is less than m). This relation considers how much a matrix can shrink vectors.

Example 1.38.
Given a matrix A and an arbitrary X vector as

$$A = \begin{bmatrix} 0 & 1 \\ -2 & -4 \end{bmatrix}, X = \begin{bmatrix} a \\ b \end{bmatrix}$$

Following the procedure of example 1.37, we can determine the two singular values as

$$\overline{\sigma}(A) = 4.5615528$$

$$\underline{\sigma}(A) = 0.4384471$$

$$AX = \begin{bmatrix} 0 & 1 \\ -2 & -4 \end{bmatrix}\begin{bmatrix} a \\ b \end{bmatrix} = \begin{bmatrix} b \\ -2a - 4b \end{bmatrix}$$

$$||AX|| = \sqrt{b^2 + (-2a - 4b)^2} = \sqrt{17b^2 + 4b^2 + 16ab}$$

$$||X|| = \sqrt{a^2 + b^2}$$

$$\sup_{x \neq 0} \left\{ \frac{\sqrt{17b^2 + 4b^2 + 16ab}}{\sqrt{a^2 + b^2}} \right\}$$

Let

$$f = \frac{17b^2 + 4b^2 + 16ab}{a^2 + b^2}$$

It can be easily shown that

$$\frac{\partial f}{\partial a} = -\frac{\partial f}{\partial b} \Rightarrow b^2 - \frac{26}{16}ab - a^2 = 0$$

Thus, for the arbitrary value of a the value of b is then

$$b = \begin{cases} 2.1009705\ a \\ -0.4759705\ a \end{cases}$$

Then, $\sup_{x \neq 0}\{.\}$ is given by $b = 2.1009705\ a$, and

$$AX = \begin{bmatrix} 2.1009705\ a \\ -2a - 4(2.1009705\ a) \end{bmatrix} = \begin{bmatrix} 2.1009705a \\ 10.403882\ a \end{bmatrix}$$

$$||AX||_2 = 10.613898\ a, \qquad ||X||_2 = 2.32568164\ a$$

$$\frac{||AX||_2}{||X||_2} = 4.56155 = \overline{\sigma}(A)$$

Similarly, if we select $b = -0.4759705\, a$, then we will end with a result equal to $\underline{\sigma}(A)$.

Definition 1.36. Condition number

Consider a matrix A, then the ratio of the maximum to minimum stretching is named a condition number $\kappa(A)$ for inversion, and it is given by

$$\kappa(A) = \frac{\overline{\sigma}(A)}{\underline{\sigma}(A)} = ||A||_2||A^{-1}||_2 \tag{1.66}$$

The calculation for $\kappa(A)$ is not restricted to the 2-norm, but the norm could be any of the previously defined norms. A large or a small condition number of a matrix states an ill-conditioned or well-conditioned matrix, respectively. Equivalently, one can say that near a singular matrix is that matrix with a small condition number. The accuracy of the numerical solution of the system $AX = K$ depends on the condition number; accepted accuracy is that with a small value of the number $\kappa(A)$. The convergence of the numerical solution degrades as this number increases. In other words, matrix A is ill-conditioned if relatively small changes in matrix A can cause more considerable changes in the system solution.

Example 1.39.

Given the matrix of Example 1.24, thus

$$A = \begin{bmatrix} 2 & 0 & 0 \\ 2 & 2 & 1 \\ 1 & 1 & 2 \end{bmatrix}, A^{-1} = \begin{bmatrix} 1/2 & 0 & 0 \\ -1/2 & 2/3 & -1/3 \\ 0 & -1/3 & 2/3 \end{bmatrix}$$

Then, we have

$$A^T A = \begin{bmatrix} 9 & 5 & 4 \\ 5 & 5 & 4 \\ 4 & 4 & 5 \end{bmatrix}$$

Then we can compute

$$\lambda(A^T A) = \{15.3814, 2.7753, 0.8433\}$$

Hence, the corresponding maximum and minimum singular values are, respectively, $\overline{\sigma}(A) = 3.9219$, and $\underline{\sigma}(A) = 0.9183$

Therefore, according to equation (1.64b) and equation (1.66), we obtain

$$||A||_2 = 3.9219, \text{ and } \kappa(A) = 4.27$$

To continue, we can similarly find the singular values of the inverse A^{-1}, thus

$$||A^{-1}||_2 = 1.0889$$

Again, according to equation (1.66), the condition number will be also 4.27. However, for non-square matrices, only the first alternative can be utilized to compute the condition number. Furthermore, the reader will prefer to use the first alternative for computing the condition number even with square matrices; the matrix dimension plays the role.

Example 1.40.

In the image-processing field, consider the simple image shown in Figure 1.5a. It is only in two colors, black and white. This image can be represented by a matrix of dimensions (8×9). The zeros' entries represent the black color in the matrix, and the ones' entries represent the white color. The total number of entries that represent the image is 72. Illustrate the use of singular value decomposition to make the image compression and noise or blurring reduction.

The matrix representing the image is given as

$$A_{8\times 9} = \begin{bmatrix} 1 & 1 & 1 & 1 & 1 & 1 & 1 & 1 & 1 \\ 1 & 1 & 1 & 1 & 1 & 1 & 1 & 1 & 1 \\ 1 & 1 & 0 & 0 & 0 & 0 & 0 & 1 & 1 \\ 1 & 1 & 1 & 0 & 1 & 1 & 0 & 1 & 1 \\ 1 & 1 & 1 & 0 & 1 & 1 & 0 & 1 & 1 \\ 1 & 1 & 1 & 0 & 1 & 1 & 0 & 1 & 1 \\ 1 & 1 & 1 & 1 & 1 & 1 & 1 & 1 & 1 \\ 1 & 1 & 1 & 1 & 1 & 1 & 1 & 1 & 1 \end{bmatrix}$$

The singular value decomposition can be used by following the procedure in example 1.37. Due to the larger matrix dimensions, the numerical procedure should be incorporated. The following results are obtained by invoking the particular function (***svd***) from the MATLAB toolbox (all the results below are obtained using MATLAB environments). The decomposition, according to equation (1.60e), is

$$A = 7.4848\begin{bmatrix}-0.3935\\-0.3935\\-0.1988\\-0.3373\\-0.3373\\-0.3373\\-0.3935\\-0.3935\end{bmatrix}\begin{bmatrix}-0.3720\\-0.3720\\-0.3454\\-0.2103\\-0.3454\\-0.3454\\-0.2103\\-0.3720\\-0.3720\end{bmatrix}^T + 1.8749\begin{bmatrix}0.3028\\0.3028\\-0.4309\\-0.3863\\-0.3863\\-0.3863\\0.3028\\0.3028\end{bmatrix}\begin{bmatrix}-0.2020\\-0.2020\\0.0279\\0.6460\\0.0279\\0.0279\\0.6460\\-0.2020\\-0.2020\end{bmatrix}^T +$$

$$+1.2093\begin{bmatrix}0.0593\\0.0593\\0.8802\\-0.2653\\-0.2653\\-0.2653\\0.0593\\0.0593\end{bmatrix}\begin{bmatrix}0.2661\\0.2661\\-0.4618\\0.1963\\-0.4618\\-0.4618\\0.1963\\0.2661\\0.2661\end{bmatrix}^T + 0$$

Only three nonzero singular values are obtained, and hence only three $u's$ vectors and corresponding $v's$ vectors are needed to implement in equation (1.60e). The number of the used elements in the decomposition is $3(8+9)+3=54$ compared to the 72 elements representing the image. It represents a sort of image compression. The saving in the number of elements will be increased as the original image number of pixels increases.

Figure 1.5a. A simple black and white image (Example 1.40).

Next, let us consider the image with one single pixel non-white pixel (assumed an image noise) for which 0.8 (this value is taken but not necessarily matches the figure) value is assigned instead of 1 in the corresponding right position ($a_{9\times8} = 0.8$) as shown in Figure 1.5b. In this case, four singular values are obtained.

$$a_1 = 7.4562, a_2 = 1.8884, a_3 = 1.207, a_4 = 0.1498$$

The idea of noise reduction is to neglect the smallest singular value over the set, to compute the decomposition, to use equation (1.61e) to obtain a computing matrix A_c , and then after applying the numerical fixing of the final result as follows

$$A_c = 7.4562\begin{bmatrix}-0.3935\\-0.3935\\-0.1990\\-0.3385\\-0.3385\\-0.3385\\-0.3950\\-0.3835\end{bmatrix}\begin{bmatrix}-0.3735\\-0.3735\\-0.3454\\-0.2103\\-0.3454\\-0.3454\\-0.2103\\-0.3735\\-0.3632\end{bmatrix}^T + 1.8884\begin{bmatrix}0.2972\\0.2972\\-0.4367\\-0.3833\\-0.3833\\-0.3833\\0.2972\\0.3218\end{bmatrix}\begin{bmatrix}-0.1976\\-0.1976\\0.0366\\0.6425\\0.0366\\0.0366\\0.6425\\-0.1976\\-0.2317\end{bmatrix}^T +$$

$$+1.2070\begin{bmatrix}0.0729\\0.0729\\0.8762\\-0.2680\\-0.2680\\-0.2680\\0.0729\\0.0297\end{bmatrix}\begin{bmatrix}0.2656\\0.2656\\-0.4603\\0.2057\\-0.4603\\-0.4603\\0.2057\\0.2656\\0.2607\end{bmatrix}^T + 0$$

$$A_c = \begin{bmatrix}1.0126 & 1.0126 & 0.9999 & 0.9991 & 0.9999 & 0.9999 & 0.9991 & 1.0126 & 0.9629\\1.0126 & 1.0126 & 0.9999 & 0.9991 & 0.9999 & 0.9999 & 0.9991 & 1.0126 & 0.9629\\0.9981 & 0.9981 & 0.0000 & 0.0001 & 0.0000 & 0.0000 & 0.0001 & 0.9981 & 1.0057\\1.0000 & 1.0000 & 1.0000 & 0.0000 & 1.0000 & 1.0000 & 0.0000 & 1.0000 & 1.0001\\1.0000 & 1.0000 & 1.0000 & 0.0000 & 1.0000 & 1.0000 & 0.0000 & 1.0000 & 1.0001\\1.0000 & 1.0000 & 1.0000 & 0.0000 & 1.0000 & 1.0000 & 0.0000 & 1.0000 & 1.0001\\1.0126 & 1.0126 & 0.9999 & 0.9991 & 0.9999 & 0.9999 & 0.9991 & 1.0126 & 0.9626\\0.9624 & 0.9624 & 1.0003 & 1.0028 & 1.0003 & 1.0003 & 1.0028 & 0.9624 & 0.9119\end{bmatrix}$$

Finally, for this simple example, the result is obtained by adding a suitable small value to all elements of A_c and fixing the value to obtain the original image matrix.

$$A = \text{fix}(A_c + 0.1)$$

In other words, the picture is reproduced after eliminating the incidental pixel. In practice, there are several sophisticated methods and algorithms to handle the noise reduction process.

Figure 1.5b. A simple black and white image with one-pixel noise (Example 1.40).

1.12. MATRIX FACTORIZATIONS

Factorization is to make a product of several matrices (at least two matrices) equal to one single matrix. For example, a product of some unitary matrices and an upper triangular matrix is a type of factorization. The factorization of matrices has several advantages in simplifying computations with large matrices and some practical algorithms are developed in many fields. There are three basic types of matrix factorization, which use triangular (upper or lower) matrices. They are:

Definition 1.37. Cholesky factorizations

The Cholesky factorization expresses a symmetric matrix (or Hermitian matrix) as the triangular matrix product and its transpose. However, not every symmetric matrix can be factorized in this way. A positive definite symmetric matrix always has a Cholesky factorization. Therefore, symmetric positive definite matrix A is factorized as

$$A = U^T U \tag{1.67}$$

where U is an upper triangular matrix.

Example 1.41.

Given a positive definite real symmetric matrix A, it is desired to find its Cholesky factorized matrix U.

$$A = \begin{bmatrix} 10 & 2 & 4 & 6 \\ 2 & 30 & 8 & 9 \\ 4 & 8 & 15 & 0 \\ 6 & 9 & 0 & 20 \end{bmatrix}$$

Let the upper triangular matrix U has the form

$$U = \begin{bmatrix} a & b & c & d \\ 0 & e & f & g \\ 0 & 0 & h & i \\ 0 & 0 & 0 & j \end{bmatrix}$$

Considering the factorization in equation (1.66), thus finding the upper matrix U requires a solution of 10 quadratic algebraic equations (generally, they are $n(n+1)/2$ equations). For our example, they and their solutions are

$$a^2 = 10 \Rightarrow a = \sqrt{10} = 3.1623$$

$$ab = 2 \Rightarrow b = \frac{2}{\sqrt{10}} = 0.6325$$

$$ac = 4 \Rightarrow c = \frac{4}{\sqrt{10}} = 1.2649$$

$$ad = 6 \Rightarrow d = \frac{6}{\sqrt{10}} = 1.8974$$

$$b^2 + e^2 = 30 \Rightarrow e = \sqrt{30 - 0.6325^2} = 5.4406$$

$$bc + ef = 8 \Rightarrow f = \frac{8 - (0.6325)(1.2649)}{5.4406} = 1.3234$$

$$bd + eg = 9 \Rightarrow g = 9 - \frac{(0.6325)(1.8974)}{5.4406} = 1.4337$$

$$c^2 + f^2 + h^2 = 15 \Rightarrow h = \sqrt{15 - 1.2649^2 - 1.3234^2} = 3.4310$$

$$cd + fg + hi = 0 \Rightarrow i = -\frac{(1.2649)(1.8974) + (1.3234)(1.4337)}{3.431} = -1.2591$$

$$d^2 + g^2 + i^2 + j^2 = 20 \Rightarrow j = \sqrt{20 - 1.8974^2 - 1.4337^2 - (-1.2591)^2} = 3.572$$

The solutions for these equations can be carried out sequentially. Then the upper matrix U is

$$U = \begin{bmatrix} 3.1623 & 0.6325 & 1.2649 & 1.8974 \\ 0 & 5.4406 & 1.3234 & 1.4337 \\ 0 & 0 & 3.413 & -1.2591 \\ 0 & 0 & 0 & 3.572 \end{bmatrix}$$

For a general *n*-order real symmetric matrix, a numerical algorithm can be developed to obtain Cholesky factorization.

Definition 1.38. LU factorization

It is also called Gaussian elimination. It is to express any square matrix A as the product of a permutation of a lower triangular matrix L with ones on its diagonal and an upper triangular matrix U, i.e.,

$$A = LU \tag{1.68}$$

This permutation has theoretical as well as computational importance. It can be easily shown that the determinant and inverse of matrix A can be computed as

$$\det(A) = \det(L)\,.\det(U) = \det(U) \tag{1.69a}$$

$$A^{-1} = U^{-1}.L^{-1} \tag{1.69b}$$

These relations provide a fast computation schema for computing the determinant and inverse.

Example 1.42.

Given a real square matrix A, it is desired to find LU factorization.

$$A = \begin{bmatrix} 9 & 4 & 1 \\ 2 & 3 & 5 \\ 1 & 1 & 8 \end{bmatrix}$$

Let the lower and upper triangular matrices have the forms

$$L = \begin{bmatrix} 1 & 0 & 0 \\ a & 1 & 0 \\ b & c & 1 \end{bmatrix}, U = \begin{bmatrix} d & e & f \\ 0 & g & h \\ 0 & 0 & i \end{bmatrix}$$

There are nine unknowns to be found from nine algebraic equations, which are easily decoupled. Considering equation (1.67), we can similarly find these unknowns as the previous example. The result is

$$L = \begin{bmatrix} 1 & 0 & 0 \\ \frac{2}{9} & 1 & 0 \\ \frac{1}{9} & \frac{5}{19} & 1 \end{bmatrix}, U = \begin{bmatrix} 9 & 4 & 1 \\ 0 & \frac{19}{9} & \frac{43}{9} \\ 0 & 0 & \frac{1134}{171} \end{bmatrix}$$

Definition 1.39. Orthogonal or QR factorizations

The orthogonal or QR factorization expresses any rectangular matrix $A_{m\times n}$ as a product of an orthogonal matrix Q (see definition 1.28) and an upper triangular matrix R. Thus, a full-size variant has the product form

$$A = QR \tag{1.70}$$

There are four variants of QR factorization; full and economic size and with and without column permutation. Here we will consider only the full-size variant. In this variant, the Q matrix is a unitary matrix of order $(m \times m)$, and the matrix R is an upper triangular matrix of order $(m \times n)$. The unknown elements of the two matrices are $m(n+2)$. The required $m(n+2)$ equations are obtained from equating the products to the corresponding $(m \times n)$ known elements of the matrix A, plus n equations obtained from the orthonormal property of each row of the matrix Q, plus other n equations obtained from the orthogonal property of each row pair of the matrix Q.

All equations are quadratic (products of two unknown elements). The solution of this system of equations is not as easy as that found with the previous two factorizations. Useful utilization of the QR factorization is the transformation of an over-determined (more equations than unknowns) linear system into an equivalent triangular system. The next example illustrates the set of the required equations (not a way of obtaining the QR factorization). Since the algebraic quadratic equations are coupled, numerical methods are usually used to obtain two matrices' factorization.

Example 1.43.
Given the rectangular matrix

$$A = \begin{bmatrix} 1 & 5 \\ 3 & 4 \\ 8 & 1 \end{bmatrix}$$

Let the matrices Q and R be

$$Q = \begin{bmatrix} a & b & c \\ d & e & f \\ g & h & i \end{bmatrix}; \; R = \begin{bmatrix} j & k \\ 0 & l \\ 0 & 0 \end{bmatrix},; \; A = QR$$

The 12 equations are:

$$aj = 1, ak + bl = 5, dj = 3$$

$$dk + el = 4, gj = 8, gk + hl = 1$$

$$a^2 + d^2 + g^2 = 1, b^2 + e^2 + h^2 = 1, c^2 + f^2 + i^2 = 1$$

$$ab + de + gh = 0, ac + df + gi = 0$$

$$bc + ef + hi = 0$$

As it is said before, such coupled quadratic equations can be solved numerically using different software. For instance, with the MATLAB software, the solution is given as

$$Q = \begin{bmatrix} 0.1162 & 0.8048 & -0.582 \\ 0.3487 & 0.5156 & 0.7827 \\ 0.930 & -0.2939 & -0.2208 \end{bmatrix}; \; R = \begin{bmatrix} 8.6023 & 2.9062 \\ 0 & 5.7926 \\ 0 & 0 \end{bmatrix}$$

The reader can check that these numbers in the two matrices satisfy the above equations; small error values can be noted with the four decimals computation.

Example 1.44.

Use the QR factorization to solve the overdetermined system whose coefficient matrix is that of example 1.43, i.e., to find an equivalent solution

$$\begin{bmatrix} 1 & 5 \\ 3 & 4 \\ 8 & 1 \end{bmatrix} \begin{bmatrix} x_1 \\ x_2 \end{bmatrix} = \begin{bmatrix} 10 \\ 20 \\ 30 \end{bmatrix}$$

Let us write the system in matrix form as

$$AX = B \implies QRX = B$$

Multiply both sides from left by Q^{-1}, and delete the last zero row of the matrix R and the last row of the product $Q^{-1}B$, we get

$$\text{Min}_{3,1}(R)X = \text{Min}_3(Q^{-1}B)$$

Multiply both sides from left by $[\text{Min}_{3,1}(R)]^{-1}$, we get the variable vector X as

$$X = [\text{Min}_{3,1}(R)]^{-1} \, \text{Min}_3(Q^{-1}B)$$

$$Q^{-1}B = \begin{bmatrix} 0.1162 & 0.8048 & -0.582 \\ 0.3487 & 0.5156 & 0.7827 \\ 0.930 & -0.2939 & -0.2208 \end{bmatrix}^{-1} \begin{bmatrix} 10 \\ 20 \\ 30 \end{bmatrix} = \begin{bmatrix} 36.0365 \\ 9.5442 \\ 3.2107 \end{bmatrix}$$

$$X = \begin{bmatrix} 8.6023 & 2.9062 \\ 0 & 5.7827 \end{bmatrix}^{-1} \begin{bmatrix} 36.0365 \\ 9.5442 \end{bmatrix}$$

Substituting the values obtained in example 1.34 yields $x_1 = 3.6316, x_2 = 1.6505$. If compared with the original system, we find

$$\begin{bmatrix} 1 & 5 \\ 3 & 4 \\ 8 & 1 \end{bmatrix} \begin{bmatrix} 3.6327 \\ 1.6472 \end{bmatrix} = \begin{bmatrix} 11.8687 \\ 17.487 \\ 30.7089 \end{bmatrix} \Leftrightarrow \begin{bmatrix} 10 \\ 20 \\ 30 \end{bmatrix}$$

Therefore, generally, for an overdetermined system $A_{m\times n}\, X = B$, the solution of the equivalent system is

$$X = [\mathrm{Min}_{m,1}(R)]^{-1}\,\mathrm{Min}_m(Q^{-1}B)$$

Definition 1.40. Positive definite matrices and Quadratic functions

A symmetric matrix A_n is positive definite if and only if $a_{ii} > 0$, and for arbitrary nonzero real vector, $X_{n\times 1}$ it holds the inequality $X^T AX > 0$. The function $f(x) = X^T AX$ is a positive quadratic function. Consequently, it can be shown that the following propositions.

1. The inverse A^{-1} exists.
2. Principle leading sub-matrices $A_k, k = 1,2 \ldots n$ are also positive definite and have positive determinants $d_1, d_2, \ldots d_n;\ d_1 = a_{11}$.
3. If S has a full column rank, then $S^T AS$ is positive definite.
4. All eigenvalues of A are real positive values.
5. The decomposition $A = LDL^T$ is valid, where L is a lower triangular matrix, and D is a diagonal matrix of elements $d_1, d_2/d_1\,, d_3/d_2 \ldots d_n/d_{n-1}$.

Simi-positive definite is defined when it holds $X^T AX \geq 0$, and in an analog way, the negative definite matrix can be defined. Such matrices play a significant role in different applications.

Example 1.45.

Given the real symmetric matrix

$$P = \begin{bmatrix} 2 & -1 & 0 \\ -1 & 2 & -1 \\ 0 & -1 & 2 \end{bmatrix}$$

It is a positive definite matrix, and

1. Since $\det(P) \neq 0$, then P^{-1} exists.
2. $d_1 = 2, d_2 = \det\left(\begin{bmatrix} 2 & -1 \\ -1 & 2 \end{bmatrix}\right) = 3, d_3 = \det(P) = 4.$
3. Let $S = \begin{bmatrix} 1 & 2 \\ 0 & 1 \\ 1 & 0 \end{bmatrix}$, which is a full column rank, then $S^T PS = \begin{bmatrix} 4 & 2 \\ 2 & 6 \end{bmatrix} > 0.$
4. $c(\lambda) = \lambda^3 - 6\lambda^2 + 10\lambda - 4 = 0 \Longrightarrow \lambda_1 = 2, \lambda_2 = 2 + \sqrt{2}, \lambda_3 = 2 - \sqrt{2};$

   all eigenvalues are positive.

5. The decomposition is

$$p = \begin{bmatrix} 1 & 0 & 0 \\ -1/2 & 1 & 0 \\ 0 & -2/3 & 1 \end{bmatrix} \begin{bmatrix} 2 & 0 & 0 \\ 0 & 3/2 & 0 \\ 0 & 0 & 4/3 \end{bmatrix} \begin{bmatrix} 1 & -1/2 & 0 \\ 0 & 1 & -2/3 \\ 0 & 0 & 1 \end{bmatrix}$$

6. Let an arbitrary vector $X = [x_1 \quad x_2 \quad x_3]^T$, then the quadratic function $X^T PX$ is

$$X^T PX = 2x_1^2 + 2x_2^2 + 2x_3^2 - 2x_1x_2 - 2x_2x_3$$

$$X^T PX = 2(x_1 - \frac{1}{2}x_2)^2 + \frac{3}{2}(x_2 - \frac{3}{2}x_3)^2 + \frac{4}{3}x_3^2 > 0$$

As a final note, there are still several important related topics, which are not included in this chapter. The representation theory for linear transformation and some computational techniques for polynomials and matrices are examples.

EXERCISES

Sections 1.1-1.5

1.1. Consider the following matrices

$$A = \begin{bmatrix} 1 & 0 \\ 4 & -3 \\ 2 & 1 \end{bmatrix}, B = \begin{bmatrix} 2 & 7 & 5 \\ 1 & 0 & 3 \\ -1 & -2 & -3 \end{bmatrix}, C = \begin{bmatrix} 2 & 4 & 8 \\ 3 & 6 & 9 \end{bmatrix}, D = \begin{bmatrix} 5 & 0 \\ 0 & 3 \end{bmatrix}, K = \begin{bmatrix} 1 \\ 3 \\ 2 \end{bmatrix}$$

$$H = \begin{bmatrix} 1 & 1+j & -1 \\ 2-j & 4 & 3j \\ 0 & 3+3j & 9 \end{bmatrix}$$

i. Determine $L(B), U(B), Dg(AC), and\ B^T$
ii. Compute the following expressions $CA + 2D$, $A + 5C^T$, $DC + \mathrm{tr}(AC + 2B)C$, $(B - AC)^T$, and $3D - \mathrm{Min}_{23}(B)$.
iii. Use the formulas (1.17 and 1.21a) to compute the determinant, adjoint, and inverse of the B matrix, respectively.
iv. Write the linear system defined by the matrix equation $BX = K$, where X is the variable vector.
v. Write the conjugate and tranjugate or Hermitian conjugate.

1.2. Use the coding table in example 1.8 to send the message S "LETTER IS SENT" that is manipulated by the formula

$$S = C(M + M^T + 3\, I_{4\times 4}), \text{where } C = \begin{bmatrix} 1 & 0 & 1 & 2 \\ 2 & 3 & 0 & 1 \\ 3 & 0 & 0 & 1 \\ 5 & 1 & 3 & 1 \end{bmatrix}$$

1.3. Given the matrix

$$A = \begin{bmatrix} 1 & 2 & 5 \\ -2 & -1 & 2 \\ -1 & 0 & 4 \end{bmatrix}$$

If $A = B + C$, where B is symmetric and C is singular skew-symmetric, determine one possible set of B and C.

1.4. Use the elementary operations to find the determinants and of the following matrices.

$$A = \begin{bmatrix} 1 & 2 & 3 \\ 4 & 5 & 6 \\ 7 & 8 & 9 \end{bmatrix}, B = \begin{bmatrix} 1 & 2 & 3 & 4 \\ 5 & 6 & 7 & 8 \\ 9 & 10 & 11 & 12 \\ 13 & 14 & 15 & 16 \end{bmatrix}$$

1.5. For a matrix, A_n and a constant c, show the property

$$\det(cA) = c^n \det(A)$$

1.6. Apply the procedure described by formula (1.23) to determine the inverse of the matrix A; check your result by applying $AA^{-1} = I_3$.

$$A = \begin{bmatrix} 1 & 2 & 6 \\ 2 & 5 & 15 \\ 6 & 15 & 46 \end{bmatrix}$$

1.7. If matrices B and C are invertible show that

$$\begin{bmatrix} A & B \\ C & 0 \end{bmatrix}^{-1} = \begin{bmatrix} 0 & C^{-1} \\ B^{-1} & -B^{-1}AC^{-1} \end{bmatrix}$$

Apply this result to find the inverse of the matrix

$$\begin{bmatrix} 2 & -1 & 3 & -1 \\ 1 & -1 & 5 & -2 \\ 3 & 2 & 0 & 0 \\ 4 & 3 & 0 & 0 \end{bmatrix}$$

1.8. Use elementary row reduced operations to find a nonsingular matrix P such that $P = A^{-1}B$, where

$$A = \begin{bmatrix} 1 & 0 & 1 \\ 3 & 1 & 0 \\ 1 & 0 & 0 \end{bmatrix}, B = \begin{bmatrix} 2 & 0 & 1 \\ -1 & 3 & -1 \\ 4 & 1 & 1 \end{bmatrix}$$

1.9. Determine the row-reduced echelon matrix for

$$E = \begin{bmatrix} 2 & 2 & 4 & 4 & 6 \\ 4 & 1 & 5 & 1 & 3 \\ 1 & 3 & 4 & 3 & 0 \\ 0 & 2 & 2 & 1 & -3 \end{bmatrix}$$

1.10. Given the matrix

$$A = \begin{bmatrix} 1 & 2 & -1 & 2 \\ -2 & -5 & 3 & 0 \\ 1 & 0 & 1 & 10 \end{bmatrix}$$

Determine the inverse matrix of the matrix $[\mathrm{Min}_{11}(A^T)\ \mathrm{Min}_{34}(A)]$

1.11. Determine the rank of the following matrices

$$M = \begin{bmatrix} 1 & 2 & 3 \\ 10 & 14 & 18 \\ 4 & 5 & 6 \end{bmatrix}, N = \begin{bmatrix} 1 & 1 & 1 & 1 \\ 1 & 3 & 2 & 4 \\ 2 & 0 & 1 & -1 \end{bmatrix}, P = \begin{bmatrix} 2 & 4 & 0 & 8 & 0 & 0 & 14 \\ 1 & 2 & 1 & 6 & 0 & 0 & 5 \\ 0 & 0 & 0 & 0 & 0 & 1 & 3 \\ 0 & 0 & 0 & 0 & 1 & 0 & 4 \\ 3 & 6 & 0 & 12 & 0 & 0 & 21 \end{bmatrix}$$

1.12. Consider the linear system $AX = K$ given by

$$x_1 + 2x_2 + x_3 = 9$$
$$x_1 + x_2 + 2x_3 = 6$$
$$2x_1 + x_2 + x_3 = 8$$

Show that the coefficient matrix A is nonsingular. Solve for the vector X by applying $X = A^{-1}K$, and by applying the augmented matrix $[A \vdots K]$ to its echelon matrix.

1.13. Solve the four-variable matrix equation

$$\begin{bmatrix} 2 & -1 & 3 & -1 \\ 1 & -1 & 4 & -2 \\ 3 & 2 & 1 & 4 \\ 4 & -3 & 3 & -3 \end{bmatrix} X = \begin{bmatrix} 7 \\ 5 \\ 31 \\ -5 \end{bmatrix}$$

1.14. Consider the linear system $AX = K$ given by

$$3x_1 + 7x_2 + Px_3 = 4$$
$$2x_1 + 3x_2 + 4x_3 = 5$$
$$x_1 + x_2 = 1$$

Find all values of P that make the system consistent. What value of P gives $x_3 = 1$?

1.15. Find the value of the constant k that makes the following system consistent.

$$\begin{bmatrix} 11 & -4 & -7 \\ 7 & -2 & -5 \\ 10 & -4 & -6 \end{bmatrix} X = \begin{bmatrix} 1 \\ 1 \\ k \end{bmatrix}$$

1.16. Find a general solution for each of the following homogenous systems.

$$i. \begin{bmatrix} 0 & 2 & 10 & 8 \\ 1 & 1 & 7 & 3 \\ 3 & 2 & 16 & 5 \end{bmatrix} \begin{bmatrix} x_1 \\ x_2 \\ x_3 \\ x_4 \end{bmatrix} = \begin{bmatrix} 0 \\ 0 \\ 0 \end{bmatrix}, ii. \begin{bmatrix} 8 & 6 & 2 \\ 4 & 3 & 1 \\ 2 & 3 & 4 \end{bmatrix} \begin{bmatrix} x_1 \\ x_2 \\ x_3 \end{bmatrix} = \begin{bmatrix} 0 \\ 0 \\ 0 \end{bmatrix}$$

1.17. The ethane is burned in the presence of oxygen to produce water and carbon dioxide as given by the chemical equation:

$$aC_2H_6 + bO_2 \rightarrow cH_2O + dCO_2$$

Use the theory of solving linear algebraic systems to find the balance weights.

1.18. Determine whether the following column vectors are linearly independent or linearly dependent.

$$x_1 = \begin{bmatrix} 1 \\ 1 \\ 0 \\ 0 \end{bmatrix}, x_2 = \begin{bmatrix} 1 \\ 1 \\ 0 \\ 1 \end{bmatrix}, x_3 = \begin{bmatrix} 0 \\ 0 \\ 1 \\ 1 \end{bmatrix}, x_4 = \begin{bmatrix} 0 \\ 1 \\ 1 \\ 1 \end{bmatrix}$$

1.19. Show that the following three-row vectors are linearly independent.

$$x_1 = [1 \quad 2 \quad 3], x_2 = [4 \quad 5 \quad 6], x_3 = [7 \quad 8 \quad 9]$$

1.20. Solve the matrix equation given by

$$\begin{bmatrix} 2 & 1 & 0 \\ 1 & 1 & 2 \\ 1 & 1 & 1 \end{bmatrix} \begin{bmatrix} x_{11} & x_{12} & x_{13} \\ x_{21} & x_{22} & x_{23} \\ x_{31} & x_{32} & x_{33} \end{bmatrix} = \begin{bmatrix} 1 & 2 & 3 \\ 4 & 2 & -1 \\ 1 & 1 & 2 \end{bmatrix}$$

1.21. The following two matrices have equal determinants and traces. Show that they are not similar.

$$A = \begin{bmatrix} 1 & 1 \\ 0 & 1 \end{bmatrix}, B = \begin{bmatrix} 1 & 0 \\ 0 & 1 \end{bmatrix}$$

1.22. Are the following two matrices similar?

$$A = \begin{bmatrix} 4 & 3 \\ 2 & 1 \end{bmatrix}, B = \begin{bmatrix} 2 & 1 \\ 3 & 5 \end{bmatrix}$$

1.23. A standard technique to show that two matrices A and B are equivalent is to use the following procedure.

$$\begin{bmatrix} A & \vdots & I \\ \cdots & \vdots & \cdots \\ I & \vdots & \end{bmatrix} \widetilde{R,C} \begin{bmatrix} B & \vdots & P \\ \cdots & \vdots & \cdots \\ Q & \vdots & \end{bmatrix}$$

where P is a record of the row operations used, and Q is a record of column operations. The two matrices P and Q are nonsingular matrices that satisfy $B = PAQ$. Apply this technique for

$$A = \begin{bmatrix} 1 & 2 & 0 \\ 3 & 1 & 1 \\ 1 & 0 & 1 \end{bmatrix}, B = \begin{bmatrix} 6 & 3 & 7 \\ 7 & 8 & 6 \\ 6 & 6 & 5 \end{bmatrix}$$

Check your answer by direct multiplication.

1.24. Use row and column operations to show that

$$\begin{bmatrix} 1 & 2 & -1 & 2 \\ -2 & -5 & 3 & 0 \\ 1 & 0 & 1 & 10 \end{bmatrix} \widetilde{R,C} \begin{bmatrix} I_2 & 0 \\ 0 & 0 \end{bmatrix}$$

1.25. If it is given that $\begin{bmatrix} A \\ B \end{bmatrix}$ can be reduced to $\begin{bmatrix} I \\ M \end{bmatrix}$ by elementary column operations, then $M = BA^{-1}$. For the following two matrices, check the validity of this result.

$$A = \begin{bmatrix} 1 & 0 & 1 \\ 3 & 1 & 0 \\ 1 & 0 & 0 \end{bmatrix}, B = \begin{bmatrix} 2 & 3 & 4 \\ 4 & 3 & 1 \\ 8 & 6 & 2 \end{bmatrix}$$

1.26. It can be proved easily that if $PAQ = I$, then $A^{-1} = QP$. For

$$A = \begin{bmatrix} 1 & 1 & 2 & 3 \\ 0 & 1 & 2 & 3 \\ 2 & 2 & 2 & 3 \\ 3 & 3 & 3 & 3 \end{bmatrix}$$

Obtain the two nonsingular matrices P and Q by applying

$$\begin{bmatrix} A & \vdots & I_4 \\ \cdots & \vdots & \cdots \\ I_4 & \vdots & \end{bmatrix} \widetilde{R,C} \begin{bmatrix} I_4 & \vdots & P \\ \cdots & \vdots & \cdots \\ Q & \vdots & \end{bmatrix}$$

Sections 1.6-1.7

1.27. Given the following three-row vectors that are spanned in $\mathcal{R}_{3\times1}$

$$X_1 = [-6 \quad 5 \quad 6], \qquad X_2 = [2 \quad -1 \quad 1], \qquad X_3 = [-2 \quad 2 \quad 2]$$

Show that the vector $x = [-12 \quad 10 \quad 15]$ is in $\text{Span}\{X_1, X_2\, X_3\}$, then describe this vector in terms of the basis X_1, X_2, and X_3.

1.28. Given the following four row-vectors

$$X_1 = [2 \quad 2 \quad -2 \quad 0], X_2 = [-3 \quad -5 \quad 0 \quad -2], X_3 = [3 \quad 3 \quad 1 \quad 1], X_4 = [0 \quad 0 \quad 3 \quad 4]$$

Does the $\text{Span}\{X_1, X_2, X_3, X_4\} = \mathcal{R}_{1\times4}$?

1.29. Given the following two row vectors

$$X_1 = [1 \quad 2 \quad -1], \qquad X_2 = [2 \quad 4 \quad -2]$$

Dose the vector $x = [-1 \quad 3 \quad 6]$ belong to the subspace of $\mathcal{R}_{1\times3}$ spanned by $\{X_1, X_2\}$?

1.30. Given the following four-row vectors

$$X_1 = [1 \quad 3 \quad 0 \quad 1], X_2 = [2 \quad 2 \quad 2 \quad 1], X_3 = [3 \quad 1 \quad 4 \quad 1], X_4 = [1 \quad 4 \quad 1 \quad 1]$$

Determine the value of p that makes the vector $x = [1 \quad 7 \quad 1 \quad p]$ in $\text{Span}\{X_1, X_2, X_3, X_4\}$.

1.31. Consider the polynomial space of 3rd-degree $\mathcal{P}$ spanned by $x^3 + 2x^2 + 1, x^2 - 2, x^3 + x$. Show that the polynomial $x^2 - x + 3$ belongs to the space $\mathcal{P}$. Also, determine whether the polynomial $4x^3 - 3x + 5$ belongs to the subset of $\mathcal{P}$.

1.32. Are the following three polynomials linearly independent?

$$11x^2 + x + 2, 2x^2 + 10x + 3, x^2 + 2x - 8$$

1.33. Determine whether the following vectors from $\mathcal{R}_{4\times 1}$

$$X_1 = [1 \quad -1 \quad 2 \quad 0], \qquad X_2 = [1 \quad 1 \quad 2 \quad 0], \qquad X_3 = [3 \quad 0 \quad 0 \quad 1],$$

1.34. Show that the following vectors are linearly independent of $C[0,5]$.

$$f(t) = e^t, g(t) = e^{3t}, h(t) = e^{-2t}$$

1.35. Construct an orthogonal matrix P that its first column is the vector $x = [3/5 \quad 0 \quad 4/5]^T$. Check your answer.

1.36. For the linear transformation **M**: $\mathcal{R}_{3\times 1} \to \mathcal{R}_{3\times 1}$ given by

$$\mathbf{M} = \begin{bmatrix} x_1 \\ x_2 \\ x_3 \end{bmatrix} = \begin{bmatrix} x_1 + 2x_2 \\ x_1 + 3x_2 - 4x_3 \\ 5x_2 - x_3 \end{bmatrix}$$

Determine $2\mathbf{M}(\delta) + 3\mathbf{M}(\mathbf{M}(\delta))$, for the vector $\delta = [-1 \quad 1 \quad -1]$

1.37. Is the transform **T** linear?

$$T = \begin{bmatrix} 3x_1 - 2x_3 \\ 7x_2 \\ x_1 + 5x_2 - 3x_3 \end{bmatrix}$$

1.38. Given the transform M and a vector α, find a vector δ such that $M(\delta) = \alpha$, where

$$M = \begin{bmatrix} x_1 + x_2 \\ 3x_1 - 7x_2 + x_3 \\ x_1 - 3x_2 + 2x_3 \end{bmatrix}, \qquad \alpha = \begin{bmatrix} 1 \\ 2 \\ 3 \end{bmatrix}$$

1.39. Given **T** as a matrix transformation $\mathcal{R}_{3\times 1} \to \mathcal{R}_{3\times 1}$ with the following matrix

$$A = \begin{bmatrix} 1 & -1 & 0 \\ -1 & 2 & 3 \\ 0 & 3 & 0 \end{bmatrix}$$

For $\alpha = [1 \quad -1 \quad 1]^T$, compute $\mathbf{T}(\mathbf{T}(\alpha))$

Sections 1.8-1.9

1.40. Find all eigenvalues and the associated eigenvectors for the following two matrices.

$$A = \begin{bmatrix} 5 & 1 \\ 4 & 8 \end{bmatrix}, \qquad M = \begin{bmatrix} 1 & 0 & -4 \\ 0 & 5 & 4 \\ -4 & 4 & 3 \end{bmatrix}$$

1.41. Consider the following symmetric matrix

$$A = \begin{bmatrix} 4 & -1 & 1 \\ -1 & 4 & -1 \\ 1 & -1 & 4 \end{bmatrix}$$

Show that one possible P matrix (whose columns are the linearly independent eigenvectors of A) is given by

$$P = \begin{bmatrix} 1 & -1 & 1 \\ 1 & 0 & -1 \\ 0 & 1 & 1 \end{bmatrix}$$

Suggest another possible matrix and check that $P^{-1}AP = Dg(\lambda_1, \lambda_2, \lambda_3)$

1.42. The coefficient matrix of an aircraft longitudinal motion is given by

$$A = \begin{bmatrix} -0.017 & 0.026 & 0 & -9.81 \\ -0.0143 & -1.02 & 1 & 0 \\ 0 & 2.06 & -1.12 & 0 \\ 0 & 0 & 1 & 0 \end{bmatrix}$$

Is the aircraft dynamically stable?

1.43. Assume that the characteristic polynomial of a square matrix A is

$$c(\lambda) = c_0\lambda^n + c_1\lambda^{n-1} + \cdots + c_{n-1}\lambda + c_n; \; c_0 = 1$$

Show that

$$c_1 = -S_1,$$

$$c_2 = -(c_1 S_1 + S_2)/2$$

$$c_k = -\frac{c_{k-1}S_1 + c_{k-2}S_2 + \cdots + c_1 S_{k-1} + S_k}{k}, k = 1,2 \dots n$$

where $S_1 = \mathrm{tr}(A), S_2 = \mathrm{tr}(A^2), \dots S_n = \mathrm{tr}(A^n)$

1.44. Use the results given in exercise 1.43 to determine the characteristic polynomial of the matrix

$$A = \begin{bmatrix} 1 & 3 & 1 & 4 \\ 0 & 1 & 2 & 1 \\ 1 & 0 & 2 & 3 \\ 2 & 3 & 0 & 2 \end{bmatrix}$$

Use equation (1.40) to check your answer.

1.45. For the matrix

$$A = \begin{bmatrix} 2 & 3 \\ 5 & 4 \end{bmatrix}$$

Use the two formulas (1.54b) and (1.45c) to find the eigenvalues of the matrices A^3, and A^{-1}.

1.46. Find the eigenvalues and eigenvectors of the following Toeplitz matrix.

$$T = \begin{bmatrix} 1 & 2 & 0 & 0 \\ 3 & 1 & 2 & 0 \\ 0 & 3 & 1 & 2 \\ 0 & 0 & 3 & 1 \end{bmatrix}$$

1.47. Consider the following companion matrix.

$$C = \begin{bmatrix} 0 & 1 & 0 \\ 0 & 0 & 1 \\ -6 & -11 & -6 \end{bmatrix}$$

i. Determine the corresponding Vandermonde matrices.
ii. Use the Cayley-Hamilton theorem to compute C^{-1}.

1.48. For the singular matrix A, which is given by

$$A = \begin{bmatrix} -2 & 1 & 1 \\ 1 & -2 & 1 \\ 1 & 1 & -2 \end{bmatrix}$$

i. Find all eigenvalues and associated eigenvectors.
ii. For the system $AX = [a \quad b \quad c]^T$, find a condition for the system solution.
iii. Use the Cayley-Hamilton theorem to compute $p(A) = A^5 + 4A^4 + 28A$, compute the eigenvalues also of the matrix $p(A)$.

iv. Compute $S(A) = \cos(A) + \sin(A)$.

1.49. Consider the matrix

$$S = \begin{bmatrix} 2 & 1 & 1 \\ 1 & 4 & 3 \\ -1 & -1 & 0 \end{bmatrix}$$

Use Sylvester's identity to find the matrix polynomial

$$p(S) = S^4 + 5S^2 - 6I$$

1.50. Find the minimum polynomials of the following two matrices.

$$A = \begin{bmatrix} 1 & 0 & 1 \\ 0 & 1 & 2 \\ 1 & 2 & 5 \end{bmatrix}, B = \begin{bmatrix} 17 & 8 & 12 & 14 \\ 46 & -22 & -35 & 41 \\ -2 & 1 & 4 & -4 \\ 4 & -2 & -2 & 3 \end{bmatrix}$$

1.51. Consider the matrix

$$A = \begin{bmatrix} 10 & 8 & 1 \\ 3 & 1 & 2 \\ 4 & 5 & 3 \end{bmatrix}$$

Based on theorem 1.9, determine a companion matrix similar to A.

Sections 1.10-1.12

1.52. Given the real matrix

$$A = \begin{bmatrix} -2 & 1 & 0 \\ -4 & 0 & 3 \\ 0 & 1 & -2 \end{bmatrix}$$

Find an orthogonal matrix R such that $R^T AR$ is an upper triangular

1.53. Given the real symmetric matrix

$$S = \begin{bmatrix} 4 & -1 & 1 \\ -1 & 4 & -1 \\ 1 & -1 & 4 \end{bmatrix}$$

Find an orthogonal matrix R such that $R^T SR$ is a diagonal matrix whose diagonal elements are the eigenvalues of the matrix S.

Check the validity of Rayleigh's ratio.

1.54. Test the properties of singular values as in section 1.11 for the following complex matrix.

$$P = \begin{bmatrix} 1 & 2 \\ 2-3j & 4j \\ 1 & 2j \end{bmatrix}$$

1.55. Find the minimum and maximum singular values of the complex matrix

$$C = \begin{bmatrix} 1-j & 3j \\ 2 & 3 \\ 1+j & 2-j \end{bmatrix}$$

1.56. Given the matrix

$$A = \begin{bmatrix} 3 & 1 & 0 \\ 2 & 0 & 1 \end{bmatrix}$$

Determine the singular values and the singular vector pair associated with the minimum singular value.

1.57. Find the minimum and maximum singular values of the square matrix.

$$B = \begin{bmatrix} 2 & 1 & -1 \\ 0 & 3 & -2 \\ 1 & 0 & -3 \end{bmatrix}$$

1.58. Given the vector $X = [1 \quad 4 \quad -1 \quad -2 \quad 3]$, determine $||X||_p, p = 1,2,\text{and } \infty$.

1.59. Given the matrix

$$A = \begin{bmatrix} 3 & -1 & 4 \\ 1 & 3 & -2 \\ 0 & 2 & 3 \end{bmatrix}$$

Determine$||A||_p, p = 1, 2, \infty$, and Frobenius norm. By direct computation, verify equation (1.65).

1.60. Compute the condition number for the following two matrices.

$$A = \begin{bmatrix} 1 & 3 & 4 \\ 4 & 5 & 6 \\ -15 & 6 & 9 \end{bmatrix}, B = \begin{bmatrix} 1 & 2 \\ 4 & 1 \\ 3 & 0 \end{bmatrix}$$

1.61. Given a real square matrix A, determine the Cholesky and the LU factorizations.

$$A = \begin{bmatrix} 6 & 27 & 50 \\ 4 & 13 & 19 \\ 2 & 5 & 6 \end{bmatrix}$$

1.62. Determine all quadratic equations that are obtained from the orthogonal QR factorization of the following matrices.

$$P = \begin{bmatrix} -6 & 3 \\ -12 & 69 \\ 6 & 25 \end{bmatrix}, W = \begin{bmatrix} 10 & 3 & 12 \\ 8 & 5 & 22 \end{bmatrix}$$

1.63. Find an equivalent solution of the following over-determinant system.

$$\begin{bmatrix} 6 & 6 \\ 12 & 48 \\ -6 & 10 \end{bmatrix} \begin{bmatrix} x_1 \\ x_2 \end{bmatrix} = \begin{bmatrix} 10 \\ 25 \\ 40 \end{bmatrix}$$

Given the following orthogonal factorization

$$\begin{bmatrix} 0.4082 & -0.2909 & 0.8653 \\ 0.8165 & 0.5403 & -0.2036 \\ -0.4082 & 0.7896 & 0.4581 \end{bmatrix} \begin{bmatrix} 14.6969 & 37.5588 \\ 0 & 32.0832 \\ 0 & 0 \end{bmatrix} = \begin{bmatrix} 6 & 6 \\ 12 & 48 \\ -6 & 10 \end{bmatrix}$$

Chapter 2

DIFFERENTIAL EQUATIONS – PART I

2.1. INTRODUCTION

Differential equations are the tool to model the variation of physical quantities such as displacement, velocity, temperature, and many others with the time variation or space coordinates. As in most scientific disciplines, in most engineering problems, differential equations have particular significance in different application fields. The operation behavior of many devices or appliances such as accelerometers, gyroscopes, electric furnaces, heat chambers, etc., is obtained through the solution of particular differential equations. Furthermore, some processes like chemical distillation or concentration of dissolved materials are described mathematically also by differential equations.

Most of the physical systems are described by a dependent variable, say *y*, and its derivatives up to a specific integer value, say *n*, in the form of the equation

$$f\left(y, y', y'' \dots y^{(n)}; x\right) = 0 \tag{2.1}$$

where f is a specific mathematical function, x is the independent variable, n is the differential equation order. Mathematically, we use the notations

$$y' = \frac{dy}{dx}, y'' = \frac{d^2y}{dx^2}, y^{(n)} = \frac{d^ny}{dx^n} \tag{2.2}$$

In other textbooks, the *D* is used as a derivative operator such as

$$Dy = \frac{dy}{dx}, D^2y = \frac{d^2y}{dx^2}, D^ny = \frac{d^ny}{dx^n} \tag{2.3}$$

For example, the following equations are differential equations of second and third-order, respectively.

$$y'' + 3(x^2 + 1)y' + 5xy = 0$$

$$(D^3 + D^2 + 2D + 3)y = 0$$

The above examples are homogeneous (when no input is applied to the system; the right-hand side is equal to zero) ordinary differential equations of different types. The dependent variable y can be a current, voltage, velocity or position, temperature or pressure, etc. On the other hand, the independent variable x could be a space distance or time. Most of the physical dynamic systems are described by time as an independent variable. The physical laws and formulas are often used to model a dynamic system to obtain the differential equation that relates the physical quantities. Furthermore, if the system is subjected to a disturbance or a forced input, which is a function of the independent variable then the system is nonhomogeneous. For example, the electric current I in a particular electric circuit subjected to a sinusoidal voltage source may be described by the following second-order nonhomogeneous differential equation.

$$I'' + 5I' + 10\,I = \sin t$$

Before leaving this introduction section, we would like to say some words about the partial differential equations, which are assumed beyond this book's scope. When there are two or more independent variables in physical systems, which are not related to each other, then a partial differential equation has resulted. For example, the temperature function $T(x, y)$ (dependent variable) is distributed in a rectangular sheet of two dimensions (independent variables x and y), then the thermodynamic law states the following partial differential equation (the Laplace equation).

$$\frac{\partial^2 T}{\partial x^2} + \frac{\partial^2 T}{\partial y^2} = 0 \qquad (2.4)$$

2.2. Classification of Differential Equations

It is an essential concept to classify the differential equations, and accordingly, the solution methods can be stated. It is a fact that the solution method is different from a class to another, and consequently, the selection of a suitable method depends on the knowledge of this classification. Here in this book, for classification, we consider only the deterministic differential equation. There are two main types of differential equations:

- Ordinary differential equations
- Partial differential equations

If the system is not subjected to input, we have a homogenous equation (the right-hand side is equal to zero); otherwise, we have a nonhomogeneous equation (a function of the independent variable). Another aspect of classification is the appearance of the dependent variable into the equation:

- Linear appearance
- Nonlinear appearance

The highest power of derivatives is the order of the equation, according to which the differential equations can be classified, for instance:

- First-order, if the only first derivative is there
- Second-order if the second derivative is the highest

Furthermore, equations are classified according to the coefficients of the dependent variable and its derivatives; they are:

- Constant-coefficients (numerical values) when all coefficients must be constant.

Variable coefficients are functions of the independent variable when at least one variable coefficient appears in the equation.

Example 2.1.

1. A linear second-order constant-coefficients nonhomogeneous ordinary differential equation, such as

$$y'' + 3y' + 2y = 1$$

It is an abstract differential equation because both dependent and independent variables are not specified physically. The input is a constant value equal to 1, the highest power of derivatives is two. The coefficients are constant values equal to 3 and 2.

2. A linear second-order variable-coefficients nonhomogeneous ordinary differential equation, such as

$$y'' + 3xy' + 2y = x;\ y' = \frac{dy}{dx}$$

The independent variable is x and the coefficient of y' is a variable function $(3x)$.

3. A nonlinear third-order constant-coefficients homogeneous ordinary differential equation, such as

$$y''' + 3y'' + 2y^2 = 0$$

The nonlinearity term in the equation is y^2.

4. A linear fourth-order constant-coefficients nonhomogeneous ordinary differential equation, such as

$$y^{iv} + 3y''' + 5\,y'' + 15\,y' + 2y = t + 1;\; y' = \frac{dy}{dt}$$

The independent variable is t, and the highest power of derivatives is four.

5. A linear second-order constant-coefficients partial differential equation

$$A\frac{\partial^2 u(x,y)}{\partial x^2} + B\frac{\partial u(x,y)}{\partial x}\frac{\partial u(x,y)}{\partial y} + C\frac{\partial^2 u(x,y)}{\partial y^2} = 0$$

Or

$$A\frac{\partial^2 u(x,y)}{\partial x^2} + 2B\frac{\partial u(x,y)}{\partial x \partial y} + C\frac{\partial^2 u(x,y)}{\partial y^2} = 0$$

The dependent variable is u, and there are two independent variables x, and y. Similarly, we may have a partial differential equation with three independent variables x, y, and z. Of course, the variables x, y, and z are independent of each other.

6. A nonlinear second-order variable-coefficients nonhomogeneous ordinary differential equation, such as

$$y'' + 3y^2y'' + 2xy^2 = \cos x$$

7. Table 2.1 lists several famous differential equations that are named after scientists.

Table 2.1. Some famous differential equations

No.	Name of differential equation	General form
1	Bernoulli's equation	$y' + p(x)y + q(x)y^n = 0;$ n is an integer
2	Bessel's equation of order v	$x^2y'' + xy' + (x^2 - v^2)y = 0;\ v \in \mathcal{R}$
3	Euler-Cauchy equation	$x^n y^{(n)} + a_{n-1}x^{n-1}y^{(n-1)} + a_{n-2}x^{n-2}y^{(n-2)} + \cdots + a_0 y = 0$ a_i, are constants
4	Hermit's equation	$y'' - 2xy' + 2vy = 0;\ v \in \mathcal{R}$
5	Legendre's equation of order λ	$(1 - x^2)y'' - 2xy' + \lambda(\lambda + 1)y = 0;\ \lambda \in \mathcal{R}$
6	Leaguer's equation	$xy'' + (1 - x)y' + ny = 0;$ n is a positive integer
7	Riccati equation	$y' + ay^2 + by + c = 0;$ a, b, c are either constants or function of x
8	Laplace equation	$\frac{\partial^2\varphi}{\partial x^2} + \frac{\partial^2\varphi}{\partial y^2} = 0$
9	Wave equation	$\frac{\partial^2\varphi}{\partial x^2} = k\frac{\partial^2\varphi}{\partial y^2}$
10	Lagrange's equations	$\frac{\partial L}{\delta \partial} - \frac{d}{dt}\left(\frac{\partial L}{\partial q_i}\right) = 0;\ i = 1,2 \dots n$

Due to the large content of the differential equations subject, we prefer to divide the content into two chapters. We will consider only the ordinary differential equations in both parts. We will limit ourselves in this chapter only to linear first, second, and higher-order constant-coefficients homogeneous and nonhomogeneous differential equations. In the next chapter, we will consider linear and nonlinear first and second-order variable-coefficients differential equations. However, both chapters do not cover all problems in this vital discipline.

2.3. Ordinary Differential Equations in Engineering Problems

So many engineering problems are described mathematically by ordinary differential equations. In all engineering disciplines, one or more similar, or different types, can be found in a specific mathematical model problem. In most engineering problems, the independent variable is the time t, and the dependent variable is a physical quantity such as electric current or voltage, displacement or velocity of a moving body, etc. Therefore, the behavior of these physical quantities during an interval of time is obtained by solving the derived differential equation. Often, it can be found that many dependent variables are related to one independent variable. This section will consider only a few engineering problems encountered in an electric circuit, mechanical motion of bodies, chemical processes, and thermal transfer.

Usually, simple electric circuits consist of energy sources (voltage or current), passive non-storage elements (resistors), active storage elements (inductances and capacitances), and power-derived components such as amplifiers. The physical laws of Ohm, Faraday, Coulomb, etc., and Kirchhoff's laws of closed-loop of voltage or current are mostly used to derive the differential equations that model the specific electric circuit. The significance of using electric circuits in modeling many other mechanical systems or chemical processes is the availability and low cost of the electric components compared with other elements. Furthermore, electric circuits are often a part of many electromechanical systems. Electric circuits can model algebraic systems by pure resistance circuits and dynamic systems by passive and storage elements such as capacitors and inductors.

The following example illustrates the methodology of obtaining the differential equation that models a simple electric circuit, the series circuit of a resistor-inductance-capacitor RLC connected to a voltage source.

Example 2.2.

Consider the RLC electrical circuit shown in Figure 2.1, where $I = I(t)$ represents the circuit current at time t, measured in seconds starting from the time the switch is closed. Let $Q = Q(t)$ is the charge on the plates of the capacitor at time t.

Without loss of generality, we will assume that at $t = 0, Q(0) = 0,$ and $I(0) = 0,$ i.e., initially, the capacitor is not charged.

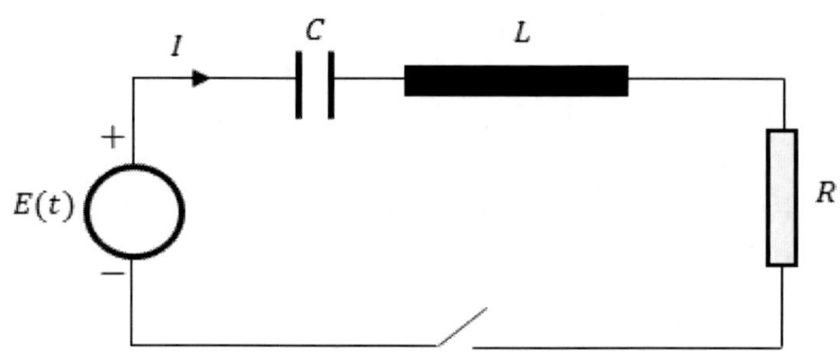

Figure 2.1. An RLC Circuit with AC power supply (Example 2.2).

From the theory of electricity, the following facts are presented:

1. The current I is the charge rate of change $I = \frac{dQ}{dt}$ or $Q = \int I\,dt$.
2. The potential drop across the resistance R (in Ohm) is RI.
3. The potential drop across the coil L (in Henry) is $\frac{dI}{dt}$.
4. The potential drop across the capacitor C (in farad) is $\frac{1}{C}Q$.
5. The sum of the potential drops across the three components within the circuit is equal to the applied voltage (Kirchhoff's laws).

Thus, (for simple writing, we drop the independent variable t from the current I, charge Q, and voltage E)

$$L\frac{dI}{dt} + RI + \frac{1}{C}Q = E$$

Or

$$L\frac{d^2Q}{dt^2}+R\frac{dQ}{dt}+\frac{1}{C}Q=E$$

$$LQ''+RQ'+\frac{1}{C}Q=E$$

Since the elements have fixed values, then the last equation is a linear second-order constant-coefficients nonhomogeneous ordinary differential equation. The equation should also satisfy the assumed initial condition *Q* (0) = 0. In this equation, the charge *Q* is the dependent variable, and the time *t* is the independent variable.

If the current *I* is desired to be the dependent variable, then

$$L\frac{dI}{dt}+RI+\frac{1}{C}\int I\,dt=E$$

The integral can be removed by differentiating both sides of the equation then we obtain

$$\frac{d^2I}{dt^2}+\frac{R}{L}\frac{dI}{dt}+\frac{1}{LC}I=\frac{1}{L}\frac{dE(t)}{dt},I(0)=0$$

Usually, we write in standard form

$$I''+2\xi w_nI'+w_n^2I=\frac{1}{L}E',\xi w_n=\frac{R}{2L},w_n=\frac{1}{\sqrt{LC}}$$

where w_n is the natural frequency, and ξ is named as a damping ratio. It is worth mentioning that the order of the derived equation is equal to the number of storage components (capacitors and coils) in the circuit

In our real world, the motion of physical bodies is surrounding us in numerous applications. One fundamental law that governs precisely the movement is Newton's second law. Second-order ordinary nonhomogeneous differential equations describe the Newton laws of transition and rotational motions, respectively as

$$my''=\sum_{i=1}^{n}F_i\,;\;y(0)=d,y'(0)=v_0 \tag{2.5}$$

$$J\,\theta''=\sum_{i=1}^{n}T_i\,;\;\theta(0)=\theta_0\,,\;y'(0)=w_0 \tag{2.6}$$

where:

m is the mass in Kilograms of the moving body.

y is the displacement of the body (in meter); the second derivative y'' is the body acceleration (in meter/second^2).

d is the initial distance (in meter).

v_0 is the initial velocity (in meter/second); if the motion starts from rest, then v_0 equals zero.

F_i is the ith input force that is applied on the body (in Newton).

J is the moment of inertia of the rotating body (in Newton-meter-second^2/radian).

θ is the angular displacement of the body (in radian).

θ' is the body angular velocity or w in(radian/seconds) θ'' is the body angular acceleration (in radian/second^2).

θ_0 and w_0 are initial angular position and velocity (in radian) and (radian/second), respectively.

T_i is the ith input torque that causes body rotation (in Newton-meter).

Note that for a particular problem, it is necessary to use the same system of units.

Like the electric circuits, the mechanical complement structures represent a rich set of mechanical and civil engineers models. The basic structure is called the spring-damper-mass problem, which models different problems like the vibration absorption problem, bending of a structural beam, or the vibration of an *n*-story shear building under the earthquake excitation. The next example considers one of these problems.

Example 2.3.

Consider the complement mechanical structure shown in Figure 2.2. It consists of two different masses m_1 and m_2. Mass m_1 is attached to a wall by a spring and a damper of coefficients k_1 and f_1, respectively. Similarly, mass m_2 is connected to mass m_1 by another spring and damper of coefficients k_2 and f_2, respectively. A pulling force $F(t)$ is applied to move the two masses on a frictionless surface. The task is to find the differential equations that govern the displacements of the two masses $y_1(t)$ and $y_2(t)$.

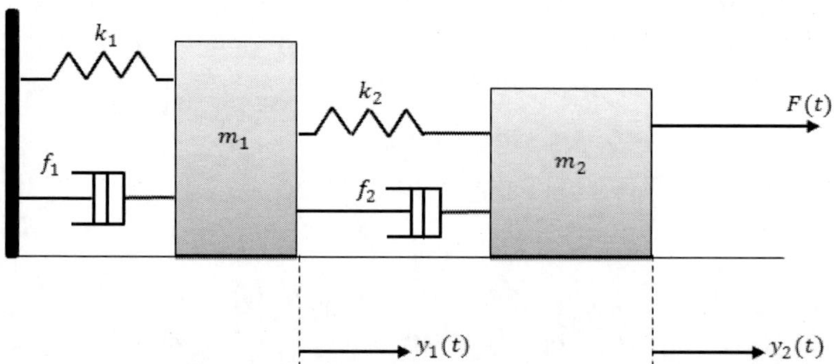

Figure 2.2. A complement mechanical structure (Example 2.3).

Newton's second law is applied to express the motion of each mass. Forces working on the system are the applied force $F(t)$, the force caused by the spring, and the force generated by the damper. It is assumed that the spring force is linearly proportional to the spring stretch. The damper force is linearly proportional to the velocity of movement.

- For mass m_1, Newton's second law states the following second-order differential equation.

$$m_1 y_1'' = k_2(y_2 - y_1) - k_1 y_1 + f_2(y_2' - y_1') - f_1 y_1'$$

- For the mass m_2, Newton's second law states the following second-order differential equation.

$$m_2 y_2'' = F(t) - k_2(y_2 - y_1) - f_2(y_2' - y_1')$$

Rewrite the two coupled second-order differential equations in the form

$$y_1'' - \frac{f_2}{m_1}(y_2' - y_1') + \frac{f_1}{m_1} y_1' - \frac{k_2}{m_1}(y_2 - y_1) + \frac{k_1}{m_1} y_1 = 0$$

$$y_2'' + \frac{f_2}{m_2}(y_2' - y_1') + \frac{k_2}{m_2}(y_2 - y_1) = F(t)$$

These last two equations can be reduced to one single fourth-order equation that relates the output displacement y_1 (or eventually y_2) to the applied input force $F(t)$. However, for analysis purposes, the matrix form is better to be considered for modeling these problems. For this purpose, the concept of state-space is introduced, where two matrix equations are used. These are the state equations and the output equations. We will consider this subject in section 2.8.

Example 2.4.

Consider the problem of the conservation of salt in two well-mixed tanks, as shown in Figure 2.3. The input flow rate and the input salt concentration are q_{in} and $c(t)$, respectively. Consequently, the tanks contain uniform functions of time salt concentrations of $c_1(t)$ and $c_2(t)$. The flow rates between the tanks are positive values of q_1 and q_2, where $q_1 > q_2$ and the output rate flow is q_{out}. It is assumed that the volume of liquid in each tank remains constant of v_1 *and* v_2. The task is to find the differential equation relating the salt concentration in the second tank $c_2(t)$ and the input salt concentration $c(t)$.

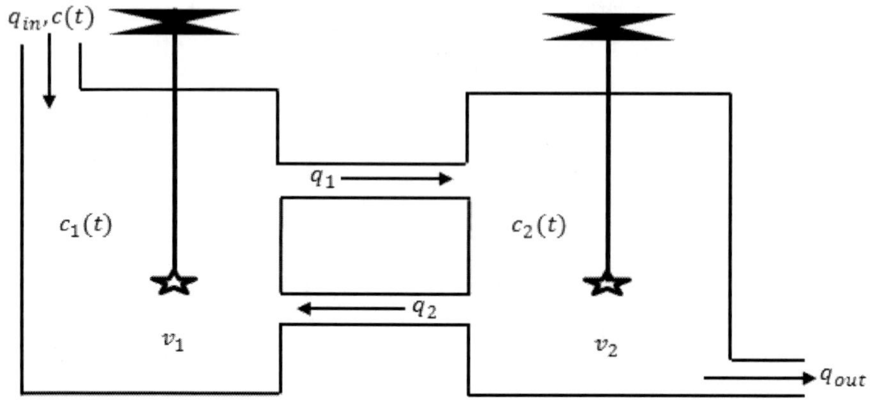

Figure 2.3. Two-mixed tanks problem (Example 2.4).

Based on the Lavoisier conservation of the mass act (the salt accumulation equation in each tank to the difference between salt rates in and out of the tanks), we obtain the following differential equations.

$$v_1 \frac{d}{dt} c_1(t) = q_{in} c + q_2 c_2 - q_1 c_1$$

$$v_2 \frac{d}{dt} c_2(t) = q_1 c_1 + q_2 c_2 - q_{out} c_2$$

Since the holdup of liquid in both tanks is constant, then it holds,

$$q_{in} = q_{out} = q_1 - q_2$$

We rewrite the differential equations in the form

$$\frac{d}{dt} c_1(t) = \frac{c}{v_1}(q_1 - q_2) + \frac{q_2 c_2}{v_1} - \frac{q_1 c_1}{v_1}$$

$$\frac{d}{dt} c_2(t) = \frac{q_1}{v_2} c_1 + \frac{q_2 c_2}{v_2} - \frac{(q_1 - q_2)}{v_2} c_2$$

Using an algebraic manipulation, the differential equation, which relates the salt concentration in the second tank $c_2(t)$ and the input salt concentration $c(t)$, can be obtained as

$$c_2'' + A c_2' + B c_2 = C$$

where,

$$A = \frac{q_1(v_1 + v_2)}{v_1 v_2}, B = \frac{q_1(q_1 + q_2)}{v_1 v_2}, C = Bc$$

The result is a linear second-order constant-coefficients nonhomogeneous ordinary differential equations.

Example 2.5.

The problem of heat flows and temperature variations via walls of an adjacent two-room building is considered; see a simplified model in Figure 2.4. The outside temperature, the temperature of the first room, and the temperature of the second room are functions of time denoted as $T_0(t)$, $T_1(t)$, and $T_2(t)$, respectively. A fireplace in the first room supplies the heat $Q(t)$. Assume the heat from the first room transfer to the outside are $Q_1(t)$ and to the second room be $Q_{12}(t)$. Also assume, the heat transfers from the second room to the outside as $Q_2(t)$. The thermal lumped parameters (resistances and capacitances) are R_1, R_2, R_{12}, C_1, and C_2; due to the difference in room volumes and the different materials of walls, the parameters' values are different too. Write the differential equations that govern the temperatures in the two rooms.

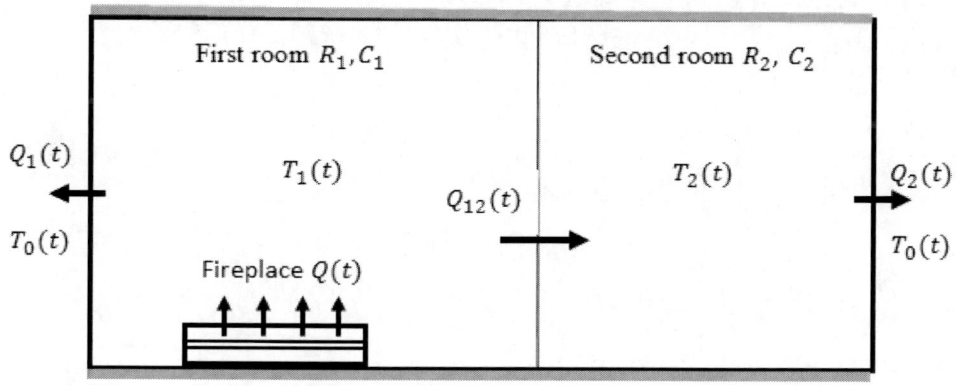

Figure 2.4. A two-room thermal problem (Example 2.5).

Based on the conservation of energy, the room-temperature rates are related to the heat as follows.

$$\frac{d}{dt}T_1(t) = \frac{1}{C_1}[Q(t) - Q_1(t) - Q_{12}(t)]$$

$$\frac{d}{dt}T_2(t) = \frac{1}{C_2}[Q_{12}(t) - Q_2(t)]$$

Also, we have the following physical relations.

$$Q_1(t) = \frac{T_1(t) - T_0(t)}{R_1}$$

$$Q_2(t) = \frac{T_2(t) - T_0(t)}{R_2}$$

$$Q_{12}(t) = \frac{T_1(t) - T_2(t)}{R_{12}}$$

Substituting the last three equations in the first two equations, we arrange the following two differential equations; the dependent variables are T_1 and T_2, the independent variable is the time t, and the input force functions are the fireplace heat Q and the outside temperature T_0.

$$T_1' + \frac{R_{12} + R_1}{R_1 R_{12} C_1} T_1 - \frac{1}{R_{12} C_1} T_2 = \frac{Q}{C_1} + \frac{1}{R_1 C_1} T_0$$

$$T_2' - \frac{1}{R_{12} C_2} T_1 + \frac{R_{12} + R_1}{R_2 R_{12} C_2} T_2 = \frac{1}{R_2 C_2} T_0$$

In matrix form,

$$\begin{bmatrix} T_1' \\ T_2' \end{bmatrix} = \begin{bmatrix} \frac{R_{12} + R_1}{R_1 R_{12} C_1} & -\frac{1}{R_{12} C_1} \\ -\frac{1}{R_{12} C_2} & \frac{R_{12} + R_1}{R_2 R_{12} C_2} \end{bmatrix} \begin{bmatrix} T_1 \\ T_2 \end{bmatrix} + \begin{bmatrix} \frac{1}{C_1} & \frac{1}{R_1 C_1} \\ 0 & \frac{1}{R_2 C_2} \end{bmatrix} \begin{bmatrix} Q \\ T_0 \end{bmatrix}$$

It is worth mentioning that different physical problems could have the same general form of the differential equation. For instance, the derived differential equations of the electric circuit of example 2.2 and the salt concentration process of example 2.4 are similar, however, with different parameters' values. This fact is very considerable in practice since we can simulate a complicated system or process by a simple system (usually electric circuits) with suitable parameters' values. To shorten the length throughout this chapter, the rest of the examples will consider mostly abstract problems. However, several conceptual problems represent the alternative version of the dimensionless physical problems, without mentioning their nature.

2.4. Solution Concepts

For the following discussion, we assume that the dependent variable is y, and the independent variable is x. The solution of a particular homogenous equation is:

The general solution is denoted by $y_h(x)$; it is the solution given by a combination of independent variable functions and several constants equal to the differential equation order. For example, for a second-order, the general solution will have the form

$$y_h(x) = c_1 g_1(x) + c_2 g_2(x) \tag{2.7}$$

where c_1 and c_2 are two arbitrary real constants, $g_1(x)$ *and* $g_2(x)$ are two different elementary functions of the independent variable x. Therefore, the general solution represents a family (or a set) of solutions.

- If the initial or final conditions (boundary) are given, then a particular solution is obtained; this is one solution in the set, which belongs to the given conditions.
- For nonhomogeneous equations, the solution consists of two parts. These are the general solution $y_h(x)$ of the homogenous equation and a particular solution $y_p(x)$ corresponding to the input force function. For example, consider the third-order constant-coefficients differential equation.

$$y''' + ay'' + ab + cy = f(x) \tag{2.8}$$

It has the solution

$$y(x) = y_h(x) + y_p(x) = c_1 g_1(x) + c_2 g_2(x) + c_3 g_3(x) + g_4(x) \tag{2.9}$$

where, $g_4(x)$ is a particular solution corresponding to the input function $f(x)$.

The initial value problem is defined by the nth-order differential equation and a set of $(n-1)$ initial conditions, for example

$$y''' + ay'' + ab + cy = f(x), y(0) = k_0, y'(0) = k_1, y''(0) = k_2 \tag{2.10}$$

It is worth mentioning that our interest in this chapter is to determine solutions rather than answer the solution's existence or uniqueness. For some equations, we cannot find an exact solution but only an approximate one with the aid of a particular numerical method (see chapter 5). The given solutions in equations (2.7) and (2.9) are in explicit forms; however, a solution could be found in implicit form as

$$g(x, y) = c \tag{2.11}$$

The integration constants can be computed if there are sufficient initial values of the dependent variable and its derivatives $y(0), y'(0)$, etc. Alternatively, when there are sufficient boundary values, as we will see later. Before going into a detailed analysis, let us consider two basic examples

Example 2.6.

Consider the second-order differential equation describing a mass movement under force influence

$$d''(t) = 4t$$

Then, the solution can be readily obtained by direct successive two integrations.

$$d'(t) = \int 4t\,dt = 2t^2 + c_1$$

$$d(t) = \int (2t^2 + c_1)dt = \frac{2}{3}t^3 + c_1 t + c_2$$

where c_1 and c_2 are the integration constants. The solution represents a family of curves in the $t - d$ plane. A particular solution can be obtained for specific initial conditions, which determine these integration constants.

Example 2.7.

Consider a harmonic system with a sinusoidal input given by

$$y'' + 4y = 8 + 6\sin t;\ y(0) = 0, y'(0) = -1, 0 \le t < \infty$$

It has the solution

$$y(t) = -2\cos 2t - \frac{3}{2}\sin 2t + 2 + 2\sin t$$

Let us check the correctness of this solution by twice differentiations of $y(t)$ and substituting in the left-hand side to check whether it is equal to the right-hand side.

$$y'(t) = 4\sin 2t - 3\cos 2t + 2\cos t$$

$$y''(t) = 8\cos 2t + 6\sin 2t - 2t$$

$$8\cos 2t + 6\sin 2t - 2\sin t + 4\left(-2\cos 2t - \frac{3}{2}\sin 2t + 2 + 2\sin t\right) = 8 + 6\sin t$$

This result is equal to the right-hand side. Moreover, if we substitute in the given solution $t = 0$, we obtain $y(0) = -2 - 0 + 2 - 0 = 0$, as it is given in the problem. Similarly, if we substitute in the first derivative of the solution, then the given initial condition $y'(0) = 0 - 3 + 2 = -1$ as it is also given. Therefore, the given solution is correct.

2.5. First-Order Ordinary Differential Equations

Despite the lowest order, these equations are found in many physical applications. The derived methods of solution represent an essential framework to understand the solution's concept for the first-order equations and high-order equations as well. These equations have many forms and are categorized into some kinds. In this section, we will consider all fundamental types.

A first-order homogeneous equation can be put in the form

$$M(x,y)dx + N(x,y)dy = 0 \tag{2.12}$$

In one case, which is easy to solve, the equation arises when $M(x,y)$ and $N(x,y)$ can be factored algebraically into the products of a function of only x and only y. Such equations are said to be of the variables separable type. If

$$M(x,y) = q(x)p(y), \text{and } N(x,y) = r(x)s(y) \tag{2.13}$$

Then we can write

$$\frac{q(x)}{r(x)}dx + \frac{s(y)}{p(y)}dy = 0 \tag{2.14}$$

Hence, the above equation can be solved by the direct integration of each term. The obtained solution either could be of the form $y = f(x)$ or $f(x,y) = c$; c is a constant. Despite the method's easy appearance, this does not mean that it will always give an exact solution. It depends on the existence of explicit solutions of both integrals. As a final note, in general, the nonhomogeneous equations' solution cannot be included in this classification. However, if the equation is linear, input functions can be included within the solution's framework.

Example 2.8.

Newton's law models the free-falling mass m (kg) under gravity g (m/sec^2) (near the Earth surface) and air resistance by the following differential equation

$$m\frac{dv}{dt} = -mg - kv, v(0) = 0$$

where v (m/sec) is the vertical velocity of the mass, and k is a positive constant (air resistance coefficient, kg/sec); note that this model approximates the air resistance by kv instead of the more actual kv^2.

Separation of variables yields

$$m\frac{dv}{mg+kv} = -dt, v(0) = 0$$

Integration on both sides, to get

$$m\int\frac{dv}{mg+kv} = -\int dt$$

$$\mathrm{m}\frac{1}{k}\ln(mg+kv) = -t + c$$

Using the initial condition $v(0) = 0$ yields

$$c = \frac{m}{k}\ln{(mg)}$$

$$\frac{m}{k}\ln\left(\frac{mg+kv}{mg}\right) = -t \Rightarrow 1+\frac{kv}{mg} = e^{-\frac{k}{m}t}$$

$$v(t) = -\frac{mg}{k}\left(1-e^{-\frac{k}{m}t}\right)$$

As the exponential term decays to zero, the velocity approaches the terminal velocity (after the air resistance balances the gravity force, and it will be equal to $(-mg/k)$. One application of this problem is parachute motion.

Example 2.9.
Find a solution to the terminal-value problem.

$$(3x+xy^2)dx-(y+x^2y)dy = 0, y(1) = -3$$

Rearranging the equation to the form

$$x(3+y^2)dx - y(1+x^2)dy = 0$$

Separation of variables yields

$$\frac{x}{1+x^2}dx = \frac{y}{3+y^2}dy$$

Integration on both sides, we get

$$\int\frac{x}{1+x^2}dx = \int\frac{y}{3+y^2}dy$$

$$\frac{1}{2}\ln(1+x^2) = \frac{1}{2}\ln(3+y^2) + c$$

To find the constant c, we substitute the initial condition, i.e.,

$$\frac{1}{2}\ln(1+1^2) = \frac{1}{2}\ln(3+(-3)^2) + c$$

$$c = \frac{1}{2}\ln(2) - \frac{1}{2}\ln(12) = \frac{1}{2}\ln\frac{2}{12} = \frac{1}{2}\ln\frac{1}{6}$$

Substitute in the general solution yields

$$\frac{1}{2}\ln(1+x^2) = \frac{1}{2}\ln(3+y^2) + \frac{1}{2}\ln\frac{1}{6}$$

Using the known rules of logarithmic yields

$$(1+x^2) = \frac{3+y^2}{6}$$

We always prefer (if it is possible) to express the solution explicitly as $y = f(x)$. We can easily do that.

$$y(x) = \pm\sqrt{6x^2+3}$$

Based on the given initial condition, the solution must be

$$y(x) = -\sqrt{6x^2+3} \Rightarrow y(1) = -\sqrt{6+3} = -3$$

Some of the first-order differential equations, which are given as in form 2.11, are called exact if there exists a function $g(x, y)$ such that

$$d[g(x,y)] = M(x,y)dx + N(x,y)dy \tag{2.15}$$

If such a function exists, then the equation reduces to

$$d[g(x,y)] = 0 \tag{2.16}$$

Consequently, the general solution is

$$g(x,y) = c \tag{2.17}$$

A sufficient condition for the existence of the function g is the equality of the mixed second-order partial derivatives.

$$\frac{\partial^2 g(x,y)}{\partial x \partial y} = \frac{\partial^2 g(x,y)}{\partial y \partial x} \tag{2.18}$$

Theorem 2.1.
Consider the first-order differential equation

$$M(x,y)dx + N(x,y)dy = 0 \tag{2.19}$$

It is exact if and only if M and N have continuous partial derivatives, which satisfy in some rectangular in the plane the equality.

$$\frac{\partial M}{\partial y} = \frac{\partial N}{\partial x} \tag{2.20}$$

Also,

$$\frac{\partial g}{\partial x} = M, \text{and } \frac{\partial g}{\partial y} = N \tag{2.21}$$

Let (x_0, y_0) be a point where M and N are defined and continuously differential, then

$$g(x,y) = \int_{x_0}^{x} M(t, y_0)dt + \int_{y_0}^{y} N(x,s)ds \tag{2.22}$$

Another way to determine the function g is called partial integration. The following two formulas are used.

$$g(x,y) = \int Mdx + c(y);\ y \text{ is treated as a constant} \tag{2.23}$$

$$g(x,y) = \int Ndy + c(x);\ x \text{ is treated as a constant} \tag{2.24}$$

Sometimes, it is possible to find the function $g(x,y)$ by cleaver inspection. The following examples use these three methods.

Example 2.10.
Use theorem 2.1 to check whether the following equation is exact

$$(x^2 + y^2 + x)dx + xydy = 0$$

We have

$$M = x^2 + y^2 + x, N = xy$$

$$\frac{\partial M}{\partial y} = 2y; \frac{\partial N}{\partial x} = y \Rightarrow \frac{\partial M}{\partial y} \neq \frac{\partial N}{\partial x}$$

Therefore, the equation is not exact.

Example 2.11.
Find the general solution of

$$(y + \sin x)dy = (x - y\cos x)dx$$

First, to write the equation in standard form,

$$(y\cos x - x)dx + (y + \sin x)dy = 0$$

Thus, we assign $M = (y\cos x - x)$, $N = (y + \sin x)$ and we test the condition

$$\frac{\partial M}{\partial y} = \cos x, \frac{\partial N}{\partial x} = \cos x$$

Therefore, the given equation is an exact one. We shall try to find the function g by clever inspection. We write the exact equation as

$$y\cos x\, dx + \sin x\, dy - xdx + ydy = 0$$

$$d(y\sin x) - d\left(\frac{x^2}{2}\right) + d\left(\frac{y^2}{2}\right) = 0$$

$$d\left(y\sin x - \frac{x^2}{2} + \frac{y^2}{2}\right) = 0 \Rightarrow g(x, y) = y\sin x - \frac{x^2}{2} + \frac{y^2}{2}$$

It follows that the general solution is

$$y\sin x - \frac{x^2}{2} + \frac{y^2}{2} = c$$

For $c = 2, 4, \text{and } 6$, Figure 2.5 shows the relations between the dependent variable y and the independent variable x. The reader can check that the initial condition is $y(0) = 2$ gives $c = 2$, as shown in the plot.

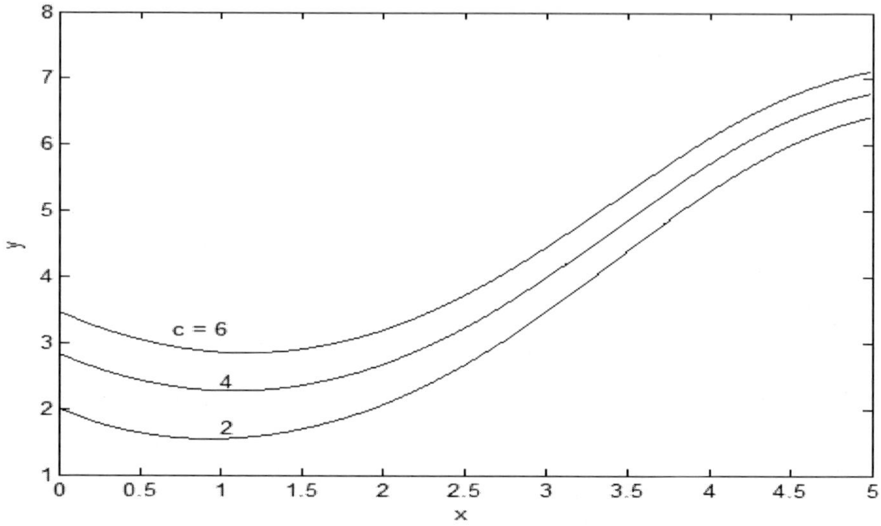

Figure 2.5. Solutions for different constant c (example 2.11).

Example 2.12.

Find the general solution of

$$\left(x^2 + \frac{y}{x}\right) dx + (\ln x + 2y)dy = 0$$

The equation is exact because

$$\frac{\partial M}{\partial y} = \frac{\partial N}{\partial x} = \frac{1}{x}$$

Let us consider the partial integration

$$g(x, y) = \int M dx = \int (x^2 + \frac{y}{x})\, dx = \frac{x^3}{3} + y \ln x + k(y)$$

To determine $k(y)$, we use

$$\frac{\partial g}{\partial y} = N \Rightarrow \ln x + k'(y) = \ln x + 2y$$

$$k'(y) = 2y \Rightarrow k(y) = y^2$$

Therefore,

$$d[g(x,y)] = d\left(\frac{x^3}{3} + y\ln x + y^2\right) = 0$$

The general solution is

$$\frac{x^3}{3} + y\ln x + y^2 = c$$

Example 2.13.
Consider the initial-value problem

$$(2xy + 3x^2)dx + x^2dy = 0, 1 \leq x < \infty, y(1) = 2$$

The equation is exact since

$$\frac{\partial M}{\partial y} = \frac{\partial N}{\partial x} = 2x$$

$$g(x,y) = \int_{x_0}^{x} M(t, y_0)dt + \int_{y_0}^{y} N(x,s)ds$$

If we take (x_0, y_0) be the point (1, 2), through which the solution must pass, we then have

$$g(x,y) = \int_{1}^{x} (4t + 3t^2)dt + \int_{2}^{y} x^2 ds$$

$$g(x,y) = 2t^2 + t^3|_1^x + x^2s|_2^y$$

$$g(x,y) = 2x^2 + x^3 - 2 - 1 + x^2y - 2x^2 = x^3 + x^2y - 3$$

The general solution is

$$g(x,y) = x^3 + x^2y - 3 = c$$

Since $y(1) = 2$, we have

$$g(1,2) = 1^3 + 1^22 - 3 = c \Rightarrow c = 0$$

Then the general explicit solution becomes

$$y = \frac{3 - x^3}{x^2}$$

It is a fact that every first-order differential equation that has a closed-form solution can be solved by reducing the equation to an exact one by multiplying it by some factor. It is called the integrator factor. These equations are called reducible to the exact equations.

Example 2.14.

Consider the non-exact equation of example 2.10.

$$(x^2 + y^2 + x)\, dx + xy\, dy = 0$$

By multiplying the equation by the integrator factor x

$$(x^3 + xy^2 + x^2)\, dx + x^2 y\, dy = 0$$

Now, the two functions M and N are

$$M = x^3 + xy^2 + x^2, \qquad N = x^2 y$$

$$\frac{\partial M}{\partial y} = 2xy; \frac{\partial N}{\partial x} = 2xy \Rightarrow \frac{\partial M}{\partial y} = \frac{\partial N}{\partial x}$$

Thus, the equation becomes an exact one.

How can we find the integrating factor for a specific equation? There is no general theory to find the integrating factor. There are some rules, which guide us to find the integrating factor. In some cases, it is possible to guess the integrator factor using intuitive sense and skill. However, for some other cases, the following formulation can be used to determine the integrating factor:

If M and N are continuously differentiable and N is not zero, and

$$\frac{\frac{\partial M}{\partial y} - \frac{\partial N}{\partial x}}{N} = f(x) \tag{2.25}$$

Then the integrating factor is

$$IF: e^{\int f(x)dx} \tag{2.26}$$

Example 2.15.

The following equations have the associated integrating factors:

$$(x^3 + y)dx - xdy = 0;\ IF: \frac{1}{x^2}$$

$$(2xy^2 + y)dx + (x + 2x^2y - x^4y^3)dy = 0;\ IF: \frac{1}{x^4y^4}$$

$$(2y + x^2 - y^2)dx + (x - xy)dy = 0;\ IF: x$$

For the first equation, the integrating factor is found by applying formulas (2.25 and 2.26). For the second equation, these formulas are not applicable and can be determined by intuitive guessing.

For a particular class of equations, another procedure exists to find the solution. The equations of this class are called homogenous first-order equations, which are reduced to a solvable form. In particular, such an equation is reduced to an equation that has separated variables.

Definition 2.1.

The function $M(x, y)$ is said to be homogenous of order *r* if, for *a* > 0,

$$M(ax, ay) = a^r M(x, y) \tag{2.27}$$

If $M(x, y)$ and $N(x, y)$ are homogenous of the same order, then equation (2.12) is said to be a homogenous first-order equation.

It can be shown that for a homogenous first-order equation, a substitution $y = vx$ will separate the variables as

$$\frac{dx}{x} + \frac{N(1, v)}{M(1, v) + vN(1, v)}dv = 0 \tag{2.28}$$

For the solution of this separable variable equation, we have to substitute instead of *v*, the term y/x to determine the original equation solution.

Alternatively, we can use the substitution $x = wy$ that will separate the variables as

$$\frac{dy}{y} + \frac{M(w, 1)}{M(w, 1) + w\,N(w, 1)}dw = 0 \tag{2.29}$$

Similarly, we substitute instead of w, the term x/y to determine the solution of the original equation. As a rule, if the function $N(x, y)$ is more straightforward than the function $M(x, y)$, then we use the substitution $y = vx$; otherwise, if $M(x, y)$ is the more straightforward, we use the substitution $x = wy$.

Example 2.16.

Find a solution to the equation

$$(3x^2 + 9xy + 5y^2)dx - (6x^2 + 4xy)dy = 0$$

Both $M(x,y)$ and $N(x,y)$ of degree equal two, and *N* is a more straightforward function than *M*. Therefore; we use the substitution $y = vx$.

The separable variable equation becomes

$$\frac{dx}{x} + \frac{(-6 - 4v)\,dv}{(3 + 9v + 5v^2) + v(-6 - 4v)} = 0$$

$$\frac{dx}{x} = \frac{2(3 + 2v)\,dv}{3 + 3v + v^2}$$

Integration on both sides yields

$$\ln x + \ln c = 2\ln(3 + 3v + v^2)$$

$$\ln cx = \ln(3 + 3v + v^2)^2$$

$$cx = \left(3 + 3\frac{y}{x} + \frac{y^2}{x^2}\right)^2$$

The general solution is

$$cx^5 = (3x^2 + 3xy + y^2)^2$$

One of the most considerable first-order differential equations is the linear equation, which has the normal form

$$y' = f(x,y) \tag{2.30}$$

This equation is called linear in y if it can be written in the form

$$\frac{dy}{dx} + p(x)y = q(x) \tag{2.31}$$

If $q(x) = 0$, then it is a homogeneous linear equation. The general solution is given by

$$y(x) = e^{-\int p(x)dx}\left[c + \int e^{\int p(x)dx}\, q(x)dx\right] \tag{2.32}$$

It is worth mentioning that the term

$$m(x) = e^{\int p(x)dx} \tag{2.33}$$

represents an integrating factor of the equation, thus

$$e^{\int p(x)dx}\left[\frac{dy}{dx}+p(x)y\right]=\frac{d}{dx}[ym(x)] \tag{2.34}$$

Example 2.17.
Solve the initial-value problem

$$\frac{dy}{dx}+3y=e^{5x}, y(0)=5$$

Therefore,

$$p(x)=3, q(x)=e^{5x}$$
$$m(x)=e^{\int 3dx}=e^{3x}$$

Multiply both sides of the equation by $m(x)$.

$$e^{3x}\left[\frac{dy}{dx}+3y\right]=e^{5x}e^{3x}=e^{8x}$$
$$\frac{d}{dx}[ye^{3x}]=e^{8x}$$

Integrating yields

$$ye^{3x}=\int e^{8x}dx \Rightarrow y=e^{-3x}\int e^{8x}dx$$

The general solution is

$$y=e^{-3x}\left[\frac{1}{8}e^{8x}+c\right]=\frac{1}{8}e^{5x}+ce^{-3x}$$

To determine the constant of integration c, we substitute the given initial conditions, i.e.,

$$5=\frac{1}{8}e^{0}+ce^{0} \Rightarrow c=\frac{39}{8}$$

The general solution of the initial-value problem is

$$y(x)=\frac{1}{8}e^{5x}+\frac{39}{8}e^{-3x}$$

Example 2.18.
Find the general solution of

$$\frac{dy}{dx}+\frac{2}{x+1}y=3x$$

Therefore,

$$p(x)=\frac{2}{x+1},q(x)=3x$$

$$m(x)=e^{\int\frac{2}{x+1}dx}=e^{2\ln(x+1)}=e^{\ln(x+1)^2}=(x+1)^2$$

Multiplying the given equation by $m(x)$ yields

$$(x+1)^2\left[\frac{dy}{dx}+\frac{2}{x+1}y\right]=3x(x+1)^2$$

$$\frac{d}{dx}[y(x+1)^2]=3x(x+1)^2$$

Integrating yields

$$y(x+1)^2=\int 3x\,(x+1)^2dx=\int(3x^3+6x^2+3x)dx$$

The general solution is

$$y(x)=\frac{\frac{3}{4}x^4+2x^3+\frac{3}{2}x^2+c}{(x+1)^2}$$

Some first-order differential equations are not linear in the dependent variable y, but it is linear in the independent variable x. For such equations, we can change the role of x and y and solve using the same method.

Example 2.19.
Find the general solution of

$$y+(3+3x-y)y'=0$$

Here because of the product yy' , the equation is not linear in y, while it is linear in x. We rewrite the equation such that x is the dependent variable and y is the independent variable as follows:

$$ydx + (3 + 3x - y)dy = 0 \;\div ydy$$

$$\frac{dx}{dy} + \frac{3 + 3x - y}{y} = 0$$

$$\frac{dx}{dy} + \frac{3}{y}x = \frac{y - 3}{y}$$

Therefore,

$$p(y) = \frac{3}{y}, \qquad q(y) = \frac{y - 3}{y}$$

$$m(y) = e^{\int \frac{3}{y}dy} = e^{3\ln y} = e^{\ln y^3} = y^3$$

$$y^3\left[\frac{dx}{dy} + \frac{3}{y}x\right] = y^3\frac{y - 3}{y}$$

$$\frac{d}{dy}(xy^3) = y^3 - 3y^2$$

Integrating yields

$$xy^3 = \int (y^3 - 3y^2)dy = \frac{1}{4}y^4 - y^3 + c$$

The general solution is

$$xy^3 - \frac{1}{4}y^4 + y^3 = c$$

Furthermore, two cases of the second-order differential equation can be solved by the technique called reduction-of-order. Consider the general second-order differential equation, where x is the independent variable and y dependent variable.

$$f(x, y, y', y'') = 0 \tag{2.35}$$

Case 1: Missing dependent variable; in this case, the equation is of the form

$$f(x, y', y'') = 0 \tag{2.36}$$

The substitution

$$p = y' = \frac{dx}{dy}, \text{and } p'' = y'' = \frac{d^2y}{dx^2} \tag{2.37}$$

will reduce the given equation to

$$f(x,p,p') = 0 \tag{2.38}$$

Thus, we have reduced the second-order equation to a system of two first-order equations.

$$y' = p \tag{2.39}$$

$$f(x,p,p') = 0 \tag{2.40}$$

Example 2.20.
Find a solution to the initial-value problem

$$y'' + 3y' = e^{2x};\ y(0) = 4, y'(0) = 0$$

Using the substituting $p = y'$ we get

$$p' + 3p = e^{2x};\ p(0) = y'(0) = 0$$

It is a first-order linear equation, and the general solution is

$$p(x) = e^{-3x}\int e^{3x}e^{2x}dx = e^{-3x}\left(\frac{1}{5}e^{5x} + c\right) = \frac{1}{5}e^{2x} + ce^{-3x}$$

$$0 = \frac{1}{5}e^0 + ce^0 \Rightarrow c = -\frac{1}{5}$$

The general solution in terms of *p* is

$$p(x) = \frac{1}{5}e^{2x} - \frac{1}{5}e^{-3x} = \frac{dy}{dx}$$

$$dy = \left(\frac{1}{5}e^{2x} - \frac{1}{5}e^{-3x}\right)dx$$

Integrating gives the general solution of the original second-order equation.

$$y(x) = \int\left(\frac{1}{5}e^{2x} - \frac{1}{5}e^{-3x}\right)dx = \frac{1}{10}e^{2x} + \frac{1}{15}e^{-3x} + k$$

To determine the constant of integration k, we use the initial condition $y(0) = 4$, i.e.,

$$y(0) = 4 = \frac{1}{10}e^0 + \frac{1}{15}e^0 + k \Rightarrow k = \frac{115}{30} = \frac{23}{6}$$

Therefore, the complete solution of the given second-order equation becomes

$$y(x) = \frac{1}{10}e^{2x} + \frac{1}{15}e^{-3x} + \frac{23}{6}$$

Case 2: Missing independent variable; in this case, the equation is of the form

$$f(y, y', y'') = 0 \tag{2.41}$$

It can be shown that equation (2.41) can be reduced to the following first-order equation.

$$f\left(y, p, p\frac{dp}{dy}\right) = 0 \tag{2.42}$$

Equation (2.42) is a first-order nonlinear equation. For the reduced equation, if the variables are separable, then the solution is easily obtained. However, sometimes difficulties in determining the general solution of the original equation may arise. We will not prefer to solve using this technique unless it is the only possible one; however, this technique will be elaborated on in chapter three.

The reader may note that the second-order equation is reduced to two first-order differential equations; in example 2.20, these are

$$y' = p$$

$$p' + 3p = e^{2x}$$

2.6. Second-Order Ordinary Differential Equations

There are several classes of second-order differential equations, but there is no general theory to solve all these classes. One of the most important classes is where the equations have a constant-coefficients of the form

$$y'' + ay' + by = f(x);\ a, b \text{ are constants} \tag{2.43}$$

First, we explore the solution of the homogenous equations in this class.

Theorem 2.2.

The general solution of the homogenous constant-coefficients second-order differential equation depends on the roots of a corresponding second-order algebraic equation (called characteristic equation).

$$r^2 + ar + b = 0 \quad (2.44)$$

The roots are r_1, and r_2 defining only the following three cases, in which the general solution has a specific form.

Case 1: If the roots are real and distinct, then

$$y_h = c_1 e^{r_1 x} + c_2 e^{r_2 x} \quad (2.45)$$

When one root is equal to zero, then the general solution is

$$y_h = c_1 + c_2 e^{r_2 x} \quad (2.46)$$

Case 2: If the roots are real and equal to r, then

$$y_h = (c_1 + c_2 x)e^{rx} \quad (2.47)$$

Case 3: If the roots are complex conjugates of the form

$$r_1 = \alpha + j\beta, r_2 = \alpha - j\beta \quad (2.48)$$

Then the general solution will be in the form

$$y_h = e^{\alpha x}(c_1 \cos \beta x + c_2 \sin \beta x) \quad (2.49)$$

Or equivalently

$$y_h = c_3 e^{\alpha x} \sin(\beta x + c_4) \quad (2.50)$$

With minor modifications, the solution method given in theorem 2.2 is valid for higher-order equations of this class (see next section). The following three examples consider the three mentioned cases in the same sequence.

Example 2.21.

Find the general solution of the equation

$$y'' + 7y' + 6y = 0$$

The characteristic equation is

$$r^2 + 7r + 6 = 0 \Rightarrow (r+6)(r+1) = 0 \Rightarrow r_1 = -6, r_2 = -1$$

According to case 1, the general solution is

$$y_h = c_1 e^{-6x} + c_2 e^{-x}$$

Example 2.22.
Find the general solution of the equation

$$y'' + 4y' + 4y = 0$$

The characteristic equation is

$$r^2 + 4r + 4 = 0 \Rightarrow (r+2)^2 = 0 \Rightarrow r_1 = r_2 = -2$$

According to case 2, the general solution is

$$y_h = (c_1 + c_2 x)e^{-2x}$$

Example 2.23.
Find the general solution of the equation

$$y'' + 4y' + 10y = 0$$

The characteristic equation is

$$r^2 + 4r + 10 = 0$$

$$r_{1,2} = \frac{-4 \pm \sqrt{16 - 4 \times 10}}{2}$$

$$r_{1,2} = -2 \pm j\sqrt{6}$$

According to case 3, the general solution is

$$y_h = e^{-2x}(c_1 \cos\sqrt{6}x + c_2 \sin\sqrt{6}x)$$

Or equivalently

$$y_h = k_1 e^{-2x}\sin(\sqrt{6}x + k_2)$$

This last expression is most suitable for finding the integration constants. The solution describes a damped harmonic motion of period $2\pi/\sqrt{6}$.

A technique called a method of undetermined coefficients is utilized to determine the solution for nonhomogeneous equations. This method is used when the input function is only a polynomial, exponential, sine, or cosine, or some combination of these functions. For a quick understanding of this method, let us first introduce an elementary example.

Example 2.24.

Find the solution to the following initial-value problem

$$y'' + 2y' - 3y = e^{-2x};\ y(0) = 1, y'(0) = -1$$

The characteristic equation is

$$r^2 + 2r\ - 3 = 0 \Rightarrow (r+3)(r-1) = 0$$

$$r_1 = -3, r_2 = 1$$

According to case 1, the general solution is

$$y_h = c_1e^{-3x} + c_2e^x$$

The general solution is

$$y(x) = y_h(x) + y_p(x)$$

where the particular solution that satisfies the given equation is given by

$$y_p(x) = ke^{-2x}$$

The derivatives of this particular solution are

$$y_p'(x) = -2ke^{-2x}, y_p''(x) = 4ke^{-2x}$$

Substitute in the given equation

$$4ke^{-2x} + 2(-2ke^{-2x}) - 3ke^{-2x} = e^{-2x}$$

$$(4k - 4k - 3k)e^{-2x} = e^{-2x}$$

Equating the coefficients of the same terms (in this example, only one term) on both sides yields

$$(4k - 4k - 3k) = 1 \Rightarrow\ k = -\frac{1}{3}$$

Thus, the particular solution becomes

$$y_p(x) = -\frac{1}{3}e^{-2x}$$

The complete solution is

$$y(x) = c_1 e^{-3x} + c_2 e^x - \frac{1}{3}e^{-2x}$$

Finally, to determine the integration constants c_1 and c_2, we use the given initial conditions as follows. First, we differentiate the solution.

$$y'(x) = -3c_1 e^{-3x} + c_2 e^x + \frac{2}{3}e^{-2x}$$

Substitute the initial conditions to obtain two algebraic equations as

$$1 = c_1 e^0 + c_2 e^0 - \frac{1}{3}e^0 \Rightarrow \frac{4}{3} = c_1 + c_2$$

$$-1 = -3c_1 e^0 + c_2 e^0 + \frac{2}{3}e^0 \Rightarrow -\frac{5}{3} = -3c_1 + c_2$$

Solving these equations yields

$$c_1 = \frac{3}{4}, c_2 = \frac{7}{12}$$

Finally, the particular solution is

$$y(x) = \frac{3}{4}e^{-3x} + \frac{7}{12}e^x - \frac{1}{3}e^{-2x}$$

Note that it is necessary to find a particular solution, and only then, the integration constants are determined.

For completing the explanation of using the method of undetermined coefficients, the following steps have to proceed.

Step 1: Find the solution to the homogeneous part of the equation.

Step 2: For each term of $f(x)$, prepare the following two points:

a) Make a list of all possible functions that can be obtained by differentiating the given terms.

b) If any term above is a solution of the homogenous part, then multiply each term by the lowest power of x that will make every term distinct from the solutions of the homogenous part.

Step 3: Assume y_p is a linear combination (with coefficients to be determined) of all the distinct terms in step 2.

Step 4: Substitute the assumed form for y_p and its derivatives into the equation.

Then, compare coefficients of similar functions on both sides of the equation. Finally, solve the resulting linear algebraic system to determine the unknown coefficients. We introduce first an example to explain how step 2 can be performed.

Example 2.25.

Consider the equation

$$y'' + 2y' - 3y = 5x^2 - x + e^{-3x} + e^{3x} + \sin\ x$$

Table 2.2. Terms in the method of undetermined coefficients

Terms in $f(x)$	x^2	x	e^{-3x}	e^{3x}	$\sin\ x$
Derivatives terms (without coefficients)	$x, 1$	1	e^{-3x}	e^{3x}	$\sin\ x, \cos x$
Do these terms appear in $y_h(x)$?	No	No	Yes	No	No
Adjust as in step 2 b	$x^2, x, 1$	$x, 1$	xe^{-3x}	e^{3x}	$\sin\ x, \cos x$

The terms of the solution of the homogenous part are e^{-3x} and e^{x}. The implementation of step 2 is contained in Table 2.2. Therefore, the distinct terms are: $x^2, x, 1, xe^{-3x}, e^{3x}, \sin\ x, \cos x$. The proper assumption of for $y_p(x)$ is

$$y_p(x) = k_1x^2 + k_2x + k_3 + k_4xe^{-3x} + k_5e^{3x} + k_6\sin\ x + k_7\cos x$$

The method of undetermined coefficients has to be utilized to find these seven coefficients. For hand calculation, this could be somehow tedious; thus, it is preferable to use the linear systems theory.

Example 2.26.

Find the general solution of the equation

$$y'' + 2y' + 5y = 3\cos 2x + 7e^{x}$$

The characteristic equation is

$$r^2 + 2r + 5 = 0 \Rightarrow (r+1)^2 + 4 = 0 \Longrightarrow [(r+1) + 2j][(r+1) - 2j] = 0$$

The roots are: $r_1 = -1 + 2j, r_2 = -1 - 2j$

The general solution of the homogeneous part is

$$y_h = e^{-x}(c_1 \cos 2x + c_2 \sin 2x)$$

The terms that should be included in $y_p(x)$ are: $\cos 2x,\ \sin 2x,\ e^x$
The particular $y_p(x)$ is

$$y_p(x) = \quad k_1 \cos 2x + k_2 \sin 2x + k_3 e^x$$

The first and second derivatives are

$$y_p'(x) = -2k_1 \sin 2x + 2k_2 \cos 2x + k_3 e^x$$

$$y_p''(x) = -4\, k_1 \cos 2x - 4k_2 \sin 2x + k_3 e^x$$

Substituting in the given equation yields

$$-4k_1 \cos 2x - 4k_2 \sin 2x + k_3 e^x + 2(-2k_1 \sin 2x + 2k_2 \cos 2x + k_3 e^x)$$
$$+5(\ \ k_1 \cos 2x + k_4 \sin 2x + k_3 e^x) = 3 \cos 2x + 7e^x$$

Collecting similar terms

$$(-4k_1 + 4k_2 + 5k_3) \cos 2x + (-4k_2 - 4k_1 + 5k_2)\sin\ 2x + (k_1 + 2k_1 + 5k_3)e^x$$
$$= 3 \cos 2x + 7e^x$$
$$(4k_2 + k_1) \cos 2x + (k_2 - 4k_1)\sin\ 2x + 8k_3 e^x = 3 \cos 2x + 7e^x$$

Equating coefficients yields

$$k_3 = \frac{7}{8}, (4k_2 + k_1) = 3, (\ k_4 - 4k_1) = 0$$

Solving for k_1 and k_2 gives

$$k_1 = \frac{3}{17}, k_2 = \frac{12}{17}$$

The general solution of the given equation is

$$y(x) = e^{-x}(c_1 \cos 2x + c_2 \sin 2x) + \frac{3}{17}\cos 2x + \frac{12}{17}\sin 2x + \frac{7}{8}e^x$$

Example 2.27.

The displacement y of one-kilogram mass m, which is attached to a spring, is determined by applying Newton's second law of motion $my'' = F_t$, where F_t is the sum of three forces. The tension force is given by $(-13\,y)$ N. The damping force proportional to motion velocity is given by $(-4\,y')$ N. The third force is a disturbing force, which is given by $(t^2 + 0.5)$ N. The minus signs with the first two forces are given because they are in the opposite effect of the disturbance force; see Figure 2.6. The mass is moved on a frictionless surface, and initially, it is in rest mode. It is required to determine the displacement response for $t \in [0, 0.5]$ seconds. Therefore, we can write the motion differential equation as

$$my'' = -4y' - 13y + (t^2 + 0.5)$$

First, we write the motion differential equation in standard form, i.e.,

$$y'' + 4y' + 13y = t^2 + 0.5, y(0) = 0, y'(0) = 0$$

The characteristic equation is

$$r^2 + 4r + 13 = 0 \Rightarrow (r + 2)^2 + 9 = 0 \Longrightarrow [(r + 2) + 3j][(r + 2) - 3j] = 0$$

The roots are

$$r_1 = -2 + 3j, r_2 = -2 - 3j$$

The general solution of the homogeneous part is

$$y_h(t) = e^{-2t}(c_1 \cos 3t + c_2 \sin 3t)$$

The terms that should be included in $y_p(t)$ are the terms $t^2, t, 1$, which do not appear in the general solution.

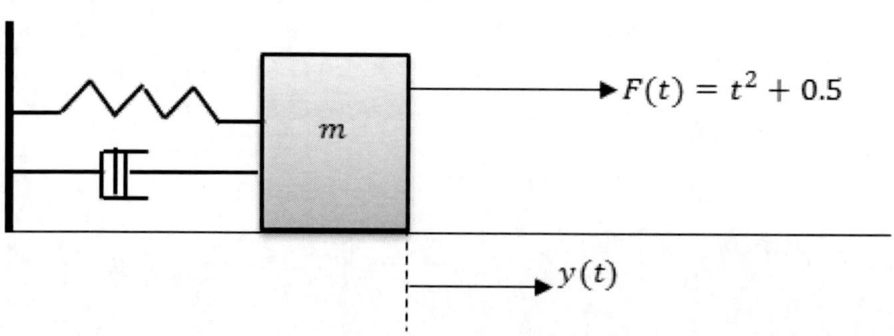

Figure 2.6. A simple dynamic system (Example 2.27).

The particular $y_p(t)$ is

$$y_p(t) = \quad k_1 t^2 + k_2 t + k_3$$

The first and second derivatives are

$$y_p'(t) = \ 2 \ k_1 t + k_2$$

$$y_p''(t) = \ 2 \ k_1$$

Substituting in the given equation gives

$$2k_1 + 4(2k_1 t + k_2) + 13(k_1 t^2 + k_2 t + k_3) = t^2 + 0.5$$

Collecting similar terms

$$13k_1 t^2 + (8k_1 + 13k_2)t + 2k_1 + 8k_2 + 13k_3 = t^2 + 0.5$$

Upon equating coefficients of the same power and solving for the coefficients $k_1, k_2,$ and k_3, we obtain

$$k_1 = \frac{1}{13}, k_2 = -\frac{8}{169}, k_3 = \frac{181}{4394}$$

The general solution of the given equation is

$$y(t) = e^{-2t}(c_1 \cos 3t + c_2 \sin 3t) + \frac{1}{13}t^2 - \frac{8}{169}x + \frac{181}{4394}$$

Upon substituting the initial conditions, we can obtain the two integration constants as

$$c_1 = -\frac{181}{4394}, c_2 = -\frac{77}{6591}$$

Then, the displacement motion is given by

$$y(t) = e^{-2t}(-\frac{181}{4394}\cos 3t - \frac{77}{6591}\sin 3t) + \frac{1}{13}t^2 - \frac{8}{169}x + \frac{181}{4394}$$

The response for $t \in [0, 0.5]$ is shown in Figure 2.7. The displacement at the end of the given interval is 3.14 cm, which could be the last value before the spring's deformation.

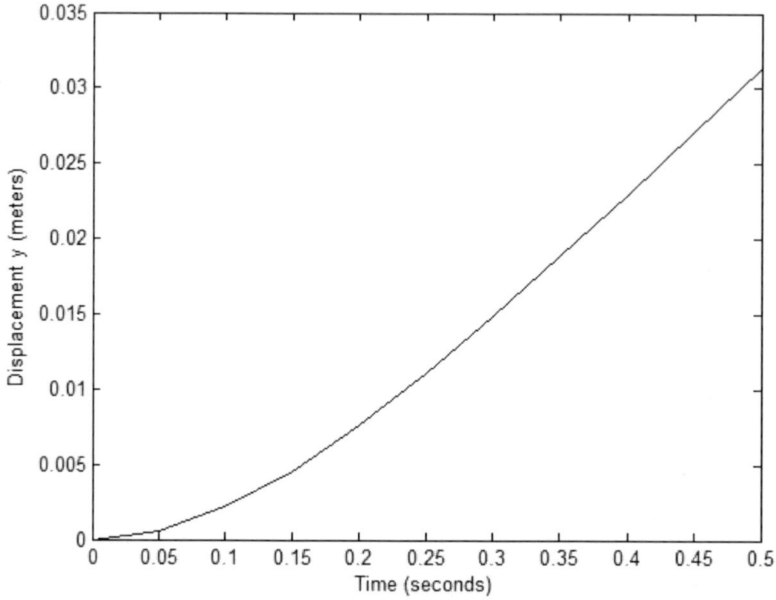

Figure 2.7. The displacement response (Example 2.27).

When the input function $f(x)$ is not of the types that the undetermined coefficients method is suitable, another approach is used. A method named a variation of parameters is a proper technique. However, although this method can be used equally for functions suitable for the method of undetermined coefficients, it is preferred only for unsuitable functions. The following theorem elucidates this method.

Theorem 2.3.

Consider the general solution of the homogenous n^{th}-order differential equation, which is given as

$$y_h(x) = c_1 g_1(x) + c_2 g_2(x) + \cdots + c_n g_n(x) \tag{2.51}$$

Then, the general solution of the nonhomogeneous n^{th}-order differential equation will be given by

$$y(x) = y_h(x) + y_p(x) \tag{2.52}$$

where

$$y_p(x) = w_1(x) g_1(x) + w_2(x) g_2(x) + \cdots w_n(x) g_n(x) \tag{2.53}$$

The functions $w_i(x)$ must satisfy

$$W(x)\begin{bmatrix} w_1'(x) \\ w_2'(x) \\ \vdots \\ w_n'(x) \end{bmatrix} = \begin{bmatrix} 0 \\ 0 \\ \vdots \\ f(x) \end{bmatrix} \tag{2.54}$$

where $W(x)$ is the Wronskian matrix, which is given in the next discussion; see equation (2.56).

Example 2.28.
Solve the second-order equation

$$y'' + y = \csc x$$

The forcing function is unsuitable for the method of undetermined method. The solution of the homogenous part of the equation is given by (as we worked previously),

$$y_h(x) = c_1 \cos x + c_2 \sin x$$

Therefore, the particular solution, as in equation (2.53), is

$$y_p(x) = w_1(x) \cos x + w_2(x) \sin x$$

The functions $w_i(x)$ must satisfy the system

$$\begin{bmatrix} \cos x & \sin x \\ -\sin x & \cos x \end{bmatrix} \begin{bmatrix} w_1'(x) \\ w_2'(x) \end{bmatrix} = \begin{bmatrix} 0 \\ \csc x \end{bmatrix}$$

The two first-order differential equations are

$$w_1'(x) \cos x + w_2'(x) \sin x = 0$$

$$-w_1'(x) \sin x + w_2'(x) \cos x = \csc x$$

From the first equation, we obtain

$$w_1'(x) = -w_2'(x) \tan x$$

Substituting in the second equation and using the trigonometric identities will give

$$w_2'(x) = \cot x$$

Thus,

$$+5(\ k_1 \cos 2x + k_4 \sin 2x + k_3 e^x) = 3\cos 2x + 7e^x$$

$$w_1'(x) = -\cot x \tan x \ = -1 \Rightarrow w_1(x) = -x$$

$$w_2(x) = \int \frac{\cos x}{\sin x}\, dx = \ln(\sin x)$$

The general solution becomes

$$y(x) = c_1 \cos x + c_2 \sin x - x\cos x + \ln(\sin x)\sin x$$

2.7. High-Order Ordinary Differential Equations

In this section, we will consider the linear constant-coefficients n^{th} –order ordinary differential equation of the form

$$y^{(n)} + a_{n-1}y^{(n-1)} + a_{n-2}y^{(n-2)} + \cdots + a_1 y' + a_0 y = f(x) \qquad (2.55)$$

where $y^{(n)}$ is the nth derivative, i.e., $y^{(n)} = \frac{d^n y}{dx^n}$, a_i are real constants; some of these constants may be equal to zero. There are two possible techniques to determine the general solution:

1. To use the roots of the characteristic equation and the method of undetermined coefficients as it was done previously with the second-order equation.
2. To reduce the nth-order equation to a system of n first-order equations: this technique is used mainly when the force function is not of the type for which the method of undetermined coefficients can be applied. This case will be discussed in the next section.

Let us start with an example for illustrating the first technique; all information gained previously was valid, and we will consider the equations with distinct roots. However, the exact solution of the characteristic equation may not be the case, and hence numerical methods should be invoked. Note that throughout the whole chapter, in all used examples, the characteristic equation had an exact form obtained by algebraic manipulation.

Example 2.29.
Find the general solution of the following equation.

$$y''' + 6y'' + 25y' = 23e^{-2x}$$

Note that in this nonhomogeneous equation, the coefficient of y is zero. The characteristic equation is

$$r^3+6r^2+25r=0 \Rightarrow r[(r+3)^2+16]=0 \Longrightarrow r[(r+3)+4j][(r+3)-4j]=0$$

The roots are: $r_1=0, r_2=-3+4j, r_3=-3-4j$

The roots are distinct, and hence the general solution of the homogeneous part is

$$y_h=c_1+e^{-3x}(c_2\cos 4x+c_3\sin 4x)$$

The particular solution $y_p(x)$ has to be of the form

$$y_p(x)=ke^{-2x}$$

Thus, $y_p'=-2ke^{-2x}, y_p''=4ke^{-2x}, y_p'''=-8ke^{-2x}$, substituting in the given equation will yield to

$$-8ke^{-2x}+24ke^{-2x}-50ke^{-2x}=23e^{-2x}$$

$$-34ke^{-2x}=23e^{-2x} \Longrightarrow k=-\frac{23}{34}$$

The general solution is

$$y(x)=y_h+y_p=c_1+e^{-3x}(c_2\cos 4x+c_3\sin 4x)-\frac{23}{34}e^{-2x}$$

Example 2.30.
Find the general solution of the equation

$$y^{iv}+4y'''+6y''+4y'+y=0$$

The characteristic equation is

$$r^4+4r^3+6r^2+4r+1=0$$

The polynomial equation is reduced to

$$(r+1)^4=0 \ \Longrightarrow \ r_{1,2,3,4}=-1$$

The general solution is

$$y(x)=c_1e^{-x}+c_2xe^{-x}+c_3x^2e^{-x}+c_4x^3e^{-x}$$

The general solution n^{th}-order linear differential equation is a linear combination of n different functions (each root gives one of these functions). The correct general solution requires a linear independent combination, i.e., none of these functions can be produced by a linear combination of two or more other functions. In a low-order case, this can be checked easily by inspection. However, for a large-order case, this could be a problem. If these functions are denoted by $y_1, y_2 \dots y_n$, then the following matrix is constructed.

$$W(x) = \begin{bmatrix} y_1 & y_2 & \cdots & y_n \\ y_1' & y_2' & \cdots & y_n' \\ \vdots & \vdots & \vdots & \vdots \\ y_1^{(n-1)} & y_2^{(n-1)} & \cdots & y_n^{(n-1)} \end{bmatrix} \tag{2.56}$$

This matrix is called the Wronskian matrix. The determinant of the matrix $\det(W)$ is called the Wronskian.

Theorem 2.4.

The functions (solutions) $y_1, y_2 \dots y_n$ are linearly independent provided their Wronskian is not identically zero.

It can be shown that for the case of distinct roots, the Wronskian is always nonzero. Then, the solution functions are linearly independent, and the delivered solution is valid.

Example 2.31.

For the equation $y''' + y'' + 4y' + 4y = 0$, the solution is

$$y(x) = c_1 e^{-x} + c_2 \sin 2x + c_3 \cos 2x$$

The solution is valid if the functions $e^{-x}, \sin 2x, \cos 2x$ are linearly independent or if the Wronskian is not equal to zero,

$$\det\left(\begin{bmatrix} e^{-x} & \sin 2x & \cos 2x \\ -e^{-x} & 2\cos 2x & -2\sin 2x \\ e^{-x} & -4\sin 2x & -4\cos 2x \end{bmatrix}\right) = -10e^{-x} \neq 0$$

Therefore, the linear combination of these functions is independent. Note that the rule of selecting the terms (see the table of example 2.25) of the particular solution, which is introduced within the undetermined coefficients method, follows the above theorem. Due to the burden of computation with the Wronskian, it is better to use the rule of delivering the general solution (see example 2.30). The rules of selecting the terms (see the table given in example 2.25) of a particular solution are issued in the method of undetermined coefficients. The reader can check the Wronskian of example 2.25 to see the correctness of the rule used in delivering the general solution.

2.8. THE SOLUTION OF A FIRST-ORDER SYSTEM OF DIFFERENTIAL EQUATIONS

As mentioned previously, one technique to solve a higher-order differential equation is to reduce it to a system of first-order differential equations. As we have seen, in modeling physical systems, a set of differential equations of different orders arises, and the solution for a specific physical variable requires converting all the elements of the set to a first-order system. Furthermore, this approach is essential for handling numerical methods of solution.

The idea of reduction has been considered in section 2.5, where two types of second-order are reduced to an equivalent system of first-order equations. The matrix theory is usually invoked to represent and solve the second (higher)-order system. An illustrative example will be introduced first to motivate the explanation of the subject.

Example 2.32.
Consider the following homogenous third-order equation

$$y''' + 5y'' + 4y' + 3y = 0$$

Let us introduce three variables x_1, x_2, x_3, which are defined as follows.

$$x_1 = y, \qquad x_2 = y', \qquad x_3 = y''$$

Then accordingly, we can write

$$x_1' = y' \Rightarrow x_1' = x_2$$

$$x_2' = y'' \Rightarrow x_2' = x_3$$

$$x_3' = y''' \Rightarrow x_3' = -3y - 4y' - 5y'' \Rightarrow x_3' = -3x_1 - 4x_2 - 5x_3$$

In matrix form, we rewrite the system of first-order equations.

$$\begin{bmatrix} x_1' \\ x_2' \\ x_3' \end{bmatrix} = \begin{bmatrix} 0 & 1 & 0 \\ 0 & 0 & 1 \\ -3 & -4 & -5 \end{bmatrix} \begin{bmatrix} x_1 \\ x_2 \\ x_3 \end{bmatrix}; \; X' = CX$$

$$y = [1 \quad 0 \quad 0] \begin{bmatrix} x_1 \\ x_2 \\ x_3 \end{bmatrix}$$

The matrix C is called a companion matrix, whose characteristic polynomial is the same as the characteristic polynomial of the differential equation, i.e.,

$$c(r) = r^3 + 5r^2 + 4r + 3$$

In engineering applications, these equations are called the canonical state-space representation of the given third-order differential equation. Based on example 2.32, the following theorem generalizes the reduction of nth-order constant-coefficients differential equations to *n* first-order system.

Theorem 2.5.
Consider the *n*th-order linear constant-coefficient differential equation

$$y^{(n-1)} + a_1 y^{(n-1)} + \cdots + a_{n-1}y' + a_0 y = f(x)$$

Similarly, the previous example procedure can be used to obtain an equivalent first-order linear differential system of *n* variables (states). The dependent variable y (the output) is obtained from the solution of the first variable x_1.

$$\begin{bmatrix} x_1' \\ x_2' \\ \vdots \\ x_{n-1}' \\ x_n' \end{bmatrix} = \begin{bmatrix} 0 & 1 & 0 & \cdots & 0 \\ 0 & 0 & 1 & \cdots & 0 \\ \vdots & \vdots & \vdots & \cdots & \vdots \\ 0 & 0 & 0 & \cdots & 1 \\ -a_0 & -a_{n-1} & -a_{n-2} & \cdots & -a_1 \end{bmatrix} \begin{bmatrix} x_1 \\ x_2 \\ \vdots \\ x_{n-1} \\ x_n \end{bmatrix} + \begin{bmatrix} 0 \\ 0 \\ \vdots \\ 0 \\ f(x) \end{bmatrix} \tag{2.57}$$

$$X' = CX + \mathrm{Col}_n(I)f(x) \tag{2.58}$$

The vector X is called the state vector, C is the companion matrix, and $\mathrm{Col}_n(I)$ is the last column of the identity matrix.

This system is a coupled one, and the technique used to solve this system is to make it uncoupled first. The ways to uncouple this first-order system depend on the type of eigenvalues of the companion matrix. The eigenvalues are either:

- All are distinct
- Some are repeated, and others are distinct
- All are equal

Next, we will consider only the case of distinct eigenvalues. For this case, it is known that

$$V^{-1}CV = \mathrm{Dg}[\lambda_1\ \lambda_2 \ldots \lambda_n] \tag{2.59}$$

where λ_i, *i* =1, 2…*n* are the eigenvalues, and V is the Vandermonde matrix given by the following nonsingular matrix.

$$V = \begin{bmatrix} 1 & 1 & \dots & 1 \\ \lambda_1 & \lambda_2 & \dots & \lambda_n \\ \lambda_1^2 & \lambda_2^2 & \cdots & \lambda_n^2 \\ \vdots & \vdots & \vdots & \vdots \\ \lambda_1^{n-1} & \lambda_2^{n-1} & \dots & \lambda_n^{n-1} \end{bmatrix} \tag{2.60}$$

Making the substitution

$$X = VZ \tag{2.61}$$

We can write

$$VZ' = CVZ + \mathrm{Col}_n(I)f(x) \tag{2.62}$$

Multiply both sides from left by V^{-1}, we obtain

$$V^{-1}VZ' = V^{-1}CVZ + V^{-1}\mathrm{Col}_n(I)f(x) \tag{2.63}$$

Hence, we obtain the uncoupled system in the Z state vector

$$Z' = V^{-1}CVZ + V^{-1}\mathrm{Col}_n(I)f(x) \tag{2.64}$$

Therefore, we obtain an uncoupled system whose equations can be solved one by one; if we write the term $V^{-1}\mathrm{Col}_n(I)$ as the last column of V^{-1} by a column vector

$$V^{-1}\mathrm{Col}_n(I) = [\alpha_1 \quad \alpha_2 \quad \cdots \quad \alpha_n]^T \tag{2.65}$$

Then we write the uncoupled system as

$$\begin{aligned} z_1'(x) &= \lambda_1 z_1(x) + \alpha_1 f(x) \\ z_2'(x) &= \lambda_2 z_2(x) + \alpha_2 f(x) \\ &\vdots \\ z_n'(x) &= \lambda_n z_n(x) + \alpha_n f(x) \end{aligned} \tag{2.66}$$

A general solution for the ith equation is

$$z_i(x) = c_i e^{-\lambda_i x} + w_i(x) \tag{2.67}$$

where, $w_i(x)$ is the ith particular solution of the nonhomogeneous part $\alpha_i f(x)$.

The next step of the solution is to determine the state vector X, i.e.,

$$X = \begin{bmatrix} x_1 \\ x_2 \\ \vdots \\ x_{n-1} \\ x_n \end{bmatrix} = V \begin{bmatrix} z_1 \\ z_2 \\ \vdots \\ z_{n-1} \\ z_n \end{bmatrix} \quad (2.68)$$

The general solution of the given equation is

$$y(x) = x_1 \quad (2.69)$$

Note that this technique also determines the derivatives up to power $(n-1)$ of the solution directly, i.e., $y', y'' \dots y^{(n-1)}$ as they are $x_2, x_3 \dots x_n$.

Two examples will be introduced to illustrate the above technique of solution for homogeneous and nonhomogeneous equations.

Example 2.33.

Solve the following fourth-order linear constant-coefficients differential equation.

$$y^{iv} + 2y''' - 16y'' - 2y' + 15y = 0$$

First, we write the companion matrix *C* as

$$C = \begin{bmatrix} 0 & 1 & 0 & 0 \\ 0 & 0 & 1 & 0 \\ 0 & 0 & 0 & 1 \\ -15 & 2 & 16 & -2 \end{bmatrix}$$

The characteristic equation is $\det([\lambda I - C]) = 0$

$$\lambda^4 + 2\lambda^3 - 16\lambda^2 - 2\lambda + 15 = 0$$

The four eigenvalues are $\lambda_1 = -5$, $\lambda_2 = -1, \lambda_3 = 1$, and $\lambda_4 = 3$, which are distinct. Therefore, the Vandermonde matrix is

$$V = \begin{bmatrix} 1 & 1 & 1 & 1 \\ -5 & -1 & 1 & 3 \\ 25 & 1 & 1 & 9 \\ -125 & -1 & 1 & 27 \end{bmatrix}$$

In terms of the Z state vector, the state equation is

$$\begin{bmatrix} z_1' \\ z_2' \\ z_3' \\ z_4' \end{bmatrix} = \begin{bmatrix} -5 & 0 & 0 & 0 \\ 0 & -1 & 0 & 0 \\ 0 & 0 & 1 & 0 \\ 0 & 0 & 0 & 3 \end{bmatrix} \begin{bmatrix} z_1 \\ z_2 \\ z_3 \\ z_4 \end{bmatrix} = \begin{bmatrix} -5z_1 \\ -z_2 \\ z_3 \\ 3z_4 \end{bmatrix}$$

In terms of the Z state vector, the solution is

$$\begin{bmatrix} z_1 \\ z_2 \\ z_3 \\ z_4 \end{bmatrix} = \begin{bmatrix} c_1e^{-5x} \\ c_2e^{-x} \\ c_3e^{x} \\ c_4e^{3x} \end{bmatrix}$$

In terms of the *X* state vector, the solution is

$$\begin{bmatrix} x_1 \\ x_2 \\ x_3 \\ x_4 \end{bmatrix} = \begin{bmatrix} 1 & 1 & 1 & 1 \\ -5 & -1 & 1 & 3 \\ 25 & 1 & 1 & 9 \\ -125 & -1 & 1 & 27 \end{bmatrix} \begin{bmatrix} c_1e^{-5x} \\ c_2e^{-x} \\ c_3e^{x} \\ c_4e^{3x} \end{bmatrix}$$

$$x_1 = c_1e^{-5x} + c_2e^{-x} + c_3e^{x} + c_4e^{3x}$$
$$x_2 = -5c_1e^{-5x} - c_2e^{-x} + c_3e^{x} + 3c_4e^{3x}$$
$$x_3 = 25c_1e^{-5x} + c_2e^{-x} + c_3e^{x} + 9c_4e^{3x}$$
$$x_4 = -125c_1e^{-5x} - c_2e^{-x} + c_3e^{x} + 27c_4e^{3x}$$

The general solution and its three derivatives are

$$y(x) = x_1 = c_1e^{-5x} + c_2e^{-x} + c_3e^{x} + c_4e^{3x}$$
$$y'(x) = x_2 = -5c_1e^{-5x} - c_2e^{-x} + c_3e^{x} + 3c_4e^{3x}$$
$$y''(x) = x_3 = 25c_1e^{-5x} + c_2e^{-x} + c_3e^{x} + 9c_4e^{3x}$$
$$y'''(x) = x_4 = -125c_1e^{-5x} - c_2e^{-x} + c_3e^{x} + 27c_4e^{3x}$$

Note that for this homogeneous case, the inverse of the Vandermonde matrix is not computed. The reader is surly noted the similarity of the matrix technique to that obtained from solving the characteristic equation in the *r* parameter. In particular, the solution of the characteristic equation in terms of the eigenvalue parameter λ is the same as the solution in terms of the *r* parameter used previously.

Example 2.34.
Solve the following third-order nonhomogeneous equation.

$$y''' + 3y'' + 2y' = 1 + x$$

The state equation in terms of the *X* state vector

$$\begin{bmatrix} x_1' \\ x_2' \\ x_3' \end{bmatrix} = \begin{bmatrix} 0 & 1 & 0 \\ 0 & 0 & 1 \\ 0 & -2 & -3 \end{bmatrix} + \begin{bmatrix} 0 \\ 0 \\ 1+x \end{bmatrix}$$

The characteristic equation of the companion matrix is

$$\lambda^3 + 3\lambda^2 + 2\lambda = 0$$

Eigenvalues are $\lambda_1 = -2,\ \lambda_2 = -1,\ \lambda_3 = 0$. The Vandermonde matrix is

$$V = \begin{bmatrix} 1 & 1 & 1 \\ -2 & -1 & 0 \\ 4 & 1 & 0 \end{bmatrix}, \qquad V^{-1} = \frac{1}{2}\begin{bmatrix} 0 & 1 & 1 \\ 0 & -4 & -2 \\ 2 & 3 & 1 \end{bmatrix}$$

In terms of the *Z* state vector, the state equation is

$$\begin{bmatrix} z_1' \\ z_2' \\ z_3' \end{bmatrix} = \begin{bmatrix} -2 & 0 & 0 \\ 0 & -1 & 0 \\ 0 & 0 & 0 \end{bmatrix}\begin{bmatrix} z_1 \\ z_2 \\ z_2 \end{bmatrix} + \begin{bmatrix} 1/2 \\ -1 \\ 1/2 \end{bmatrix}(1+x)$$

$$z_1' = -2z_1 + \frac{1}{2} + \frac{1}{2}x$$

$$z_2' = -z_2 - 1 - x$$

$$z_3' = \frac{1}{2} + \frac{1}{2}x$$

Then, the solutions of the above three linear first-order differential equations are respectively.

$$z_1 = c_1 e^{-2x} + \frac{1}{8} + \frac{1}{4}x$$

$$z_2 = c_2 e^{-x} - x$$

$$z_3 = \frac{1}{2}x + \frac{1}{4}x^2 + c_3$$

In terms of the *X* state vector, the solution is

$$\begin{bmatrix} x_1 \\ x_2 \\ x_3 \end{bmatrix} = \begin{bmatrix} 1 & 1 & 1 \\ -2 & -1 & 0 \\ 4 & 1 & 0 \end{bmatrix}\begin{bmatrix} z_1 \\ z_2 \\ z_3 \end{bmatrix}$$

The general solution

$$y(x) = x_1 = z_1 + z_2 + z_3 = +c_1 e^{-2x} + c_2 e^{-x} + \frac{1}{4}x^2 - \frac{1}{4}x + (c_3 + \frac{1}{8})$$

In the next example, we will show the computation complications when there is at least one complex conjugate pair of eigenvalues.

Example 2.35.
Solve the linear third-order nonhomogeneous equation

$$y''' + y = 1 + 2\sin x$$

In terms of the *X* state vector, the state equation is

$$\begin{bmatrix} x_1' \\ x_2' \\ x_3' \end{bmatrix} = \begin{bmatrix} 0 & 1 & 0 \\ 0 & 0 & 1 \\ -1 & 0 & 0 \end{bmatrix} + \begin{bmatrix} 0 \\ 0 \\ 1 + 2\sin x \end{bmatrix}$$

The characteristic equation of the companion matrix is

$$\lambda^3 + 1 = 0$$

The three Eigenvalues are $\lambda_1 = -1$, $\lambda_{1,2} = \frac{1}{2} \pm \frac{\sqrt{3}}{2}j$. The Vandermonde matrix is

$$V = \begin{bmatrix} 1 & 1 & 1 \\ -1 & \frac{1}{2} + \frac{\sqrt{3}}{2}j & \frac{1}{2} - \frac{\sqrt{3}}{2}j \\ 1 & -\frac{1}{2} + \frac{\sqrt{3}}{2}j & -\frac{1}{2} - \frac{\sqrt{3}}{2}j \end{bmatrix}$$

In terms of the Z state vector, the state equation is

$$\begin{bmatrix} z_1' \\ z_2' \\ z_3' \end{bmatrix} = \begin{bmatrix} -1 & 0 & 0 \\ 0 & \frac{1}{2} + \frac{\sqrt{3}}{2}j & 0 \\ 0 & 0 & \frac{1}{2} - \frac{\sqrt{3}}{2}j \end{bmatrix} \begin{bmatrix} z_1 \\ z_2 \\ z_2 \end{bmatrix} + \begin{bmatrix} 1 \\ \frac{1}{2} - \frac{\sqrt{3}}{2}j \\ -\frac{1}{2} - \frac{\sqrt{3}}{2}j \end{bmatrix} (1 + 2\sin x)$$

$$z_1' = -z_1 + 1 + 2\sin x$$

$$z_2' = (\frac{1}{2} + \frac{\sqrt{3}}{2}j)z_2 + (\frac{1}{2} - \frac{\sqrt{3}}{2}j)(1 + 2\sin x)$$

$$z_3' = (\frac{1}{2} - \frac{\sqrt{3}}{2}j)z_3 + (-\frac{1}{2} - \frac{\sqrt{3}}{2}j)(1 + 2\sin x)$$

As we can see, although the theorem is applied in this case of complex eigenvalues, the solutions of the three equations are tedious to continue. Therefore, we prefer to use the technique of undetermined coefficients in such cases.

In practice, the dynamic system may be described initially by a set of first-order differential equations, as shown in examples 2.3-2.5. Therefore, the canonical state-space may not be available, and the coefficient matrix is not a companion matrix. The illustrated solution procedure, in previous examples, can still be used, but the transformation of the variable must use the general eigenvector matrix P; see chapter one. The next example illustrates this different case.

Example 2.36.

Figure 2.8 shows the system of two interacting tanks of areas A_1 and A_2. The liquid rate flows into the two tanks are $w_1(t)$ and $w_2(t)$, with liquid heads $h_1(t)$ and $h_2(t)$ and they are initially empty. The valve orifice outlet from the first to the second tank is O_{12} and the valve orifice outlet from the second tank is O_2. Assume that the volumes of liquid are A_1h_1 and A_2h_2. After passing units of time, we have to find the heads at that value of time in each tank.

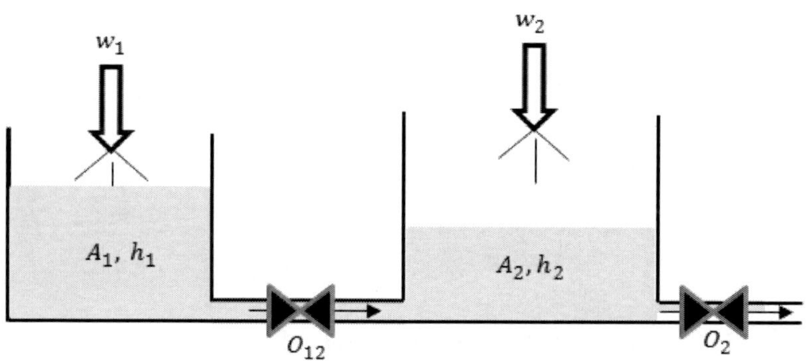

Figure 2.8. Two interacting tanks system (Example 2.36).

Assuming a constant liquid rat flows, the heads in the two tanks are governed by the following two first-order differential equations.

$$h_1' + \frac{1}{A_1 O_{12}}(h_1 - h_2) = \frac{1}{A_1} w_1$$

$$h_2' + \frac{1}{A_2 O_{12}}(-h_1 + h_2) + \frac{1}{A_2 O_2} h_2 = \frac{1}{A_2} w_2$$

In matrix form

$$\begin{bmatrix} h_1' \\ h_2' \end{bmatrix} = \begin{bmatrix} -\dfrac{1}{A_1 O_{12}} & \dfrac{1}{A_1 O_{12}} \\ \dfrac{1}{A_2 O_{12}} & -\dfrac{1}{A_2 O_{12}} - \dfrac{1}{A_2 O_2} \end{bmatrix} \begin{bmatrix} h_1 \\ h_2 \end{bmatrix} + \begin{bmatrix} \dfrac{w_1}{A_1} \\ \dfrac{w_2}{A_2} \end{bmatrix}$$

To continue, let us consider the following parameters: $A_1 = 2.5 \text{ m}^2$, $A_2 = 5 \text{ m}^2, O_{12} = 0.75 \text{ m/m}^3\text{/seconds}$, and $O_2 = 2 \text{ m/m}^3\text{/seconds}$.

The liquid rate flows are: $w_1 = 0.06 \text{ m}^3\text{/seconds}$, and $w_2 = 0.04 \text{ m}^3\text{/seconds}$. The matrix system becomes

$$\begin{bmatrix} h_1' \\ h_2' \end{bmatrix} = \begin{bmatrix} -0.5333 & 0.5333 \\ 0.2667 & -0.3667 \end{bmatrix} \begin{bmatrix} h_1 \\ h_2 \end{bmatrix} + \begin{bmatrix} 0.0024 \\ 0.0004 \end{bmatrix}$$

The eigenvalues are $\lambda_1 = -0.8362, \lambda_2 = -0.0638$, and $P = \begin{bmatrix} -0.8696 & -0.7506 \\ 0.4938 & -0.6608 \end{bmatrix}$.

Using the transformation $H = PZ$, and noting that we have zero initial conditions, the equivalent system becomes

$$\begin{bmatrix} z_1' \\ z_2' \end{bmatrix} = \begin{bmatrix} -0.8362 & 0 \\ 0 & -0.0638 \end{bmatrix} \begin{bmatrix} z_1 \\ z_2 \end{bmatrix} + \begin{bmatrix} -0.6991 & 0.794 \\ -0.5224 & -0.9199 \end{bmatrix} \begin{bmatrix} 0.024 \\ 0.008 \end{bmatrix}$$

Then,

$$z_1 = 0.0104e^{-0.8362\,t} - 0.0104$$

$$z_2 = 0.0199e^{-0.0638\,t} - 0.0199$$

The solution is

$$\begin{bmatrix} h_1 \\ h_2 \end{bmatrix} = \begin{bmatrix} -0.8695 & -0.7506 \\ 0.4939 & -0.6608 \end{bmatrix} \begin{bmatrix} 0.0104e^{-0.8362\,t} - 0.0104 \\ 0.0199e^{-0.0638\,t} - 0.0199 \end{bmatrix}$$

Therefore,

$$h_1(t) = 0.2449 - 0.2341e^{-0.0638\,t} - 0.0108e^{-0.8362\,t}$$

$$h_2(t) = 0.2 - 0.2061e^{-0.0638\,t} + 0.0061e^{-0.8362\,t}$$

The results indicate that the maximum heads will not exceed 24.49 cm and 20 cm in the first and second tanks, respectively. Consequently, the maximum volumes will be 0.61225 m^3 and 1.0 m^3. Also, these volumes are reached approximately after one minute.

EXERCISES

Sections 2.1-2.4

2.1. Classify the following differential equations:

i. $y'' - 4y = \cos x$

ii. $I'' + 5I' + 6I = 0$

iii. $y''' + xy'' - y' + (x - 4)y = 10$

iv. $y''y + 4y' + 3y = e^x$

2.2. Derive the differential equations that describe the following electrical circuits:

i. The electrical circuit that is shown in Figure 2.9

ii. The parallel RLC circuit with a current source $I(t)$; put your answer in the standard form in terms of the natural frequency and damping ratio.

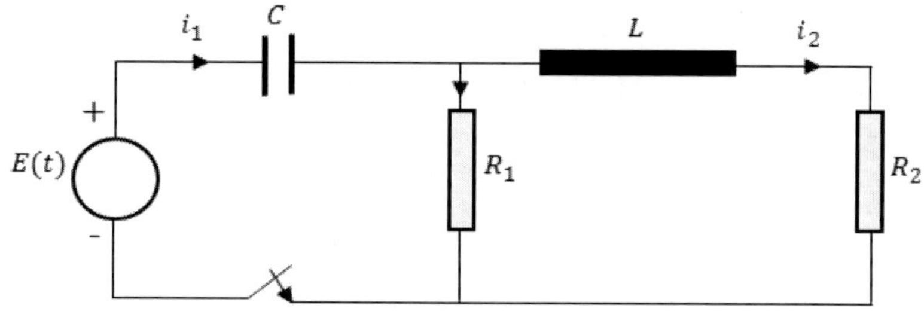

Figure 2.9. An electric circuit (exercise 2.2 i and exercise 2.3).

2.3. In the electric circuit shown in Figure 2.9, the input source is $10 \sin 0.075t$ volts, the capacitance is $0.1\ \mu\text{F}$, the inductance is $50\ \text{mH}$, and the resistors R_1 and R_2 are $10\ \text{K}\Omega$ and $100\ \Omega$ respectively. Determine a single differential equation that calculates the second current.

2.4. Consider the complement structure shown in Figure 2.10. Determine the differential equations that govern the behavior of the system. Then, determine a single differential equation that relates the displacement $y_1(t)$ to the applied force $F(t)$.

2.5. A 1000 liters tank contains a 500 liters solution in which 20 kg of salt is dissolved entirely (the mixture is perfectly stirred at all the time). A solution containing 1 kg/liter of salt is pumped into the tank at a 2.5 kg/min flow rate. The mixture is allowed to flow out at a flow rate of 0.5 kg/min. Write the differential equation that describes this concentration problem.

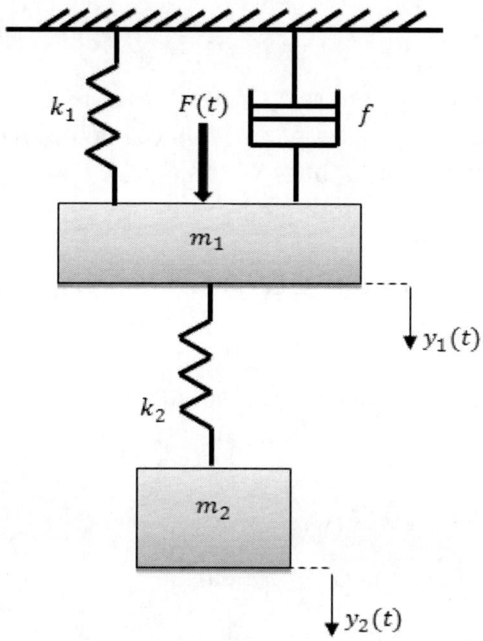

Figure 2.10. A complement structure (exercise 2.4).

2.6. Consider a satellite moves in a plane about the Earth. The mass of the satellite and Earth is m, and M respectively. It is given that the separation distance between them is r. Use the attractive gravitational force concepts between the two masses and the conservation of angular momentum to write the differential equations that govern the satellite motion.

2.7. Solve the following differential equations by direct integration.

i. $y' = xe^{2x}$

ii. $y'' = \sin 2x;\ y(0) = 1, y'(0) = -1$

iii. $w' = t \cos 5t;\ w(0) = 20$

iv. $y' = (x + 1)/(x^2 + 2x)$

2.8. Check the correctness of the given solution for the corresponding differential equation. That is to show that the circuit current satisfies the given differential equation describing the electric circuit

$$I(t) = \frac{5}{6}(2t + 5) - \frac{625000}{48(2t + 5)^5}$$

$$\frac{dI}{dt} + \frac{10I}{2t + 5} = 10;\ I(0) = 0$$

Section 2.5

2.9. Consider a chemical reaction of converting substance S_1 into substance S_2 at a rate that is proportional to the square of the amount of substance S_1. Initially $(t = 0)$, 100 grams of S_1 are there, and after 75 minutes, only 25 grams of S_1 remain unconverted. How many grams remain of S_1 after 90 minutes?

2.10. For concentration exercise 2.5, find the amount of salt in the tank after one hour.

2.11. Solve the following differential equations using the method of variable separation:

i. $4xydx + (x^2 + 1)dy = 0;\ y(0) = 2$

ii. $(x^3y + x^2y)dx + (x^2y^3 + 2x^2y)dy = 0;\ y(2) = 0$

iii. $\sqrt{xy}\frac{dy}{dx} = 1$

iv. $(x + 9)dx - (x^2 - 3x - 4)dy = 0;\ y(5) = 10$

2.12. Solve the following exact equations.

i. $(4x^3y^3 - 2xy)dx + (3x^4y^2 - x^2)dy = 0$

ii. $(x^3 + xy^2 + x^2)dx + x^2ydy = 0$

iii. $(\sin y - y\sin x)dx + (\cos x + x\cos y)dy = 0$

2.13. Solve the following homogenous first-order differential equations.

i. $(2xy + 3y^2)dx - (2xy + x^2)dy = 0, y(1) = 1$

ii. $\left(x^3 + y^2\sqrt{x^2 + y^2}\right)dx - \left(xy\sqrt{x^2 + y^2}\right)dy = 0$

2.14. Consider the following non-exact equation

$$y(\cos^3 x + y\sin x)\ dx + \cos x(\sin x\cos x + 2y)\ dy = 0$$

Show that the term $(1/\cos^2 x)$ is a suitable integrating factor, and then solve the equation.

2.15. Solve the following equations. Note that the considered equations are not given in the same sequence as in the text. Some of these equations need a suitable integrating factor. If possible, write the solution in explicit form.

i. $\left(\frac{1}{x}+\frac{y^4}{x^5}\right)dx-\frac{y^3}{x^4}dy=0$

ii. $\frac{2x}{y}dy+(2\ln 5y+\frac{1}{x})dx=0$

iii. $xy^2dy=(x^2+y^2)dx;\ y(1)=0$

iv. $xdy-ydx=(x^2+y^2)dx$

v. $y'+y\tan x=2\sin x\cos x;\ y(0)=5$

vi. $y+3x\frac{dy}{dx}=y^2\frac{dy}{dx};\ y(0)=1$

vii. $\frac{dr}{d\theta}=\theta-\frac{r}{30};\ r(1)=1$

viii. $ty'-y=t^3-t;\ y(1)=3$

ix. $xy''+3y'=0$

x. $I'+5I=\cos t;\ I(0)=0$

xi. $y''+2y'=3e^x$

xii. $y''+4y=1$

Section 2.6

2.16. Solve the following nonhomogeneous second-order equations using the method of undetermined coefficients.

i. $y''+2y'+7y=t^2+3t+5$

ii. $y''-4y'+3y=3x^3+65\sin 2x$

iii. $y''+6y'+5y=10e^x;\ y(0)=2, y'(0)=-2$

iv. $y''+4y'+5y=3\cos 2x+7e^x$

v. $y''+4y'+13y=x\,e^{-2x}+3\cos 3x$

vi. $y''-3y'+2y=x\,e^x;\ y(0)=1, y'(0)=0$

vii. $y''+y=\sin t;\ y(0)=1, y'(0)=-1$

viii. $y''-2y'=1+x$

2.17. Use the method of variation of parameters to solve the following equations.

i. $y'' + y = \csc x$

ii. $y'' - 2y' - 3y = x \sin x$

iii. $y'' + 4y' = 2\cos 2x$

iv. $y'' + 3y' + 2y = e^{-2x} + e^{-x}$

Sections 2.7-2.8

2.18. Solve the following high-order linear nonhomogeneous equations.

i. $y''' + y = 1 + 2\sin x$

ii. $y''' + 4y' = e^{x} + \cos x$

iii. $y^{iv} + 3y'' + 2y = x, y(0) = y'(0) = 1, y''(0) = y'''(0) = 0$

2.19. Solve the following high-order equations; use the matrix formulation of a first-order system.

i. $y''' + 5y'' + 4y' = e^{-x}$

ii. $y''' + 4y'' + 9y' = \cos 3x;\ y(0) = 2, y'(0) = -1, y''(0) = 0$

iii. $y''' + 8y = e^{-2x}\cos x$

2.20. Apply the solution procedure used in example 2.36 to determine the output voltage v_0 and the voltage across the capacitor C_1 of the bridged-T network shown in Figure 2.11, when an input voltage of 10 volts is applied, $t \geq 0$.

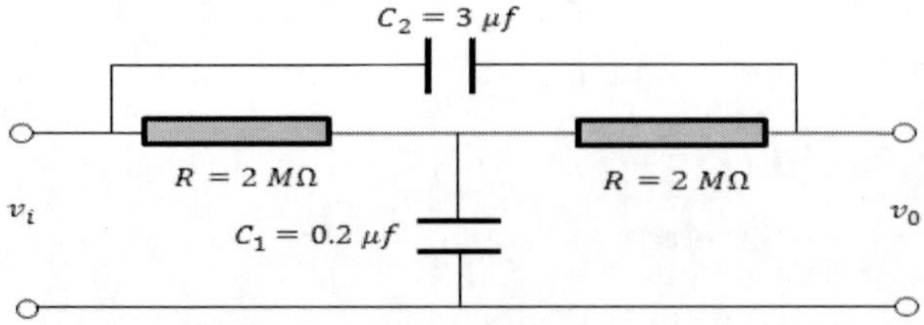

Figure 2.11. A bridged-T network (exercise 2.20)

2.21. Consider the unstable dynamic system that is described by the following state equations.

i. $x_1'(t) = 2x_1(t) + x_2(t) + x_3(t);\ x_1(t) = 1$

ii. $x_2'(t) = 2x_1(t) + 3x_2(t) + 2x_3(t);\ x_2(0) = 2$

iii. $x_3'(t) = x_1(t) + x_2(t) + 2x_3(t);\ x_3(0) = 2$

iv. $y(t) = [1 \quad 0 \quad 0]\, X(t)$

Find the unstable output $y(t)$.

2.22. Resolve the problem in example 2.36 with the following parameters: $A_1 = 1.5\, m^2,\ A_2 = 2m^2, O_{12} = 0.5\ \mathrm{m/m^3/seconds}, O_2 = 1\ \mathrm{m/m^3/sec},\ w_1 = 0.075\ \mathrm{m^3/seconds},$ and $w_2 = 0.05\ \mathrm{m^3/seconds.}$

2.23 For the complement mechanical structure of example 2.3, Consider that the required output is the displacement y_2. Determine the fourth-order differential equation that describes the system then set state equation when a sinusoidal force is applied as $F(t) = \sin wt$.

2.24 A general solution of a coupled first-order linear homogenous system $X'(t) = AX$ consists of n equations is given by

$$X(t) = \sum_{j=1}^{n} C_j v_j e^{\lambda_j t}, \qquad X(t) = \begin{bmatrix} x_1(t) \\ x_2(t) \\ \vdots \\ x_n(t) \end{bmatrix}$$

Where

- C_j are constants of integration
- v_j are the eigenvectors of the coefficients matrix A
- λ_j are the eigenvalues of the coefficients matrix A

Determine the solution of the system

$$X'(t) = \begin{bmatrix} 0 & 1 & 0 & 0 \\ 0 & -2 & 0 & -18 \\ -2 & 0 & 0 & 0 \\ 0 & 0 & -1 & 2 \end{bmatrix} X(t)$$

Chapter 3

Differential Equations – Part II

3.1. Introduction

In this second part, we will study additional topics from the differential equation discipline. We will mainly consider the second-order linear homogeneous variable-coefficients equations and the types of nonlinear equations, which have exact solutions through suitable substitutions. Some named equations, such as Bessel's equations that have significant engineering applications, are solved. The exact solution, the approximated solution's concepts, and the closed-form polynomial solution will be all introduced. A related topic of solving difference equations for the discrete-time system will be given in chapter six. However, we do not claim that we elaborate on this rich topic's complete treatment in this book.

3.2. Second-Order Variable-Coefficients Ordinary Differential Equations

A general linear differential equation of nth-order is

$$a_n(x)y^{(n)} + a_{n-1}(x)y^{(n-1)} + \cdots + a_1(x)y' + a_0(x)y = f(x) \tag{3.1}$$

These equations have coefficients *a*'s and the force function as continuous functions of the independent variable *x*. Only a few of these equations can be solved using elementary functions to give exact closed-form solutions. One of the famous n^{th}-order differential equations with variable coefficients, which have exact closed-form solutions, is the Euler-Cauchy equation (named after the French mathematician Louis Cauchy, see Table 2.1). It has the form (3.1) in which the variable coefficients are:

$$a_j(x) = a_j x^j, j = 0,1,2 \ldots n, a_j \text{ are constants} \tag{3.2}$$

The solution of the Euler-Cauchy equation can be obtained by introducing the following logarithmic substitution.

$$w = \ln|x|, or\ |x| = e^{w} \tag{3.3}$$

Therefore, using the chain rule yields

$$y' = \frac{dy}{dx} = \frac{dy}{dw}\frac{dw}{dx} = \frac{1}{x}\frac{dy}{dw} \tag{3.4}$$

$$y'' = \frac{dy'}{dx} = \frac{d}{dx}\left[\frac{1}{x}\frac{dy}{dw}\right] = -\frac{1}{x^2}\frac{dy}{dw} + \frac{1}{x^2}\frac{d^2y}{dw^2} \tag{3.5}$$

In general, one can easily prove that

$$x^n D^n y = \Delta(\Delta - 1)(\Delta - 2) \dots (\Delta - n + 1)y;\ n \geq 1 \tag{3.6}$$

where the operational symbol D is the first derivative of y with respect to x, and the derivative operator Δ is defined for y with respect to w such that $\Delta y = xDy$.

The Euler-Cauchy equation will be converted to a constant-coefficient differential equation using these substitutions. Then the converted equation can be solved according to the methods of chapter two. The next example solves a third-order Cauchy-Euler equation.

Example 3.1.
Find a complete solution of the equation

$$x^3y''' - 2y'' + 4xy' - 4y = 5x$$

Using equation (3.6) yields

$$\Delta(\Delta - 1)(\Delta - 2)y - 2\Delta(\Delta - 1)y + 5\Delta y - 5y = 5e^{w}$$

Rearranging to have

$$(\Delta^3 - 5\Delta^2 + 8\Delta\ - 4)y = 20e^{w}$$

The characteristic equation is

$$r^3 - 5r^2 + 8r - 4 = (r - 1)(r - 2)^2 = 0$$

The general solution is

$$y_h(w) = c_1e^{w} + c_2e^{2w} + c_3we^{2w}$$

To find the particular solution $y_p(w)$, we proceed as in section 2.7; we let

$$y_p(w) = kwe^w, y_p'(w) = kwe^w + ke^w$$

$$y_p''(w) = kwe^w + 2ke^w, y_p'''(w) = kwe^w + 3ke^w$$

Substituting in the equation gives

$$kwe^w + 3ke^w - 5kwe^w - 10ke^w + 8kwe^w + 8ke^w - 4kwe^w = 5e^w$$

$$ke^w = 5e^w \Rightarrow k = 5$$

Thus, the complete solution in terms of the independent variable w is

$$y(w) = c_1e^w + c_2e^{2w} + c_3we^{3w} + 5we^w$$

Therefore, the complete solution in terms of the independent variable x is

$$y(x) = c_1e^{\ln|x|} + c_2e^{2\ln|x|} + c_3\ln|x|\,e^{3\ln|x|} + 5\ln|x|e^{\ln|x|}$$

$$y(x) = c_1x + c_2x^2 + c_3x^3\ln|x| + 5x\ln|x|$$

For different cases of root values, the reader will find additional results in exercises 3.1 and 3.2.

No general method exists for solving variable-coefficients differential equations. However, the nearest thing to a general solution procedure is the use of infinite series; in some cases, the series have closed-form in terms of elementary functions. Further, in this section, we shall consider a particular type of equation, which is the second-order homogenous with variable coefficients. The considered form is

$$y'' + P(x)y' + Q(x)y = 0 \tag{3.7}$$

The importance of this type is because the modeling of many physical and engineering problems leads to such equations. Many of the nonlinear equations are converted to a variable-coefficients equation via suitable transformations. Moreover, many partial differential equations may be solved using equivalent ordinary differential equations of this type. Before we start the solution procedure, the following definitions are introduced.

Definition 3.1: Analytic functions

They are functions with the continuous derivatives of all orders and a convergent Taylor's series representation.

Definition 3.2: Ordinary Point OP

The point $x = a$ is said to be an ordinary point of the differential equation when both of $xP(x)$ and $x^2Q(x)$ functions are analytic at $x = a$.

Definition 3.3: Singular Point SP

The point $x = a$ is said to be a singular point of the differential equation when it is not an ordinary point.

Definition 3.4: Regular singular point RSP

The point $x = a$ is said to be a regular singular point of the differential equation when both $(x - a)P(x)$ and $(x - a)^2Q(x)$ functions are analytic at $x = a$.

Definition 3.5: Irregular singular point ISP

It is any point $x = a$ of the differential equation, which is singular but not regular.

Example 3.2.

Find the OP, SP, RSP, and ISP of the differential equation

$$x^2(1 - x)y'' + (1 - x)y' + y = 0$$

$$P(x) = \frac{1 - x}{x^2(1 - x)}; \ Q(x) = \frac{1}{x^2(1 - x)}$$

1. All points, except the two points $x = 0$, and $x = 1$, are ordinary points because, at these two points, both $xP(x)$ and $x^2Q(x)$ functions are undefined, and hence they are not analytic. Therefore, they are singular.

2. The point $x = 1$ is a regular singular point because

$$(1 - x)P(x)|_{x=1} = 0, \text{and } (1 - x)^2Q(x)|_{x=1} = 0$$

3. The point $x = 0$ is irregular because

$$xP(x)|_{x=0} = \frac{1 - x}{x\,(1 - x)}|_{x=0} = \infty$$

Although $x^2Q(x)|_{x=0} = 1$

Definition 3.6: Solution form

The general solution of this type of equation can be in the form of infinite series as

$$y(x) = (x-a)^r[a_0 + a_1(x-a) + a_2(x-a)^2 + \cdots + a_n(x-a)^n + \cdots]$$

We can write in a compact form using the series notation as

$$y(x) = (x-a)^r \sum_{k=0}^{\infty} a_k(x-a)^k$$

Or simply

$$y(x) = \sum_{k=0}^{\infty} a_k(x-a)^{k+r}, r \in z \text{ complex} \quad (3.8)$$

In our text, we will consider only the real values of the exponent r.

Theorem 3.1.

The following are valid:

1. At an ordinary point $x = a$ of the equation $y'' + P(x)y' + Q(x)y = 0$, every solution is analytic and can be represented by the infinite series.

$$y(x) = \sum_{k=0}^{\infty} a_k(x-a)^k \quad (3.9)$$

The radius of convergence R is not less than the distance between $x = a$, and the nearest singular point.

2. Around a regular singular point $x = a$, of the considered equation, there is at least one solution, which can be represented by

$$y(x) = \sum_{k=0}^{\infty} a_k(x-a)^{k+r} \quad (3.10)$$

The solution converges for $0 < |x-a| < R$

3. There are, in general, no solutions with expansions consisting solely of powers of $(x-a)$, when the point $x = a$, is an irregular point.

The algebraic handling of the infinite series requires the knowledge of some basic rules for addition or multiplication, expansion, the series of elementary functions, etc., for instance,

1.
$$(x-a)^r \sum_{k=0}^{\infty} a_k (x-a)^k = \sum_{k=0}^{\infty} a_k (x-a)^{k+r}$$
$$= a_0 (x-a)^r + \sum_{k=1}^{\infty} a_k (x-a)^{k+r}$$
$$= a_0 (x-a)^r + \sum_{k=0}^{\infty} a_{k+1} (x-a)^{k+r+1}$$

2.
$$\sum_{k=0}^{\infty} a_k (x-a)^{k+r} - \sum_{k=0}^{\infty} a_{k+2} (x-a)^{k+r+2}$$
$$= a_0 (x-a)^r + a_1 (x-a)^{r+1} + \sum_{k=0}^{\infty} a_{k+2} (x-a)^{k+r+2}$$
$$- \sum_{k=0}^{\infty} a_{k+2} (x-a)^{k+r+2} = a_0 (x-a)^r + a_1 (x-a)^{r+1}$$

3.
$$\sum_{m=0}^{\infty} x^m . \sum_{k=0}^{\infty} a_k x^{k+r} = \sum_{k=0}^{\infty} [\sum_{n=0}^{n-k} a_n] x^{k+r}$$

4.
$$\frac{d}{dx}\left[\sum_{k=0}^{\infty} a_k (x-a)^{r+k}\right] = \sum_{k=0}^{\infty} a_k (r+k)(x-a)^{r+k-1}$$

The method of solution is named after the German mathematician Ferdinand Georg Frobenius as the "method of Frobenius." It is used to find the solution of linear second-order variable-coefficients differential equations at an ordinary point or near a regular singular point in the form of an infinite series. The method procedure will be explored using a nontrivial example.

Example 3.3.
Consider the equation

$$x(x-4)y'' + (x-2)y' - 4y = 0$$

The following steps are the procedure of solution:

Step 1:

To find out the ordinary and regular singular points

$$P(x) = \frac{x-2}{x(x-4)}, Q(x) = -\frac{4}{x(x-4)}$$

Therefore, all points are ordinary except the singular points $x = 0$ and $x = 4$, which are both regular because

$$(x-0)P(x)|_{x=0} = \frac{1}{2}, (x-0)^2Q(x)|_{x=0} = 0$$

$$(x-4)P(x)|_{x=4} = \frac{1}{2}, (x-4)^2Q(x)|_{x=4} = 0$$

The solutions can be obtained around $x = 0$, or around $x = 4$, or any other ordinary point like $x = 1$. These solutions are different, and they are valid at different intervals. We continue to obtain the solution around $x = 0$ (i.e., $a = 0$).

Step 2:

In this step, if the point around which the solution is desired is not the origin $x = 0$, then we have to perform a linear transformation of axes. This case will be discussed later.

Step 3:

In this step, we look for the form of the functions $xP(x)$ and $x^2Q(x)$. Because they are both analytic at the point around which the solution is required, we can write them in polynomial forms as

$$xP(x) = \frac{x-2}{(x-4)} = \sum_{i=0}^{\infty} p_i x^i = p_0 + p_1 x + p_2 x^2 + \cdots$$

$$x^2Q(x) = -\frac{4x}{(x-4)} = \sum_{i=0}^{\infty} q_i x^i = q_0 + q_1 x + q_2 x^2 + \cdots$$

Next, we will show how the values, p_0, p_1, q_0, q_1 are useful for computation and delivering a correct solution. These values can be obtained directly or by an algebraic manipulation or by expanding the function of elementary functions such as $\sin x$ by an infinite series.

Step 4:

We write the solution and its two derivatives. For our example $a = 0$, they are

$$y=\sum_{k=0}^{\infty} a_k x^{k+r}\,,y'=\sum_{k=0}^{\infty} a_k(k+r)x^{k+r-1}\,,y''=\sum_{k=0}^{\infty} a_k(k+r-1)(k+r)x^{k+r-2}$$

Upon substituting these series of the assumed solution and its derivatives in the given equation (note that, in this step, if necessary, we also substitute the polynomial expansion of $xP(x)$ and $x^2Q(x)$), we obtain

$$x(x-4)\sum_{k=0}^{\infty} a_k(k+r-1)(k+r)x^{k+r-2}+(x-2)\sum_{k=0}^{\infty} a_k(k+r)x^{k+r-1}-4\sum_{k=0}^{\infty} a_k x^{k+r}=0$$

$$\sum_{k=0}^{\infty} a_k(k+r-1)(k+r)x^{k+r}-4\sum_{k=0}^{\infty} a_k(k+r-1)(k+r)x^{k+r-1}+\sum_{k=0}^{\infty} a_k(k+r)x^{k+r}$$

$$-2\sum_{k=0}^{\infty} a_k(k+r)x^{k+r-1}-4\sum_{k=0}^{\infty} a_k x^{k+r}=0$$

Collecting summations of the same power of x yields

$$\sum_{k=0}^{\infty} a_k[-4(k+r-1)(k+r)-2(k+r)]x^{k+r-1}$$
$$+\sum_{k=0}^{\infty} a_k[(k+r-1)(k+r)+(k+r)-4]x^{k+r}=0$$

$$a_0[-4(r-1)(r)-2r]x^{r-1}$$
$$+\sum_{k=0}^{\infty} a_{k+1}[-4(k+r-1+1)(k+r+1)-2(k+r+1)]x^{k+r-1+1}$$
$$+\sum_{k=0}^{\infty} a_k[(k+r-1)(k+r)+(k+r)-4]x^{k+r}=0$$

We shift the first series up by one and collect it with the second series of the highest power of x to obtain with one term outside the following series.

$$a_0[-4r^2+2r]x^{r-1}+\sum_{k=0}^{\infty}[\,a_{k+1}[-4(k+r)(k+r+1)-2(k+r+1)]$$
$$+a_k[(k+r-1)(k+r)+(k+r)-4]]x^{k+r}=0$$

Step 5:

If the last equation is to be satisfied, then each coefficient in the series must vanish. Therefore,

$$a_0(-4r^2 + 2r) = 0$$

Without loss of generality, we assume $a_0 \neq 0$, and then we obtain the indicial equation, whose roots are called exponents.

$$(-4r^2 + 2r) = 0 \Longrightarrow r_1 = 0, r_2 = \frac{1}{2}; \text{ distinct values}$$

The other coefficients in the series will vanish if and only if, for each k, it is satisfied with the following recurrent formula.

$$a_{k+1} = \frac{(k + r - 1)(k + r) + (k + r) - 4}{4(k + r)(k + r + 1) + 2(k + r + 1)} a_k, k = 1,2,3 \ldots.$$

Or simply

$$a_{k+1} = \frac{(k + r)^2 - 4}{4(k + r)^2 + 6(k + r) + 2} a_k;\ k = 1,2,3 \ldots.$$

Step 6:

In this step, we construct a complete solution. We expect to obtain two linearly independent solutions corresponding to the distinct values of r.

For $r_1 = 0$, we find the solution y_1 as follows:

$$a_{k+1} = \frac{k^2 - 4}{4k^2 + 6k + 2} a_k;\ k = 1,2,3 \ldots.$$

Hence,

$$a_1 = -2a_0, a_2 = -\frac{3}{12} a_1 = \frac{1}{2} a_0,\ a_3 = 0a_2 = 0, a_j = 0; j = 4,5 \ldots$$

Assuming $a_0 = 1$, we get a closed-form solution of y_1 as

$$y_1 = \sum_{k=0}^{\infty} a_k x^k = a_0 + a_1 x + a_2 x^2 = 1 - 2x + \frac{1}{2} x^2$$

It is a closed-form solution; this could be the case in some problems. Since this solution has a closed form, then we can check its correctness. We have

$$y_1'(x) = -2 + x, y_1''(x) = 1$$

Substituting in the given equation gives a result of zero.

$$x(x-4)1 + (x-2)(-2+x) - 4(1 - 2x + \frac{1}{2}x^2) = 0$$

Similarly, to obtain the second solution, we compute in the recurrent formula with $r_2 = \frac{1}{2}$,

$$a_{k+1} = \frac{(2k+1)^2 - 16}{4[(k+1)^2 + 3(2k+1) + 2} a_k;\ k = 1,2,3 \dots.$$

We find y_2 as

$$y_2 = \sum_{k=0}^{\infty} a_k x^{k+\frac{1}{2}} = x^{\frac{1}{2}}[1 - 0.625x + 0.0547x^2 + 0.00293x^3 + \cdots]$$

The two solutions are linearly independent, and then the complete solution $y(x)$ is a linear combination of $y_1$and y_2, i.e.,

$$y(x) = c_1 y_1 + c_2 y_2$$

$$y(x) = c_1\left[1 - 2x + \frac{1}{2}x^2\right] + c_2 x^{\frac{1}{2}}[1 - 0.625x + 0.0547x^2 + 0.00293x^3 + \cdots]$$

Step 7:

To determine the interval of convergence of the solution (series), we use the ratio test. The series $\sum_0^\infty a_k x^k$ converges for $|x| < \frac{1}{L}$, where

$$L = \lim_{k\to\infty} |\frac{a_{k+1}}{a_k}| = \lim_{k\to\infty} |\frac{(2k+1)^2 - 16}{4[(k+1)^2 + 3(2k+1) + 2}| = \frac{1}{4}$$

Therefore, the series (solution) converges for $|x| < 4$, and the solution of the given equation has the convergence interval $0 < x < 4$. Note that the interval $-4 < x < 0$ is excluded because $x^{1/2}$ is not real on this interval. Therefore y_2 should have the form

$$y_2 = |x^{\frac{1}{2}}|[1 - 0.625x + 0.0547x^2 + 0.00293x^3 + \cdots]$$

The general solution becomes

$$y(x) = c_1\left[1 - 2x + \frac{1}{2}x^2\right] + c_2|x^{\frac{1}{2}}|[1 - 0.625x + 0.0547x^2 + 0.00293x^3 + \cdots]$$

In Figure 3.1, the solution is valid in the shaded zone, and the radiuses of convergence between any ordinary point, say $x = 1$, and the two regular singular points are $R_1 = 1$, and $R_2 = 3$. Note that we will postpone the solution around the ordinary point $x = 1$ until we introduce some additional remarks.

Remark 3.1.

Another different complete general solution, which is valid in a different interval, can be obtained around the second regular point $x = 4$, or near any ordinary point, say $x = 1$. In such cases, we have first to perform linear transformation as we will tickle next.

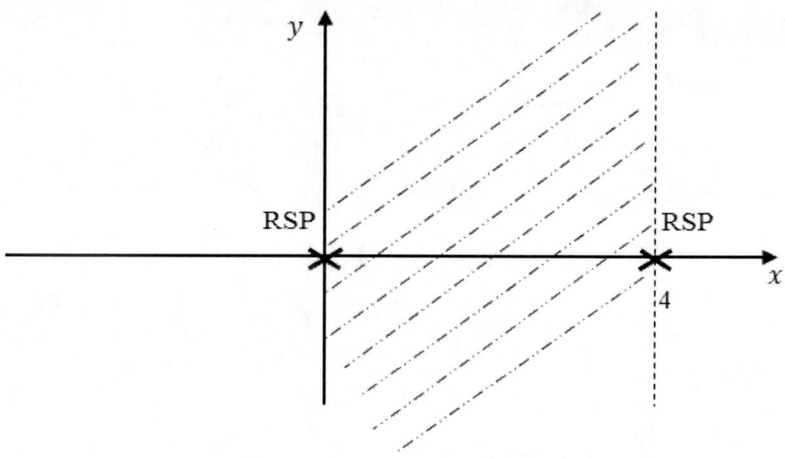

Figure 3.1. Convergence region of the solution (Example 3.3).

Remark 3.2.

For any equation, we can obtain the indicial equation directly from

$$r(r-1) + p_0 r + q_0 = 0 \tag{3.11}$$

Furthermore, for any equation, it can be shown that near any ordinary point, the indicial equation is

$$r(r-1) = 0 \tag{3.12}$$

Let us show that for the equation of example 3.3. We assume

$$xP(x) = \frac{x-2}{(x-4)} = \sum_{i=0}^{\infty} p_i x^i = p_0 + p_1 x + p_2 x^2 + \cdots$$

$$x - 2 = p_0 x + p_1 x^2 + p_2 x^3 + \cdots - 4p_0 - 4p_1 x - 4p_2 x^2 - \cdots$$

If the above equation is correct, then the coefficients of equal power of x in both sides are equal, i.e.,

$$-2 = -4p_0 \Rightarrow p_0 = \frac{1}{2}$$
$$1 = p_0 - 4p_1 \Rightarrow p_1 = -\frac{1}{8}$$

Similarly,

$$x^2 Q(x) = -\frac{4x}{(x-4)} = \sum_{i=0}^{\infty} q_i x^i = q_0 + q_1 x + q_2 x^2 + \cdots$$

$$-4x = q_0 x + q_1 x^2 + q_2 x^3 + \cdots - 4q_0 - 4q_1 x - 4q_2 x^2 - \cdots$$

$$0 = -4q_0 \Rightarrow q_0 = 0$$
$$-4 = q_0 - 4q_1 \Rightarrow q_1 = 1$$

The indicial equation

$$r(r-1) + \frac{1}{2}r + 0 = 0 \Rightarrow r^2 - r + \frac{1}{2}r = 0 \Rightarrow r^2 - \frac{1}{2}r = 0 \Rightarrow r\left(r - \frac{1}{2}\right) = 0$$
$$\Rightarrow r_1 = 0, \qquad r_2 = \frac{1}{2}$$

It is the same answer as we find in the procedure of the method of Frobenius.

Remark 3.3.
The types of exponents of the indicial equation have the following cases:

1. Distinct real exponents with a noninteger difference: as in the case of example 3.3, the complete general solution is a linear combination of the two solutions y_1, and y_2 corresponding to each exponent.
2. If the two exponents are equal, then a complete general solution is given by

$$y(x) = c_1 y_1(x) + c_2 \ln|x| y_1(x) \qquad (3.13)$$

3. Distinct real exponents with integer difference: for each difference, if a specific condition is satisfied, then it is possible to construct the complete general solution as a linear combination of the two solutions y_1, and y_2 corresponding to each exponent. Otherwise, the method of Frobenius will fail to give a complete general solution. For instance, if the difference is 1, then to obtain the complete general solution, the following condition must be satisfied:

$$p_1 r_1 + q_1 = 0, r_1 \text{ is the smallest exponent} \tag{3.14}$$

4. If condition (3.14) is not satisfied, then the two particular solutions must have the forms

$$y_1(x) = \sum_{k=0}^{\infty} a_k x^{k+r_1} \tag{3.15}$$

$$y_2(x) = \sum_{k=0}^{\infty} b_k x^{k+r_1} + c y_1(x) \ln|x| \tag{3.16}$$

With $b_0 = 1$ and *c* is a fixed unknown integer.

Example 3.4.
The equation

$$x^2 y'' - x(4-x)y' + (6-2x)y = 0$$

$$xP(x) = -x\frac{x(4-x)}{x^2} = -4 + x \Rightarrow p_0 = -4, p_1 = 1$$

$$x^2 Q(x) = x^2\frac{(6-2x)}{x^2} = 6 - 2x \Rightarrow q_0 = 6, q_1 = -2$$

The indicial equation is

$$r^2 + (p_0 - 1)r + q_0 = 0 \Rightarrow r^2 - 5r + 6 = 0 \Rightarrow r_1 = 2, r_2 = 3$$

The difference is $r_2 - r_1 = 3 - 2 = 1$. The condition is $p_1 r_1 + q_1 = 0 \Rightarrow 1(2) - 2 = 0$; therefore, we can construct two particular independent solutions, and the complete general solution is a linear combination of these solutions. It is worth mentioning that a difference value of 1 will arise when the exponents are not necessary integers; for example, $\pm\frac{1}{2}$ give a difference of 1 and so on. For other values of this difference, more complicated conditions have to be applied. However, if the condition is unknown, we can determine the two solutions and observe the two solutions for dependency.

Example 3.5.

Find whether the complete general solution (composed of two linearly independent solutions) can be obtained by the method of Frobenius. The equation is

$$x^2y'' - 2xy' + (2 - x)y = 0$$

$$xP(x) = -x\frac{2x}{x^2} = -2 \Rightarrow p_0 = -2, p_1 = 0$$

$$x^2Q(x) = x^2\frac{(2 - x)}{x^2} = 2 - x \Rightarrow q_0 = 2, q_1 = -1$$

The indicial equation is

$$r^2 + (p_0 - 1)r + q_0 = 0 \Rightarrow r^2 - 3r + 2 = 0 \Rightarrow r_1 = 1, r_2 = 2$$

The difference $r_2 - r_1$ is equal to 1, which is an integer. The condition to have two linearly independent solutions is

$$p_1r_1 + q_1 = 0 \Rightarrow \ 0(1) - 1 = -1 \neq 0$$

Therefore, the method of Frobenius fails to give two linearly independent solutions and we have to proceed as is stated in remark 3.3.

Example 3.6.

Let us solve again the equation given in example 3.3 at an ordinary point $x = x_0 = 1$

For simple calculation and by applying Theorem 3.1, we make a linear translation of the *x*-axis. Thus, we introduce another independent variable w, which satisfies

$$w = x - x_0;$$

$$x = w + 1, dx = dw$$

Therefore, a solution around the point $w = 0$ gives the required solution around $x = 1$. By substituting in the given equation, we obtain the equation in terms of the variable *w*.

$$(w^2 - 2w - 3)y'' + (w - 1)y' - 4y = 0; \ y' = \frac{dy}{dw}, y'' = \frac{d^2y}{dw^2}$$

Note now that the point $w = 0$ is also an ordinary point; therefore, it is expected to obtain the two roots equal to 0 and 1. Despite this, we will continue to get the roots as we did in example 3.3.

Assume the solution and its two derivatives as

$$y = \sum_{k=0}^{\infty} a_k w^{k+r}, y' = \sum_{k=0}^{\infty} a_k (k+r) w^{k+r-1}, y'' = \sum_{k=0}^{\infty} a_k (k+r-1)(k+r) w^{k+r-2}$$

Upon substituting in equation with independent variable w yields

$$(w^2 - 2w - 3) \sum_{k=0}^{\infty} a_k (k+r-1)(k+r) w^{k+r-2} + (w-1) \sum_{k=0}^{\infty} a_k (k+r) w^{k+r-1}$$
$$- 4 \sum_{k=0}^{\infty} a_k w^{k+r} = 0$$

Multiplying and collecting terms yields

$$-3 \sum_{k=0}^{\infty} a_k (k+r-1)(k+r) w^{k+r-2} - \sum_{k=0}^{\infty} \big[a_k [2(k+r)(k+r-1) + (k+r)]\big] w^{k+r-1}$$
$$+ \sum_{k=0}^{\infty} \big[a_k [(k+r)(k+r-1) + (k+r) - 4]\big] w^{k+r} = 0$$

To collect all series in one of the highest power, we perform two shifts up for the first series and one shift up of the second series as follows:

$$-3a_0 r(r-1) w^{r-2} - 3a_1 r(r+1) w^{r-1}$$
$$- 3 \sum_{k=0}^{\infty} a_{k+2} (k+r+1)(k+r+2) w^{k+r} - a_0 [2r(r-1) + r] w^{r-1}$$
$$- \sum_{k=0}^{\infty} a_{k+1} [2(k+r+1)(k+r) + (k+r+1)] w^{k+r}$$
$$+ \sum_{k=0}^{\infty} \big[a_k [(k+r)(k+r-1) + (k+r) - 4]\big] w^{k+r} = 0$$

The above equation will be satisfied for every w if and only if each coefficient is equal to zero.

$$w^{r-2}: -3a_0 r(r-1) = 0;\ \text{if } a_0 = 1, \text{then } r_1 = 0, r_2 = 1$$

Although the difference between the exponents is equal to 1, it can be shown that two linearly independent solutions can be obtained; applying the condition given in equation (3.14).

$$w^{r-1}: -3a_1 r(r-1) - a_0[2r(r-1) + r] = 0 \Rightarrow$$

$$a_1 = -\frac{2r(r-1) + r}{3r(r+1)} a_0;\ a_0 = 1$$

$$a_1 = \begin{cases} \text{arbitrary}, r_1 = 0 \\ -\frac{1}{6}, r_2 = 1 \end{cases}$$

Let us chose $a_1 = 0 \text{ for } r_1 = 0$

$$w^{k+r}: -3a_{k+2}(k+r+1)(k+r+2) - a_{k+1}[2(k+r+1)(k+r) + (k+r+1)]$$
$$+a_k[(k+r)(k+r-1) + (k+r) - 4] = 0$$

The recurrent formula is

$$a_{k+2} = \frac{[(k+r)^2 - 4]a_k - [2(k+r)^2 + 3(k+r) + 1]a_{k+1}}{3(k+r+2)(k+r+1)}; k = 0,1,.$$

Now we construct the first particular solution y_1 with $r_1 = 0$ as follows.

$$a_{k+2} = \frac{[k^2 - 4]a_k - [2k^2 + 3k + 1]a_{k+1}}{3(k+2)(k+1)}; k = 0,1,\dots;\ a_0 = 1$$

$$a_2 = \frac{-4a_0 - a_1}{6} = -\frac{2}{3}$$

$$a_3 = \frac{-3a_1 - 6a_2}{18} = \frac{0 - 6(-\frac{2}{3})}{18} = \frac{2}{9}$$

$$a_4 = \frac{(4-4)a_2 - 15a_3}{36} = -\frac{5}{54}$$

$$y_1(w) = \sum_{k=0}^{\infty} a_k w^k = a_0 + a_1 w + a_2 w^2 + \cdots$$

$$y_1(w) = 1 + 0w - \frac{2}{3}w^2 + \frac{2}{9}w^3 - \frac{5}{54}w^4 + \cdots$$

$$y_1(x) = 1 - \frac{2}{3}(x-1)^2 + \frac{2}{9}(x-1)^3 - \frac{5}{54}(x-1)^4 + \cdots$$

Similarly, the second particular solution y_2 with $r_2 = 1$ will be determined.

$$y_2(w) = \sum_{k=0}^{\infty} a_k w^{k+1} = a_0 w + a_1 w^2 + a_2 w^3 + \cdots$$

$$y_2(w) = w - \frac{1}{6}w^2 - \frac{1}{9}w^3 + \frac{5}{108}w^4 + \cdots$$

$$y_2(x) = (x-1) - \frac{1}{6}(x-1)^2 - \frac{1}{9}(x-1)^3 + \frac{5}{108}(x-1)^4 + \cdots$$

$$y(x) = c_1 y_1(x) + c_2 y_2(x)$$

Example 3.7.
Solve the equation

$$y'' + y' + \frac{1}{x-1}y = 0$$

At the ordinary point $x = 0$
First, we write the term $1/(x-1)$as infinite series, since $\frac{1}{x-1} = \sum_{m=0}^{\infty} x^m$

$$y = \sum_{k=0}^{\infty} a_k x^{k+r}, y' = \sum_{k=0}^{\infty} a_k (k+r) x^{k+r-1}, y'' = \sum_{k=0}^{\infty} a_k (k+r-1)(k+r) x^{k+r-2}$$

Substitute in the given equation

$$\sum_{k=0}^{\infty} a_k (k+r-1)(k+r) x^{k+r-2} + \sum_{k=0}^{\infty} a_k (k+r) x^{k+r-1} + \sum_{m=0}^{\infty} x^m . \sum_{k=0}^{\infty} a_k x^{k+r} = 0$$

$$\sum_{k=0}^{\infty} a_k (k+r-1)(k+r) x^{k+r-2} + \sum_{k=0}^{\infty} a_k (k+r) x^{k+r-1} + \sum_{k=0}^{\infty} [\sum_{n=0}^{n-k} a_n] x^{k+r} = 0$$

As we did in example 3.3, we shift the series up to the highest power of x, the $(k + r)$ power, and collect terms from the lowest to the highest power of x. In this example, we get three terms outside the single series

$$a_0[r(r-1)]x^{r-2} + [a_1 r(r+1) +$$

$$+a_0 r]x^{r-1} \sum_{k=0}^{\infty} [a_{k+2}(k+r+2)(k+r+1) + a_{k+1}(k+r+1) + \sum_{n=0}^{n=k} a_n]x^{k+r} = 0$$

Equating the coefficients of similar power to zero yields

$$a_0[r(r-1)] = 0 \Rightarrow \text{with } a_0 = 1,\ r_1 = 0,\ r_2 = 1$$

$$a_1[r(r+1) + a_0 r] = 0 \Rightarrow a_1 = -\frac{a_0 r}{r(r+1)}$$

Therefore, when $r = 0$, the coefficient a_1 can be chosen arbitrarily. The simple choice $a_1 = 0$, thus

$$a_1 = \begin{cases} 0, r = r_1 = 0 \\ -\frac{1}{2}, r = r_2 = 1 \end{cases}$$

$$[a_{k+2}(k+r+2)(k+r+1) + a_{k+1}(k+r+1) + \sum_{n=0}^{n=k} a_n = 0$$

Thus, the recurrent formula is

$$a_{k+2} = -\frac{\sum_{n=0}^{n=k} a_n + (k+r+1)a_{k+1}}{(k+r+2)(k+r+1)}$$

For $r = 0$, we find y_1

$$a_2 = -\frac{a_0 + a_1}{(2)(1)} = -\frac{1+0}{2} = -\frac{1}{2}$$

$$a_3 = -\frac{a_0 + a_1 + a_2(1+1)}{(3)(2)} = -\frac{1 + 0 + \left(\frac{1}{2}\right)2}{6} = 0$$

$$a_4 = -\frac{a_0 + a_1 + a_2 + a_3(2+1)}{(4)(3)} = -\frac{1 + 0 - \frac{1}{2} + 0}{12} = -\frac{1}{24}$$

The first particular solution is

$$y_1(x) = \sum_{k=0}^{\infty} a_k x^k = 1 + (0)x + \left(-\frac{1}{2}\right)x^2 + (0)x^3 + \left(-\frac{1}{24}\right)x^4 + \cdots$$

$$y_1(x) = 1 - \frac{1}{2}x^2 - \frac{1}{24}x^4 + \cdots$$

For $r = 1$, the recurrent formula and the second solution are

$$a_{k+2} = -\frac{\sum_{n=0}^{n=k} a_n + (k+2)a_{k+1}}{(k+3)(k+2)}$$

$$y_2(x) = x - \frac{x^2}{2} - \frac{x^4}{24} + \cdots$$

The complete general solution is

$$y(x) = c_1\left[1 - \frac{1}{2}x^2 - \frac{1}{24}x^4 + \cdots\right] + c_2\left[x - \frac{x^2}{2} - \frac{x^4}{24} + \cdots\right]$$

3.3. Special Second-Order Ordinary Differential Equations

There are many special second-order variable-coefficients differential equations that most of them are named after mathematicians or scientists. The solutions to some of these equations are manipulated mathematically in a similar manner as the elementary functions. The solutions to some other equations are given in polynomial forms. We will consider in this section one of the most important and known equations; it has the form

$$t^2y'' + ty' + (\lambda^2t^2 - \nu^2)y = 0 \tag{3.17a}$$

It is called Bessel's (German astronomer, mathematician, physicist, and geodesist) equation of order ν and parameter λ. If we make the substitution $x = \lambda t$, then we obtain

$$x^2y'' + xy' + (x^2 - \nu^2)y = 0 \tag{3.17b}$$

Equation (3.16b) is called simply a Bessel's equation of order ν.

The solution of Bessel's equation of order ν can be obtained by the method of Frobenius. It can be shown that its indicial equation is $r^2 - \nu^2 = 0$, and thus the exponents are $\pm\nu$. For the positive exponent $r = \nu$, we can obtain, using an algebraic manipulation, the solution in the form

$$y_1(x) = \sum_{m=0}^{\infty} \frac{(-1)^m x^{\nu+2m}}{2^{\nu+2m}\, m!\, \Gamma(\nu+m+1)} \tag{3.18}$$

where Γ is the gamma function (the generalized factorial function).

The solution $y_1(x)$ is known as Bessel's function of the first kind of order ν, and it is denoted by the symbol $J_\nu(x)$. The function $J_\nu(x)$ converges for all values of t if $\nu \geq 0$, and the equation $J_\nu(x) = 0$ has an infinite number of real roots. Figure 3.2 shows a plot of $J_0(x)$, $J_1(x)$ and $J_2(x)$. The reader might note the similarity with the sinusoidal functions. Numerically, the computation of these functions for noninteger ν requires an available table of the Γ function.

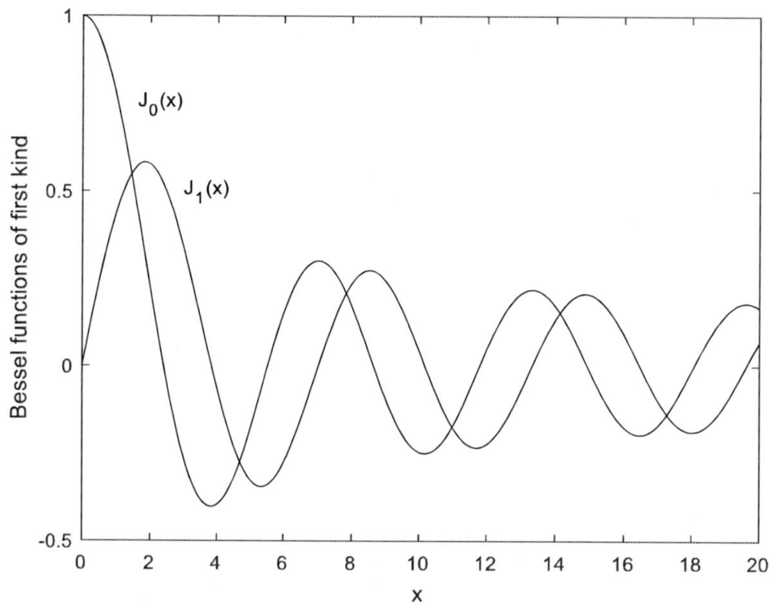

Figure 3.2. Bessel's function of the first kind.

Similarly, we can determine the second solution associated with the negative exponent $r = -\nu$ directly by substituting in $y_1(x)$, $-\nu$ instead of $+\nu$ provided that the Γ-function in all series terms are defined (or ν should not be an integer). The second solution is

$$J_{-\nu}(x) = \sum_{m=0}^{\infty} \frac{(-1)^m x^{-\nu+2m}}{2^{-\nu+2m}\, m!\, \Gamma(-\nu + m + 1)} \tag{3.19}$$

Because the function $J_{-\nu}(x)$ has a negative power of x at $m = 0$, then $J_{-\nu}(x)$ is unbounded at the $x = 0$, while the function $J_\nu(x)$ is finite at $x = 0$. Therefore, they are linearly independent, and a complete solution of the Bessel's equation can be obtained, i.e.,

$$y(x) = c_1 J_\nu(x) + c_2 J_{-\nu}(x) \tag{3.20}$$

For many reasons concerning the applications, another form of solution is considered. The following linear combination is taken as a second independent solution, see Figure 3.3.

$$Y_\nu(x) = \frac{\cos \nu\pi \; J_\nu(x) - J_{-\nu}(x)}{\sin \nu\pi} \tag{3.21}$$

The complete solution is

$$y(x) = c_1 J_\nu(x) + c_2 Y_\nu(x); \; \nu \text{ is not an integer} \tag{3.22}$$

The solution $Y_\nu(x)$ is called the second kind of order ν. The previous two complete solutions are correct even if the value of ν is an odd multiple of 0.5, and the exponents of the indicial equation are differed by an odd integer. However, when ν is an integer, these solutions are not valid (the linear independence will be broken). Figure 3.3 depicts different second-order kinds of Bessel functions. By comparing the plots of Figures 3.3 and 3.2, it is evident that the two solutions $J_\nu(x)$ and $Y_\nu(x)$ are linearly independent.

Another equation named modified Bessel equation of order ν is found in many applications. It has the form

$$y'' + \frac{1}{x}y' - \left(\lambda^2 + \frac{\nu^2}{x^2}\right)y = 0 \tag{3.23}$$

Since we can write

$$y'' + \frac{1}{x}y' + \left(j^2\lambda^2 - \frac{\nu^2}{x^2}\right)y = 0; \; j = \sqrt{-1} \tag{3.24}$$

It is a Bessel equation of order $\nu = j\lambda$; the complete solution is expressed by the other Bessel functions as

$$y(x) = c_1 I_\nu(x) + c_2 I_{-\nu}(x) \tag{3.25}$$

where I_ν and $I_{-\nu}$ are called the modified first-kind Bessel functions.

Similarly, we replace the function $I_{-\nu}$ by the modified second kind Bessel function K_ν, which is given by

$$K_\nu(x) = \frac{\pi}{2} \frac{I_{-\nu}(x) - I_\nu(x)}{\sin \; \nu x} \tag{3.26}$$

Figures 3.4 and 3.5 illustrate these modified Bessel functions, which are linearly independent. Therefore, the complete solution can be expressed similarly by the other Bessel functions as

$$y(x) = c_1 I_\nu(x) + c_2 K_\nu(x) \tag{3.27}$$

If ν is an integer equal n, then to obtain a complete solution, we compute

$$Y_n(x) = \lim_{\nu \to n} \frac{\cos \nu\pi \; J_\nu(x) - J_{-\nu}(x)}{\sin \nu\pi} \tag{3.28}$$

$$y(x) = c_1 J_n(x) + c_2 Y_n(x) \tag{3.29}$$

Similarly, for the modified Bessel equation, the complete solution is

$$y(x) = c_1 I_n(x) + c_2 K_n(x) \tag{3.30}$$

where

$$K_n(x) = \lim_{\nu \to n} \frac{\pi}{2} \frac{I_{-\nu}(x) - I_\nu(x)}{\sin \nu x} \tag{3.31}$$

In practice, these functions are tabulated (or programmed) for direct use. In math books, several identities, relations, and even forms for differentiation and integration are found. Moreover, these functions constitute the basics for solving a large number of specific variable-coefficients differential equations that arise in different fields, as we will explain at the end of this section.

It is worth mentioning that there is a package of formulas and calculus rules to work with Bessel functions. For example, the following formulas are a sample from many others that are found in the literature.

$$J_\nu(x) = \frac{1}{\pi} \int_0^\pi \cos(n\varphi - x \sin \varphi)\, d\varphi, n \text{ is an integer}$$

$$J_{\frac{1}{2}}(x) = \sqrt{\frac{2}{\pi x}} \sin x, J_{-\frac{1}{2}}(x) = \sqrt{\frac{2}{\pi x}} \cos x$$

$$J_{\frac{3}{2}}(x) = \sqrt{\frac{2}{\pi x}} (\frac{\sin x}{x} - \cos x), J_{-\frac{3}{2}}(x) = -\sqrt{\frac{2}{\pi x}} (\frac{\cos x}{x} + \sin x)$$

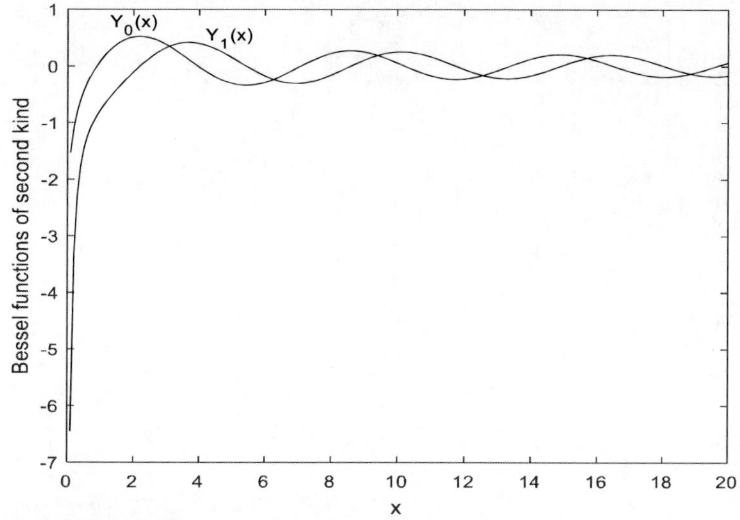

Figure 3.3. Bessel's function of the second kind.

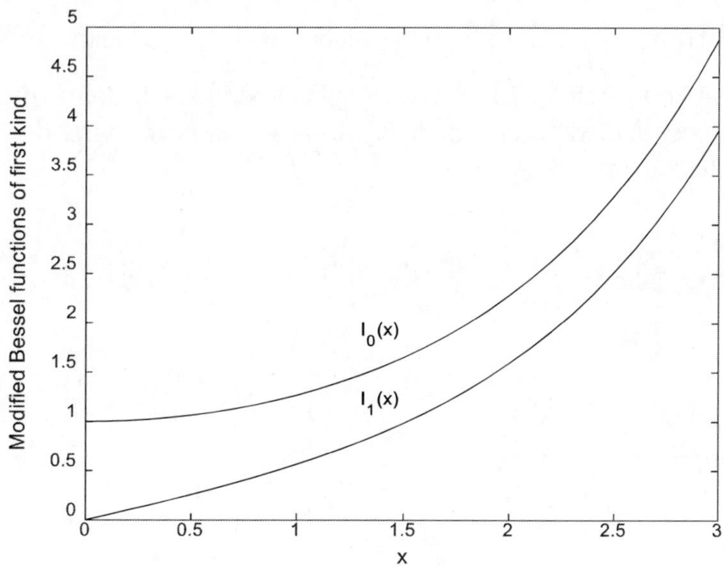

Figure 3.4. Modified Bessel's function of the first kind.

$$\frac{d}{dx}[x^m J_v(ax)] = ax^m J_{v-1}(ax) - (v-m)x^{m-1}J_v(ax)$$

$$J_{v-1}(ax) = \frac{2v}{ax}J_v(ax) - J_{v+1}(ax)$$

$$\int x^v J_{v-1}(x)dx = x^v J_v(x) + c$$

$$\int x^{-v} J_{v+1}(x)dx = -x^{-v} J_v(x) + c$$

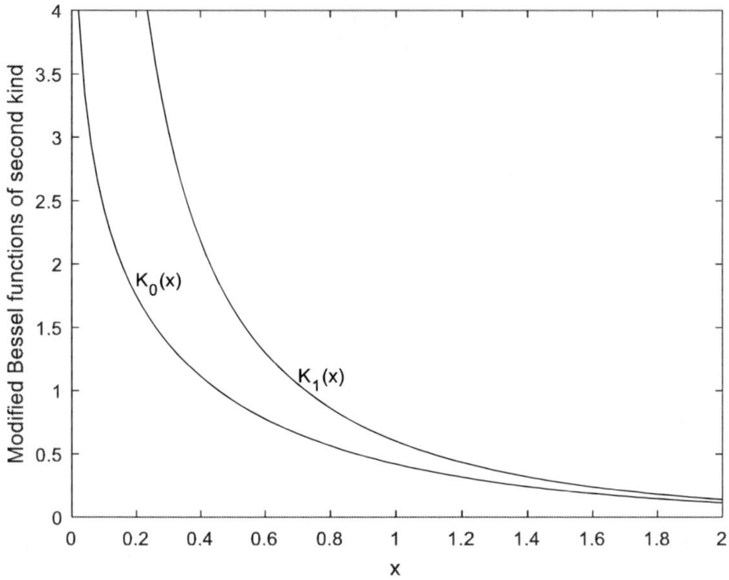

Figure 3.5. Modified Bessel's function of the second kind.

The reader may refer to the mathematic textbooks to learn more about these special functions. We will consider only the application of these functions in solving variable-coefficients differential equations.

Example 3.8.

Solve the equation

$$x^2y'' + xy' + \left(9x^2 - \frac{1}{4}\right)y = 0$$

By comparing to the Bessel equation we get

$$\nu^2 = \frac{1}{4}, \qquad \lambda^2 = 9 \;\Rightarrow \nu = \frac{1}{2}, \qquad \lambda = 3$$

Since ν is not an integer, then the complete solution is

$$y(x) = c_1 J_{0.5}(3x) + c_2 J_{-0.5}(3x)$$

where

$$J_{0.5}(3x) = \sum_{m=0}^{\infty} \frac{(-1)^m (3x)^{0.5+2m}}{2^{0.5+2m}\, m!\; \Gamma(m+1.5)}$$

$$J_{-0.5}(3x) = \sum_{m=0}^{\infty} \frac{(-1)^m (3x)^{-0.5+2m}}{2^{-0.5+2m}\, m!\, \Gamma(m+0.5)}$$

The reader can use remarks 3.2 to find that the roots of the indicial equation are ± 0.5 too, and the condition (3.14) of remark 3.3 is satisfied to have two independent solutions around the singular point $x = 0$. Obviously, despite the difference in shape, the solution obtained by the method of Frobenius is identical to that obtained by the Bessel function above; see exercise 3.15. However, Bessel functions are linearly independent without requiring any test. Further, note that the Γ function is usually found from a particular table. However, in MATLAB toolboxes, this function is ready to calculate or plot the behavior of the solution of the given differential equation.

There are many second-order variable-coefficients differential equations, whose solutions can be obtained in terms of Bessel functions. The following theorem and its corollary contribute to reducing equations of specific forms to equivalent Bessel equations, and the solutions can be reached directly using the Bessel functions.

Theorem 3.2.

The equation

$$x^2 y'' + x(a + 2bx^p)y' + [c + dx^{2q} + b(a + p - 1)x^q + b^2 x^{2p}]y = 0 \quad (3.32)$$

Has a complete solution

$$y(x) = x^{\alpha} e^{-\beta x^p} [c_1 J_\nu(\lambda x^q) + c_2 Y_\nu(\lambda x^q)] \quad (3.33)$$

Where

$$\alpha = \frac{1}{2}(1 - a), \qquad \beta = \frac{b}{p}, \qquad \lambda = \frac{\sqrt{|d|}}{q}, \qquad \nu = \frac{\sqrt{(1-a)^2 - 4c}}{2q} \quad (3.34)$$

Provided that the following conditions be satisfied

1. $(1 - a)^2 \geq 4c$
2. The values of d, p, q are not equal to zero.

Besides, if $d < 0$, then we replace J_ν and Y_ν by I_ν and K_ν, respectively. However, in the complete solution, if ν is not an integer, then we can use the functions $J_{-\nu}$ or $I_{-\nu}$.

Corollary 3.1
The equation

$$(x^r y')' + (ax^s + bx^{r-2})y = 0 \tag{3.35}$$

Has a complete solution of the form

$$y(x) = x^\alpha [c_1 J_v(\lambda x^\gamma) + c_2 Y_v(\lambda x^\gamma)] \tag{3.36}$$

where

$$\alpha = \frac{1-r}{2}, \qquad \gamma = \frac{2-r+s}{2}, \qquad \lambda = \frac{2\sqrt{|a|}}{2-r+s}, \qquad \nu = \frac{\sqrt{(1-r)^2 - 4b}}{2-r+s} \tag{3.37}$$

The following conditions should be satisfied

$$a \neq 0, \qquad (1-r)^2 \geq 4b, \qquad s > r-2 \text{ or } b = 0$$

If $a < 0$, if v is not an integer, we do the replacement as in the theorem. Furthermore, we add the following lemma

Lemma 3.1.
A complete solution of the equation $(x^m y'')'' = k^2 x^{m-2} y$ can be obtained by a linear combination of the solutions of the following two differential equations.

$$(x^{m-1} y')' = k\, x^{m-2} y \tag{3.38}$$

$$(x^{m-1} y')' = -k\, x^{m-2} y \tag{3.39}$$

Example 3.9.
Find a complete solution in terms of Bessel functions of the equation

$$x^2 y'' + 2x^2 y' + (x^4 + x^2 - 2)y = 0$$

First, we equate each of the coefficients of the given equation to the equation given in theorem 3.2, i.e.,

$$2x^2 = x(a + 2bx^p) \Rightarrow a = 0, p = 1, b = 1$$

$$(x^4 + x^2 - 2) = c + dx^{2q} + b(a + p - 1)x^q + b^2 x^{2p}$$

$$= c + dx^{2q} + 1(0 + 1 - 1)x^q + 1^2 x^{2\times 1} = c + dx^{2q} + x^2$$

$$\Rightarrow c = -2, \qquad d = 1, \qquad q = 2$$

Check the necessary conditions

1. $(1-a)^2 \geq 4c \Longrightarrow 1 \geq -8$
2. $d \neq 0, p \neq 0, q \neq 0$

Therefore, we can apply the theorem to express the complete solution. We compute

$$\alpha = \frac{1}{2}(1-0), \qquad \beta = \frac{1}{1}, \qquad \lambda = \frac{\sqrt{|1|}}{2}, \qquad \nu = \frac{\sqrt{(1-0)^2+8}}{4}$$

$$\Longrightarrow \alpha = \frac{1}{2}, \qquad \beta = 1, \qquad \lambda = \frac{1}{2}, \qquad \nu = \frac{3}{4}$$

Since ν is not an integer, we can write the complete solution as

$$y(x) = x^{\frac{1}{2}} e^{-x} \left[c_1 J_{\frac{3}{4}}\left(\frac{1}{2}x^2\right) + c_2 J_{\frac{-3}{4}}\left(\frac{1}{2}x^2\right)\right]$$

Note that we use $J_{\frac{-3}{4}}$ because $d > 0$ and ν is not an integer.

Example 3.10.
Solve in terms of Bessel function the equation

$$x^2 y'' + (2x^3 + 3x)y' - (3x^4 - 4x^2 - 1)y = 0$$

Equate each of the coefficients of the given equation to the equation given in the theorem 3.2 yields.

$$2x^3 + 3x = x(a + 2bx^p) \quad \Longrightarrow \quad a = 3, \qquad p = 2, \qquad b = 1$$

$$(-3x^4 + 4x^2 - 1) = c + dx^{2q} + b(a + p - 1)x^q + b^2 x^{2p}$$

$$= c + dx^{2q} + 4x^q + x^4$$

$$\Longrightarrow c = 1, \qquad d = -3 - 1 = -4, \qquad 2q = 4, q = 2$$

The necessary conditions are satisfied; thus, we compute

$$\alpha = \frac{1}{2}(1-3) = -1, \qquad \beta = \frac{1}{2}, \qquad \lambda = \frac{\sqrt{|-4|}}{2} = 1, \qquad \nu = \frac{\sqrt{(1-3)^2+4}}{4} = 0$$

Since $d < 0$, the complete solution will be

$$y(x) = x^{-1}e^{-\frac{1}{2}x^2}\left[c_1 I_0(x^2) + c_2 K_0(x^2)\right]$$

Example 3.11.
Solve in terms of Bessel's function the equation

$$y'' + x^m y = 0$$

According to lemma 3.1, we can rewrite the given equation as

$$(x^0 y')' + x^m y = 0$$

Using corollary 3.1, we can find that

1. $r = 0$
2. $x^m = ax^s + bx^{0-2} \Rightarrow b = 0, a = 1, s = m$

The necessary conditions are satisfied since

$$a \neq 0, (1-r)^2 \geq 4b, b = 0$$

We compute

$$\alpha = \frac{1-r}{2}, \gamma = \frac{2-r+s}{2}, \lambda = \frac{2\sqrt{|a|}}{2-r+s}, \nu = \frac{\sqrt{(1-r)^2-4b}}{2-r+s}$$

$$\alpha = \frac{1-0}{2} = \frac{1}{2}, \qquad \gamma = \frac{2-0+m}{2} = \frac{2+m}{2},$$

$$\lambda = \frac{2\sqrt{|1|}}{2-0+m} = \frac{2}{2+m}, \qquad \nu = \frac{\sqrt{(1-0)^2-4(0)}}{2-0+m} = \frac{1}{2+m}$$

Because $a > 0$ and v may have an integer value (m = -1), the complete solution should be written in the form

$$y(x) = x^{\frac{1}{2}}[c_1 J_{\frac{1}{2+m}}(\frac{2}{2+m}x^{\frac{2+m}{2}}) + c_2 Y_{\frac{1}{2+m}}(\frac{2}{2+m}x^{\frac{2+m}{2}})]$$

Example 3.12.

Solve in terms of Bessel's function the equation

$$(x^2 y'')'' = 9y$$

The complete solution is obtained by solving the following two differential equations.

$$(x^{2-1}y')' = 3y, \text{ and } (x^{2-1}y')' = -3y$$

For the first equation, using corollary gives

$$r = 1, \quad b = 0, \quad s = 0, \quad a = -3$$

$$\alpha = 0, \quad \gamma = \frac{1}{2}, \quad \lambda = 2\sqrt{3}, \quad v = 0$$

The necessary conditions are satisfied, and since $a = 0$ and $v = 0$, then the solution of the first equation becomes

$$y(x) = [c_1 I_0\left(2\sqrt{3}, x^{\frac{1}{2}}\right) + c_2 K_0\left(2\sqrt{3}, x^{\frac{1}{2}}\right)]$$

For the second equation, we can similarly determine a solution but with $a = 3$, thus

$$y(x) = [c_3 J_0\left(2\sqrt{3}, x^{\frac{1}{2}}\right) + c_4 Y_0\left(2\sqrt{3}, x^{\frac{1}{2}}\right)]$$

Since the two solutions are linearly independent, we can use their linear combination to obtain a complete fourth-order equation solution.

$$y(x) = k_1 J_0\left(2\sqrt{3}, x^{\frac{1}{2}}\right) + k_2 Y_0\left(2\sqrt{3}, x^{\frac{1}{2}}\right) + k_3 I_0\left(2\sqrt{3}, x^{\frac{1}{2}}\right) + k_4 K_0\left(2\sqrt{3}, x^{\frac{1}{2}}\right)$$

Example 3.13.

A body whose mass varies according to the law $m = m_0(1 + \delta t)$ moves along the x-axis under the influence of a force of attraction F, which varies directly as the distance from the origin changes. Determine the motion equation of the body if it starts from rest at the point $x = x_0$; see Figure 3.6.

We apply the general form of Newton second law, i.e.,

$$\frac{d}{dt}(mv) = -F = -k^2x$$

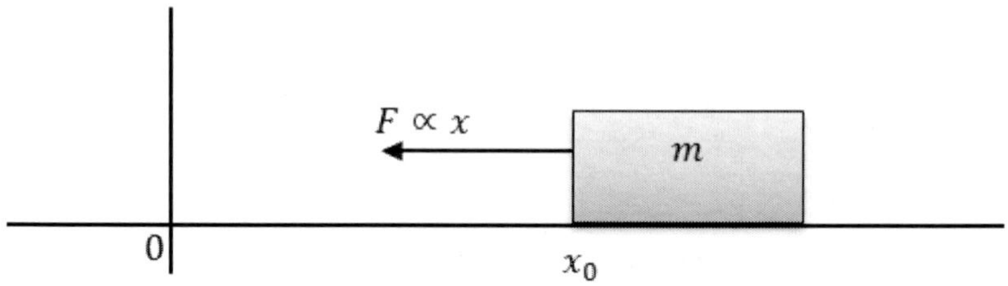

Figure 3.6. A moving body (Example 3.13).

where k^2 is a proportional constant, and v is the velocity of motion equal to x'. The minus sign is introduced because the applied force is an attraction force. Therefore, the equation that governs the motion is

$$\frac{d}{dt}(mx') + k^2x = 0$$

The above equation can be solved if we relate the position variable x to the mass variable m, i.e., we require obtaining the equation in the following form

$$l.\frac{d}{m}\left(m\frac{dx}{dm}\right) + k^2x = 0$$

where l is constant.

Since we have given $m = m_0(1 + \alpha t)$, thus

$$\frac{dm}{dt} = m_0\delta$$

$$\frac{dm}{dt}.\frac{dx}{dm} = m_0\delta\frac{dx}{dm} \Rightarrow \frac{dx}{dt} = m_0\delta\frac{dx}{dm}$$

To obtain the required equation, we successively substitute it into the equation governing the motion.

$$\frac{d}{dt}\left(mm_0\delta\frac{dx}{dm}\right) + k^2x = 0$$

$$m_0\delta\frac{d}{dt}\left(m\frac{dx}{dm}\right) + k^2x = 0$$

$$m_0\delta\left(\frac{dm}{dt}\frac{dx}{dm}+m\frac{d^2x}{dm\,dt}\right)+k^2x=0$$

$$m_0\delta\left(m_0\delta\frac{dx}{dm}+m\frac{dx}{dm}.\frac{dx}{dt}\right)+k^2x=0$$
$$m_0\delta\left(m_0\delta\frac{dx}{dm}+mm_0\delta\frac{d^2x}{dm^2}\right)+k^2x=0$$

$$m_0^2\delta^2\left(\frac{dx}{dm}+m\frac{d^2x}{dm^2}\right)+k^2x=0$$

$$m_0^2\delta^2\frac{d}{dm}\left(m\frac{d}{m}\right)+k^2x=0$$

Finally, we obtain

$$m_0^2\delta^2(mx')'+k^2x=0$$

Or

$$(mx')'+\frac{k^2}{m_0^2\delta^2}\,x=0$$

To solve this last equation, we use the corollary of theorem 2.7; thus

$$r=1,\qquad b=0,\qquad s=0,\qquad a=\frac{k^2}{m_0^2\delta^2}$$

Therefore, we compute

$$\alpha=0,\qquad \gamma=\frac{1}{2},\qquad \lambda=\frac{2k}{m_0\delta},\qquad v=0$$

The solution in terms of Bessel's functions is

$$x(t)=c_1J_0\left(\frac{2k}{m_0\delta}\,m^{\frac{1}{2}}\right)+c_2Y_0\left(\frac{2k}{m_0\delta}\,m^{\frac{1}{2}}\right)$$

$$x(t)=c_1J_0\left(\frac{2k}{\delta}\sqrt{\frac{1+\delta t}{m_0}}\right)+c_2Y_0\left(\frac{2k}{\delta}\sqrt{\frac{1+\delta t}{m_0}}\right)$$

To evaluate the constants c_1 *and* c_2, we use the initial conditions $x(0) = x_0$, and $x'(0) = 0$. To complete, we have to use the calculus of Bessel's function; therefore, we leave the solution up to this point.

Like Bessel's equations, other second-order equations variable-coefficients are found, and also they named after famous scientists and mathematicians. Most of these equations are originated from physical problems that are modeled by partial differential equations. The primary method for solving these partial differential equations is to convert them first to ordinary variable-coefficients differential equations. For example, Legendre's equation is obtained from the heat transfer (described by a second-order partial differential equation named the Laplace equation) in spherical coordinates. In principle, their solutions are obtained by the power series Frobenius method and follow the same procedure discussed in the previous section. It is discovered that some of these solutions can be put in standard forms or polynomials. Some of these equations are:

- Gauss or hypergeometric equation, which has the form

$$x(1-x)y'' + [c-(a+b+1)x]y' - aby = 0 \tag{3.40}$$

- Kummer or confluent hypergeometric equation, which has the form

$$xy'' + (c-x)y' + 2y = 0 \tag{3.41}$$

- Legendre's equation, which has the form

$$(1-x^2)y'' - 2xy' + \lambda(\lambda+1)y = 0;\ \lambda \in \mathcal{R} \tag{3.42}$$

- Chebyshev's equation, which has the form

$$(1-x^2)y'' - xy' + n^2y = 0;\ n = 0,1,2,3\ldots \tag{3.43}$$

For these equations, the solutions derived using the Frobenius method are given in Table 3.1.

Example 3.14.
Use the method of Frobenius to solve the Legendre equation about the origin.

$$(1-x^2)y'' - 2xy' + 6y = 0$$

As we refer to Table 2.1, the given Legendre's equation has the parameter λ is a positive integer equal to 2. The singular points of the equation are ± 1; therefore, the solution around the origin converges in the interval $-1 < x < 1$. First, let us analyze according to remarks 3.2 and 3.3.

$$xP(x) = -\frac{2x^2}{(1-x^2)} = p_0 + p_1 x + p_2 x^2 + \cdots \Rightarrow p_0 = 0, p_1 = 0$$

$$x^2 Q(x) = x^2 \frac{6}{(1-x^2)} = q_0 + q_1 x + q_2 x^2 + \cdots \Rightarrow q_0 = 0, q_1 = 0$$

Table 3.1. Solutions of Gauss, Kummer, and Legendre equations

Name	Solution Form
Gauss or hyper-geometric	1. Around the singular point $x = 0$ $r_1 = 0, r_2 = 1 - c$ $a_{k+1} = \frac{(k+a)(k+b)}{(k+1)(k+c)} a_k;\ k = 0,1,2 \ldots$ $y_1(x) = F(a,b,c,x) = 1 + \frac{a \times b}{1 \times c} x + \frac{ab(a+1)(b+1)}{1 \times 2 \times c(c+1)} x^2 + \frac{ab(a+1)(b+1)(a+2)(b+2)}{1 \times 2 \times 3 \times c(c+1)(c+2)} x^3 + \cdots$ $y_2(x) = G(a,b,c,x) = Ax^{1-c} F(a+1-c, b+1-c, 2-c, x)$ where $A = \frac{\Gamma(a+1-c)\Gamma(b+1-c)\Gamma(c-1)}{\Gamma(a)\Gamma(b)\Gamma(1-c)}$ $y(x) = m_1 F(a,b,c,x) + m_2 x^{1-c} F(a+1-c, b+1-c, 2-c, x)$ 2. Around the singular point $x = 1$ $y(x) = m_1 F(a,b,a+b+1-c, 1-x) + m_2 (1-x)^{c-a-b} F(c-b, c-a, c-a-b+1, 1-x)$
Kummer or confluent hyper-geometric	Around the singular point $x = 0$, for $a_0 = 1$ $y_1(x) = 1 + \frac{1}{n!} \prod_{i=0}^{n-1} \frac{a+i}{c+i} x^n + \cdots; n = 1,2,3 \ldots$ $y_2(x) = x^{1-c} \left[1 + \frac{1}{n!} \prod_{i=0}^{n-1} \frac{a+i+1-c}{2+i-c} x^n + \cdots \right]; n = 1,2,3 \ldots$ $y(x) = m_1 y_1(x) + m_2 y_2(x)$
Legendre	1. For integer λ: The solutions are called Legendre polynomial, which is described by $P_n(x) = \frac{1}{2^n\, n!} \frac{d^n}{dx^n} [(x^2 - 1)^n]\ ;\ n = \lambda$ 2. For noninteger k: For the singular point $x = -1$, the solution can be expressed by the hypergeometric function F, as $y(x) = CF(k+1, -k, 1, \frac{1}{2}(x+1))$ 3. For noninteger k: For the singular point $x = -1$, the solution can be expressed by the hypergeometric function F, as $y(x) = CF(-k, k+1, 1, \frac{1}{2}(1-x))$
Chebyshev	$y(x) = c_1 \cos(n \cos^{-1}(x)) + c_2 \sin(n \cos^{-1}(x));\ \lvert x \rvert < 1$ $y(x) = c_1 T_n(x) + c_2 U_n(x);\ \lvert x \rvert < 1$ where: $T_n(x)$ and $T_n(x)$ are first and second kinds of Chebyshev polynomials, $T_n(x) = \frac{(-2^n)n!}{(2n)!} \sqrt{1-x^2} \frac{d^n}{dx^n} \left[(1-x^2)^{\frac{n-1}{2}} \right]$ $U_n(x) = (x \pm \sqrt{x^2 - 1})^n - T_n(x)$

Thus, the indicial equation is $r(r-1) = 0$, and condition 3.13 is satisfied to obtain two linearly independent solutions. Next, we determine the solution associated with the root $r = 0$.

Assume the solution and its two derivatives as

$$y = \sum_{k=0}^{\infty} a_k x^k, y' = \sum_{k=0}^{\infty} a_k k x^{k-1}, y'' = \sum_{k=0}^{\infty} a_k k(k-1) x^{k-2}$$

$$(1-x^2)\sum_{k=0}^{\infty} a_k k(k-1)x^{k-2} - 2x\sum_{k=0}^{\infty} a_k k x^{k-1} + 6\sum_{k=0}^{\infty} a_k x^k = 0$$

$$\sum_{k=0}^{\infty} a_k k(k-1)x^{k-2} - \sum_{k=0}^{\infty} a_k k(k-1)x^k - 2\sum_{k=0}^{\infty} a_k k x^k + 6\sum_{k=0}^{\infty} a_k x^k = 0$$

$$\sum_{k=0}^{\infty} a_k k(k-1)x^{k-2} - \sum_{k=0}^{\infty} a_k (k^2+k-6)x^k = 0$$

$$a_0(0)(-1)x^{-2} + a_1(1)(0)x^{-1} + \sum_{k=0}^{\infty} a_{k+2}(k+2)(k+1)x^k - \sum_{k=0}^{\infty} a_k(k^2+k-6)x^k = 0$$

$$a_0(0)x^{-2} + a_1(0)x^{-1} + \sum_{k=0}^{\infty} [a_{k+2}(k^2+3k+1) - a_k(k^2+k-6)]x^k = 0$$

As it can be noticed, both coefficients a_0, and a_1 can be set arbitrarily, and then by a proper choice of these coefficients, we can readily determine two linearly independent solutions. The recurrent formula is

$$a_{k+2} = \frac{(k^2+k-6)}{(k^2+3k+2)} a_k;\ k = 0,1,2 \dots$$

$$a_2 = -3a_0, a_3 = -\frac{2}{3}a_1, a_4 = 0, a_5 = -\frac{1}{5}a_1, a_6 = 0, a_7 = -\frac{4}{35}a_1$$

The general solution is

$$y(x) = a_0 + a_1 x - 3a_0 x^2 - \frac{2}{3}a_1 x^3 - \frac{1}{5}a_1 x^5 - \frac{4}{35}a_1 x^7 - \cdots$$

Alternatively, we can write the general solution as

$$y(x) = a_0(1-3x^2) + a_1(x - \frac{1}{5}x^5 - \frac{4}{35}x^7 - \cdots)$$

For the moment, let us choose $a_0 = -0.5$, and $a_1 = 0$, then the solution becomes

$$y(x) = \frac{1}{2}(3x^2 - 1)$$

In our example, the parameter $\lambda = 2$, and the choice of the constants a_0 and a_1 is taken to coincide with the results given in Legendre's polynomial (Legendre's functions of the first kind).

$$P_n(x) = \frac{1}{2^n\, n!}\frac{d^n}{dx^n}[(x^2 - 1)^n]\ ;\ n = \lambda$$

$$P_2(x) = \frac{1}{2^2\, 2!}\frac{d^2}{dx^2}[(x^2 - 1)^2] = \frac{1}{8}\frac{d}{dx}[4x(x^2 - 1)] = \frac{1}{2}(3x^2 - 1)$$

If we postulate the results above, we will have a solution for every positive integer value of λ. For instance, the Legendre equation with $\lambda = 3$; i.e., $(1 - x^2)y'' - 2xy' + 12y = 0$, has the solution

$$P_3(x) = \frac{1}{2^3\, 3!}\frac{d^3}{dx^3}[(x^2 - 1)^3] = \frac{1}{2}(5x^3 - 3x)$$

The Frobenius solution that coincides with the polynomial $P_3(x)$, requires another choice of the coefficients a_0 and a_1. The proper choice is $a_0 = 0$ and $a_1 = -3/2$. In fact, for even value of λ, the Frobenius solution is obtained by assigning a specific value to a_0 and $a_1 = 0$, and for the odd value of λ by assigned $a_0 = 0$ and a particular value of a_1. In mathematics literature, it is derived that to obtain the standard polynomials as solutions, the choice of these coefficients is according to

$$a_0 = (-1)^{n/2}\frac{n!}{2^n[\left(\frac{n}{2}\right)!]^2};\ n = \lambda$$

$$a_1 = (-1)^{(n-1)/2}\frac{(n+1)!}{2^n\left[\left(\frac{n-1}{2}\right)!\right]\left[\left(\frac{n+1}{2}\right)!\right]};\ n = \lambda$$

3.4. Nonlinear Second-Order Differential Equations

The real world has more nonlinear dynamic problems than linear problems. The dynamic systems that are described by differential equations are said to be nonlinear if the nonlinearity appears in the dependent variable y, as it was illustrated in the previous chapter. For example, for the motion like a pendulum, the dependent variable is the angle θ that the pendulum makes with the vertical. For the motion of a body attached to a spring, the dependent variable is the position of the body y for one end. The differential equations that govern the motion are respectively:

$$\theta'' + k\sin\theta = 0$$

$$y'' + ay + by^3 = 0$$

The practical problems with nonlinear elements face many theoretical difficulties such as the inapplicability of the superposition theorem, the complicated stability analysis, changing of the solution as the initial conditions change, the occurrence of limit cycles in the unforced systems, and so on. Table 3.2 describes the most famous nonlinear differential equations, which are studied thoroughly in math literature. Unfortunately, there is no general method to solve all types of nonlinear differential equations. However, besides the numerical approach, many nonlinear differential equations have exact closed-form solutions (in terms of elementary functions). The next discussion will be oriented to the types that have closed-form solutions in explicit or implicit forms.

Table 3.2. Some Known nonlinear differential equations

Differential Equation Name	**Mathematical Form ($y' = dy/dx$)**
Rayleigh	$y'' + ky' + my'^2 + n^2y = 0$
Van der Pol	$y'' - \epsilon(1 - y^2)y' + ay = 0$
Duffing	$y'' + ay + by^2 = k\sin wt$
Emden	$y'' + \frac{2}{x}y' + y^n = 0$
White dwarf	$\frac{1}{x^2}\frac{d}{dx}(x^2y') + (y^2 - C)^{1.5} = 0$
Generalized Blasius	$y''' + ayy'' = \beta(y'^2 - 1)$
Thomas-Fermi	$y'' = \frac{1}{\sqrt{x}}y^{1.5}$
Langmuir	$3yy'' + y'^2 + 4yy' + y^2 = 1$

Before we discuss the solution methods of second-order nonlinear differential equations, we start with traceable techniques to solve some particular first-order nonlinear differential equations. The principle of changing the independent variable by a suitable substitution will demonstrate a method for solving these equations.

Example 3.15.
Consider the first-order nonlinear nonhomogeneous equation

$$y' + y^2 + \frac{1}{x}y = \frac{4}{x^2}$$

Let us introduce a new variable z, which is defined as

$$z = e^{\int y dx},$$

$$\text{then } z' = \frac{dz}{dx} = zy, \text{and}$$

$$z'' = \frac{d^2z}{dx^2} = (y' + y^2)z$$

Thus, the given equation can be rewritten in terms of the variable z as

$$x^2z'' + xz' - 4z = 0$$

The obtained equation is the homogeneous Euler equation (a special kind of the second-order variable- coefficients equations). It can be solved using different methods (also possibly by the method of Frobenius) to get

$$z = ax^2 + \frac{b}{x^2}; \ a \text{ and } b \text{ are constants}$$

Since we have

$$\ln z = \int y dx, \Rightarrow \ y = \frac{1}{z}\frac{dz}{dx}$$

Therefore

$$y = \frac{1}{(ax^2 + \frac{b}{x^2})}\frac{d}{dx}\left[ax^2 + \frac{b}{x^2}\right] = \frac{2(x^4 - c)}{x(x^4 + c)}; \ c = \frac{b}{a}$$

Example 3.16.

Consider a famous first-order differential equation that is named Riccati equation (a fundamental equation in optimal control engineering)

$$y' + ay^2 + by + c = 0$$

We discuss and solve the following cases:

1. When the parameters $a, b,$ and c are constant values; the solution can be easily obtained by the variable separation method as follows:

$$\frac{dy}{ay^2 + by + c} + dx = 0$$

Integrating both sides yields

$$\int \frac{dy}{(y - m_1)(y - m_2)} + \int dx = 0$$

where

$$m_{1,2} = \frac{-b \mp \sqrt{b^2 - 4ac}}{2a}$$

Using the partial fraction method for the first term, we get

$$\int \frac{Q}{(y - m_1)} dy - \int \frac{Q}{(y - m_2)} dy + \int dx = 0$$

where

$$Q = \frac{1}{m_1 - m_2}$$

Solving the integrals yields

$$Q \ln|y - m_1| - Q \ln|y - m_2| + \ln x + \ln k = 0$$

where k is integration constant.

$$\ln k |\frac{y - m_1}{y - m_2}| + \ln e^{x/Q} = 0$$

$$k \frac{y - m_1}{y - m_2} e^{x/Q} = 1$$

Solving for y; the solution is

$$y(x) = \frac{m_1 k - m_2 e^{-x/Q}}{k - e^{-x/Q}}$$

2. When the parameters a is a constant, and the parameters b and c are a function of x, then, in this case, the solution is obtained by introducing the substitution

$$\mu = e^{a \int y \, dx}$$

Then, we have

$$\mu' = \frac{d\mu}{dx} = ay\mu$$

$$\mu'' = \frac{d^2\mu}{dx^2} = (a\frac{dy}{dx} + a^2y^2)\mu$$

Substituting in the Riccati equation gives

$$\frac{1}{a\mu}\frac{d^2\mu}{dx^2} + \frac{b(x)}{a\mu}\frac{d\mu}{dx} + c(x) = 0$$

Alternatively, we write

$$y'' + b(x)\mu' + ac(x)\mu = 0$$

According to the functions $b(x)$ and $c(x)$, this last equation can be solved to give an exact solution in terms of elementary functions or an infinite power series. Then, the solution of the original equation is

$$y(x) = \frac{d}{dx}\left[\ln\frac{\mu}{a}\right]$$

3. When the parameters $a, b,$ and c are a function of the independent variable x, it represents the most challenging case. For some $a(x), b(x)$ and $c(x)$, the original equation can be solved by converting the equation to a specific Bernoulli equation that can be solved analytically, and then the required solution is obtained. However, the conversion to Bernoulli's equation needs additional effort. This procedure will be shown in the next example.

Example 3.17.
Solve the following Riccati equation

$$y' - y^2 - 2xy - x^2 + 2 = 0$$

Using some trials, one can find a solution $v = 1 - x$, which satisfies the given equation, i.e.,

$$v' - v^2 - 2xv - x^2 + 2 = (-1) - (1 - x)^2 - 2x(1 - x) - x^2 + 2 = 0$$

Now let us assume that the required solution is given by

$$y(x) = 1 - x + u$$

Then the given equation is reduced to Bernoulli's equation, which has the form

$$u' + au^2 + (2av + b)u = 0$$

With $a=-1, b=-2x$, we get

$$u'-u^2+\big(2(-1)(1-x)+(-2x)\big)u=0$$

$$u'-u^2-2u=0$$

$$\frac{du}{dx}=u^2+2u$$

Separation of variables and integrating gives

$$\frac{du}{u(u+2)}=dx \Rightarrow \frac{du}{2u}-\frac{du}{2(u+2)}=dx$$

$$\frac{1}{2}\ln u-\frac{1}{2}\ln(u+2)+\ln c=x$$

$$u(x)=\frac{2e^{2x}}{c-e^{2x}}$$

Finally, the solution of the given equation is

$$y(x)=1-x+\frac{2e^{2x}}{c-e^{2x}}$$

It can be shown that the relation between the initial condition $y(0)$ and the constant c is given by

$$c=\frac{1+y(0)}{-1+y(0)}$$

For two different values of the initial conditions $y(0)=0,$ and $y(0)=2,$ the integration constant will be $c=-1,$ and $c=3,$ respectively. The corresponding solutions are shown in Figure 3.7. For the first initial value, the solution is continuous over the interval $[0,\infty)$, while for the second initial value, the solution is not continuous. It is one of the nonlinear features; the different initial conditions do not give a family set of solutions as with the linear differential equations.

For second-order nonlinear equations, the problem becomes more difficult. A few types of these equations can be solved in terms of elementary functions or special functions exactly. The possible technique is to reduce the equation to a first-order equation whose variables are separable.

Example 3.18.

Solve the following nonlinear equation

$$y'' - \frac{2y-1}{y^2+1}(y')^2 = 0$$

Let

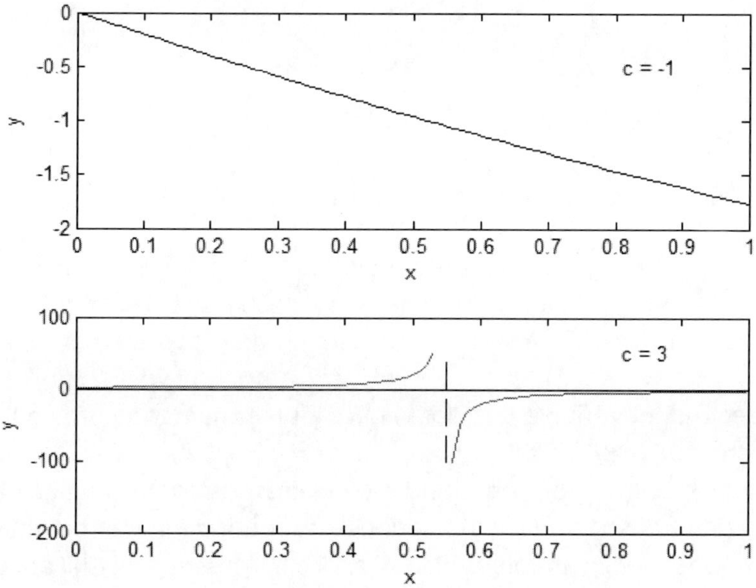

Figure 3.7. Solutions for different initial conditions (Example 3.17).

$$z = \frac{dy}{dx}\text{, then} \qquad \frac{d^2y}{dx^2} = z\frac{dz}{dy}$$

Substitute in the given equation gives

$$z\frac{dz}{dy} = \frac{2y-1}{y^2+1}\,z^2$$

$$\frac{dz}{z} = \frac{2y-1}{y^2+1}\,dy = \frac{2y}{y^2+1}dy - \frac{1}{y^2+1}dy$$

Integrating both sides yields

$$\ln\left(\frac{z}{c_1}\right) = \ln(y^2+1) - \tan^{-1}y$$

$$z = \frac{dy}{dx} = c_1(y^2+1)e^{-\tan^{-1}y}$$

Integrating again

$$\int \frac{e^{\tan^{-1}y}}{(y^2+1)}dy = c_1 \int dx$$

$$e^{\tan^{-1}y} = c_1 x + c_2$$

Therefore, the solution is

$$y(x) = \tan[\ln(c_1 x + c_2)]$$

Let us consider first the technique of using a suitable substitution (by introducing a new variable) to reduce a first-order equation, which can be solved by implementing the methods used before. There are three cases to be elaborated on.

1. The dependent variable y is not explicitly present in the equation. In this case, the technique is to use the substitution $y' = z$, which converts the second-order equation to a first-order one, and if the reduced-order equation can be solved then a general solution for the second-order equation can be obtained.
2. The independent variable x is not explicitly present in the equation. In this case, the technique is to use the substitutions $y' = z$, and $y'' = z\frac{dz}{dy}$, which reduced the equation to a first-order of the form $z = f(z, y)$, then it could hopefully be solved. This technique may lead to the exact solution only for specified initial conditions.
3. None of the above two cases. In this case, the used substitution is

$$y(x) = e^{\int z(x)dx}$$

The following three examples illustrate the above three cases, respectively.

Example 3.19.
Solve

$$y'' + 2x(1 + y')^2 = 0$$

We set $y' = z$, then the given equation becomes

$$z' + 2x(1 + z)^2 = 0$$

Using the separation of variables and integrating yield

$$\frac{dz}{(1+z)^2} = -2x\,dx$$

$$-\frac{1}{(1+z)} = -x^2 + k$$

$$z = -1 + \frac{1}{x^2 + c_1} = \frac{dy}{dx}$$

The general solution is

$$y(x) = -x + \frac{1}{c_1}\tan^{-1}\left(\frac{x}{c_1}\right) + c_2$$

Example 3.20.
Solve the initial value problem

$$y'' + 3y^2y' = 0; \quad y(1) = 1, \qquad y'(1) = -1$$

We use the substitution

$$y' = z, \text{and} \qquad y'' = z\frac{dz}{dy}$$

With these substitutions, the given equation is reduced to

$$z\frac{dz}{dy} + 3y^2z = 0$$

Divided by $z, (z \neq 0)$ and integrating, we get

$$z = -y^3 + c_1$$

$$y' = -y^3 + c_1$$

By substituting the initial conditions, we can find that $c_1 = 0$, and thus

$$\frac{dy}{dx} = -y^3 \Rightarrow -\frac{1}{2}y^{-2} = -x + c_2$$

Therefore,

$$y^2 = \frac{1}{2x - 2c_2}$$

By substituting the initial conditions, we obtain $c_2 = 0.5$, and then the solution is

$$y^2 = \frac{1}{2x-1}$$

Based on the given initial condition, we can write the solution as

$$y = \frac{1}{\sqrt{2x-1}}$$

Example 3.21.

Solve

$$x^2yy'' - (y - xy')^2 = 0$$

We use the substitution

$$y = e^{\int zdx}, \qquad \text{and hence} \quad y' = ze^{\int zdx}, \qquad y'' = (z' + z^2)e^{\int zdx}$$

With these substitutions, the given equation is reduced to

$$x^2z' + 2xz - 1 = 0$$

$$z' + \frac{2}{x}z = \frac{1}{x^2}$$

Refer to section 2.4; this last equation is a first-order linear equation, which can be solved using expression 2.28.

$$z(x) = e^{-\int \frac{2}{x}dx}\left[c_1 + \int \frac{1}{x^2}\, e^{\int \frac{2}{x}dx}dx\right] = \frac{c_1}{x^2} + \frac{1}{x}$$

Therefore, the required solution is given by

$$y(x) = e^{\int zdx} = e^{\int \frac{1}{x}dx}.e^{\int \frac{c_1}{x^2}dx} = c_2xe^{-\frac{c_1}{x}}$$

As we see from previous examples, the used technique reduces the second-order equation to a first-order one by applying a suitable substitution. An alternative approach of order reducing is when the equation's left-hand side has a complete derivative shape (concerning the independent variable) of some reduced-order expression. In general, this approach is not easy to handle; however, several equations can be solved by this technique, as shown in the next example.

Example 3.22.

Solve the following second-order nonlinear differential equation

$$yy'' - y'^2 = 0$$

Let us first solve using appropriate substitution. Since the independent variable x does not appear explicitly, the right substitution is

$$y' = z, \text{and} \quad y'' = z\frac{dz}{dy}$$

With these substitutions, the given equation is reduced to

$$yzz' - z^2 = 0 \Rightarrow yz' - z = 0, z \neq 0$$

Separation of variables yields

$$\frac{dz}{z} = \frac{dy}{y}$$

Integrating sides gives

$$\ln z = \ln y + \ln c_1 = \ln c_1 y$$
$$z = c_1 y$$

Therefore,

$$\frac{dy}{dx} = c_1 y \Rightarrow \frac{dy}{y} = c_1 dx$$

Thus

$$\ln y - \ln c_2 = c_1 x$$

$$\ln\frac{y}{c_2} = c_1 x$$

Then, the general solution is given by

$$y(x) = c_2 e^{c_1 x}$$

Now let us solve by rearranging the equation as (dividing by y^2; where $y \neq 0$)

$$\frac{yy'' - y'^2}{y^2} = 0$$

The left-hand side of the last equation is the complete derivative of the ratio (y'/y), i.e.,

$$\frac{yy'' - y'^2}{y^2} = \frac{d}{dx}\left\{\frac{y'}{y}\right\} = 0$$

Therefore,

$$\frac{y'}{y} = c_1 \Rightarrow \frac{dy}{y} = c_1 dx$$

Hence, the same answer can be obtained.

3.5. Method of a Parameter

Another set of nonlinear differential equations can be solved by following a particular procedure called the method of a parameter. This set contains the equations that have the form

$$F\left(y^{(n)}, y^{(n-1)}\right) = 0 \tag{3.44}$$

When these equations cannot be arranged as $y^{(n)} = f(y^{n-1})$, this method is invoked to perform a solution for both independent variable x and the dependent variable y in terms of a new variable (parameter), say t. Initially, we find two functions $\varphi(t)$ and $\psi(t)$ equal to the nth and $(n-1)^{th}$ derivatives respectively, and they satisfy the given differential equation, i.e.,

$$y^{(n)} = \varphi(t), y^{(n-1)} = \psi(t) \tag{3.45}$$

$$F\left(\varphi(t), \psi(t)\right) = 0 \tag{3.46}$$

Next, to have convinced synthesis, we will introduce the method of a parameter for third-order nonlinear equations $(n = 3)$; i.e.,

$$F(y''', y'') = 0, y''' = \varphi(t), y'' = \psi(t) \tag{3.47}$$

Therefore, we can write

$$dy'' = y''' dx$$

$$dx = \frac{dy''}{y'''} = \frac{\psi'(t)dt}{\varphi(t)} \tag{3.48}$$

Integrating sides yields

$$x(t) = \int \frac{\psi'(t)}{\varphi(t)}\, dt = \mu(t) + c_1 \tag{3.49}$$

Next,

$$dy' = y''dx = \frac{\psi(t)\psi'(t)dt}{\varphi(t)}$$

$$y'(t) = \int \frac{\psi(t)\psi'(t)}{\varphi(t)}\, dt = \eta(t) + c_2 \tag{3.50}$$

Finally,

$$dy = y'dx = [\eta(t) + c_2]\, \frac{\psi'(t)dt}{\varphi(t)}$$

By integrating, we obtain

$$y(t) = \int [\eta(t) + c_2] \frac{\psi'(t)}{\varphi(t)} dt = \lambda(t, c_2) + c_3 \tag{3.51}$$

Equations (3.48) and (3.50) represent the solution of equation (3.47) in terms of the single parameter t. If it is possible to have from either of the obtained solutions $x(t)$ or $y(t)$ relation of the form $t = h(x)$ or $t = g(y)$, respectively, then a substitution in the other solution will give the relation between the original variables x, and y.

With some modifications, the method of a parameter can also be used. For example, the equation

$$F(y''', y') = 0 \tag{3.52}$$

Similarly, by introducing the parametric functions that satisfy equation (3.52), we have

$$y''' = \varphi(t), y' = \psi(t) \tag{3.53}$$

We can similarly obtain,

$$\frac{1}{2}[y''(t)]^2 = \int \varphi(t)\psi'(t)\, dt \tag{3.54}$$

Once $y''(t)$ is obtained say $\alpha(t, c_1)$, we proceed again from equation (3.48); however, with a determined function of time $dy''(t)$, say equal to $\alpha'(t, c_1)\, dt$, i.e.,

$$dx = \frac{dy''}{y'''} = \frac{\alpha'(t, c_1)dt}{\varphi(t)} \tag{3.55a}$$

$$dy' = y''dx = \alpha(t, c_1)\frac{\alpha'(t, c_1)dt}{\varphi(t)} \tag{3.55b}$$

By integrating equations (3.55a, and 3.55b), we obtain the variables $x(t)$, and $y'(t)$. Next, we set

$$dy = y'dx = \psi(t)\frac{\alpha'(t, c_1)dt}{\varphi(t)} \tag{3.56}$$

Finally, we integrate to obtain the variable, $y(t)$.

Even if the functions $\varphi(t)$ and $\psi(t)$ are found, the problem may still be challenging to be solved with these found functions. In some cases, the main obstacle in applying the method of a parameter is when the integrations do not have a closed-form solution. Therefore, this obstacle reduces significantly the number of differential equations that can be solved by this method.

Example 3.23.

Find a parametric solution of the nonlinear differential equation

$$y'''^2 + y''^2 - 4 = 0$$

We choose the parametric form as

$$\varphi(t) = y''' = 2\cos t,\ \psi(t) = y'' = 2\sin t$$

Therefore, the chosen functions satisfy the given equation since

$$(2\cos t)^2 + (2\sin t)^2 - 4 = 0$$

Using equation (3.49), we find a solution for the variable x.

$$x = \int \frac{\psi'(t)}{\varphi(t)}\, dt = \int \frac{2\cos t}{2\cos t}\, dt = t + c_1$$

Using equation (3.50) gives

$$y'(t) = \int \frac{\psi(t)\psi'(t)}{\varphi(t)} dt = \int \frac{4 \sin t \cos t}{2 \cos t} dt = \int 2 \sin t \, dt = -2 \cos t + c_2$$

Using equation (3.51), we find a solution for the variable y.

$$y(t) = \int [\eta(t) + c_2] \frac{\psi'(t)}{\varphi(t)} dt = \int [-2 \cos t + c_2] \frac{2 \cos t}{2 \cos t} dt = -2 \sin t + c_2 t + c_3$$

For this example, we can eliminate the parameter t, to obtain the relation $y = f(x)$, i.e., $t = x - c_1$

$$y = -2 \sin(x - c_1) + c_2(x - c_1) + c_3$$

The complete solution is

$$y(x) = -2 \sin(x - c_1) + c_2 x + C; \; C = -c_1 c_2 + c_3$$

Example 3.24.
Find a parametric solution of the nonlinear differential equation

$$y''^2 - y'^3 = 0$$

We choose the parametric form as

$$\varphi(t) = y'' = e^{-2t}, \psi(t) = y' = e^{-\frac{4}{3}t}$$

Therefore, the chosen functions satisfy the given equation since

$$(e^{-2t})^2 - (e^{-\frac{4}{3}t})^3 = 0$$

Using equation (3.49), we find a solution for the variable x.

$$dx = \frac{dy'}{y''} = \frac{\psi'(t)dt}{\varphi(t)}$$

$$x = \int \frac{\left(-\frac{4}{3} e^{-\frac{4}{3}t}\right)}{e^{-2t}} dt = -2e^{\frac{2}{3}t} + c_1$$

Unlike in the previous example, here we do need to use equation (3.50) as it is given, but we have directly to apply

$$dy = y'dx = \frac{\psi(t)\psi'(t)dt}{\varphi(t)}$$

$$y = \int \frac{\psi(t)\psi'(t)}{\varphi(t)}\,dt = \int \frac{\left(-\frac{4}{3}e^{-\frac{4}{3}t}\right)\left(e^{-\frac{4}{3}t}\right)}{e^{-2t}}\,dt = -\frac{4}{3}\int e^{-\frac{2}{3}t}dt = 2e^{-\frac{2}{3}t} + c_2$$

If we let the variable t in terms of the variable x, we get

$$t = \frac{3}{2}\ln\frac{(x - c_1)}{2}$$

Substituting in the solution of y yields

$$y = 2e^{-\frac{2}{3}\times\frac{3}{2}\ln\frac{(x-c_1)}{2}} + c_2 = \frac{4}{(x - c_1)} + c_2$$

Example 3.25.
Find a parametric solution of the nonlinear differential equation

$$y'''^2 - 64y'^4 = 0$$

We choose the parametric form as

$$\varphi(t) = y''' = 8e^{-2t}, \qquad \psi(t) = y' = e^{-t}$$

Therefore, the chosen functions satisfy the given equation since

$$(8e^{-2t})^2 - 64(e^{-t})^4 = 0$$

First, we apply equation (3.55)

$$\frac{1}{2}[y''(t)]^2 = \int \varphi(t)\psi'(t)\,dt = \int (8e^{-2t})(-e^{-t})dt = \frac{8}{3}e^{-3t} + c_1$$

$$y''(t) = \sqrt{\frac{16}{3}e^{-3t} + c_1} = \alpha(t,, c_1)$$

$$dx = \frac{\alpha'(t, c_1)dt}{\varphi(t)} = \frac{(-\frac{1}{2}16e^{-3t})}{8e^{-2t}\sqrt{\frac{16}{3}e^{-3t} + c_1}}\,dt$$

$$x = \int \frac{-e^{-t}}{\sqrt{\frac{16}{3}e^{-3t} + c_1}} dt$$

For this integration, there is no exact closed-form solution in terms of elementary functions. Therefore, the method of parameters cannot solve this equation.

For some equations, the elimination of the t parameter cannot be performed to obtain an explicit or even implicit solution of form $f(x, y) = 0$. However, practically this is not a problem; the obtained relations $x = X(t)$ and $y = Y(t)$ are enough to find the trajectory for a specific range of values of the t parameter. For example, for the equation $y'^3 - axy' + x^3 = 0; a \neq 0$, the solution in parametric form is

$$y = \frac{a^2(4t^3 + 1)}{6(t^3 + 1)^2} + c, x = \frac{at}{t^3 + 1}$$

Even if the elimination is possible, it might be preferable to obtain the solution in a parametric form instead of a complicated solution in implicit form; see exercise 3.29.

3.6. Solution Method Based on Jacobi Elliptic Functions

The previous solvable problems represent only a minimal subset of the nonlinear equation set. Even for the same types of the discussed cases, some equations cannot be solved by the suggested techniques. Since we might meet equations that do not have a closed-form solution in many applications, the first alternative is to use approximated methods. The second alternative is to use numerical methods to obtain the solution's behavior at some intervals within acceptable accuracy. It is known that approximate solutions can see the effects of the changes in the equation parameters.

In what follows, we will discuss an approximated method, which is not so familiar in the textbooks. The method is based on using the Jacobi elliptic functions. Although these functions do not have closed-forms, they have similar properties as the elementary sinusoidal functions have. However, it is used only with a special class of second-order nonlinear differential equations. Before using these functions for solving second-order nonlinear differential equations, some features are necessary to be noted. The three Jacobi elliptic functions, which are valid for the parameter $0 < k < 1$, are:

- The alike sine function, which is denoted by $\text{sn}(u)$. It is given by the infinite series

$$\text{sn}(u) = u - \frac{1}{3!}(1 + k^2)u^3 + \frac{1}{5!}(1 + 14k^2 + k^4)u^5 - \cdots; \text{ sn}(0) = 0 \quad (3.57)$$

- The alike cosine function, which is denoted by $\mathrm{cn}(u)$. It is given by the infinite series

$$\mathrm{cn}(u) = 1 - \frac{1}{2!}u^2 + \frac{1}{4!}(1 + 4k^2)u^4 - \frac{1}{6!}(1 + 44k^2 + 16k^4)u^6 + \cdots;\ \mathrm{cn}(0) = 1 \quad (3.58)$$

- The third function, which is denoted by $\mathrm{dn}(u)$. It is given by the infinite series

$$\mathrm{dn}(u) = 1 - \frac{k^2}{2!}u^2 + \frac{1}{4!}(4k^2 + k^4)u^4 - \frac{1}{6!}(16k^2 + 44k^4 + k^6)u^6 + \cdots;\ \mathrm{dn}(0) = 1 \quad (3.59)$$

The parameter k^2 is called the elliptic modulus. These three elliptic functions are also defined by what is called elliptical integrals of the first kind:

$$u = \int_0^x \frac{dx}{\sqrt{(1 - x^2)(1 - k^2x^2)}} = \int_0^\varphi \frac{d\varphi}{\sqrt{1 - k^2\sin^2\varphi}} \quad (3.60)$$

$$\mathrm{cn}(u) = \sqrt{1 - \mathrm{sn}^2 u} = \sqrt{1 - x^2} \quad (3.61)$$

$$\mathrm{dn}(u) = \sqrt{1 - k^2\mathrm{sn}^2 u} = \sqrt{1 - k^2x^2} \quad (3.62)$$

The sn and cn functions have a period of $4P(k)$, while the dn function has a period of $2P(k)$. The value of elliptic function P depends on the k value, and it is equal to the argument value of the corresponding zero value of the elliptic function. Furthermore, the following identities are valid (more relations can be found).

$$\mathrm{sn}^2(u) + \mathrm{cn}^2(u) = 1$$

$$k^2\mathrm{sn}^2(u) + \mathrm{dn}^2(u) = 1$$

$$\mathrm{sn}(u + 2p) = -\mathrm{sn}(u);\ \mathrm{sn}(u + 4p) = \mathrm{sn}(u);\ \mathrm{sn}(-u) = -\mathrm{sn}(u)$$

$$\mathrm{cn}(u + 2p) = -\mathrm{cn}(u);\ \mathrm{cn}(u + 4p) = \mathrm{cn}(u);\ \mathrm{cn}(-u) = \mathrm{cn}(u)$$

$$\mathrm{dn}(u + 2p) = \mathrm{dn}(u);\ \mathrm{dn}(-u) = \mathrm{dn}(\mathrm{u}) \quad (3.63)$$

$$\mathrm{dn}(u + 2p) = \mathrm{dn}(u);\ \mathrm{dn}(-u) = \mathrm{dn}(u)$$

$$\frac{d}{du}\mathrm{sn}(au) = a\ \mathrm{cn}(\ u)\mathrm{dn}(\ \mathrm{u})$$

$$\frac{d}{du}\mathrm{cn}(au) = -a\ \mathrm{sn}(u)\ \mathrm{dn}(u)$$

$$\frac{d}{du}\text{dn}(au) = -k^2\text{a sn}(\text{u})\text{cn}(u) \tag{3.64}$$

$$\lim_{k\to 0}\text{sn}(u) = \sin u, \lim_{k\to 0}\text{cn}(u) = \cos u, \lim_{k\to 0}\text{dn}(u) = 1 \tag{3.65}$$

The plots of Jacobi elliptic functions with $k = 0.5$ are shown in Figure 3.8, which shows the similarity of the $\text{sn}(u)$ and $\text{cn}(u)$ functions with the elementary sine and cosine functions. However, as the parameter k approaches 1, this similarity is destroyed.

The class of second-order differential equations is given as

$$y'' + (1 + k^2)y - \frac{2k^2}{A^2}y^3 = 0 \tag{3.66}$$

where the amplitude A is a real constant. The standard form of a solution of this class is

$$y(x) = A\ \text{sn}(\text{u}) \tag{3.67}$$

With the aid of the derivative formulas (3.64), it can be easily shown that this solution satisfies equation (3.66). If instead of u (for dynamic systems), we put wt; where t represents a time variable, then equation (3.66) becomes

$$y'' + (1 + k^2)w^2y - \frac{2k^2w^2}{A^2}y^3 = 0, y' = \frac{dy}{dt} \tag{3.68}$$

Let us denote

$$a = (1 + k^2)w^2, \text{and } b = \frac{2k^2w^2}{A^2};\ a, b > 0$$

Thus, a general form of equation (3.68) is

$$y'' + ay - by^3 = 0 \tag{3.69}$$

Then solving for the frequency w in terms of $a, b,$ and A yields

$$w = \sqrt{a - \frac{1}{2}bA^2} \tag{3.70}$$

The solution is

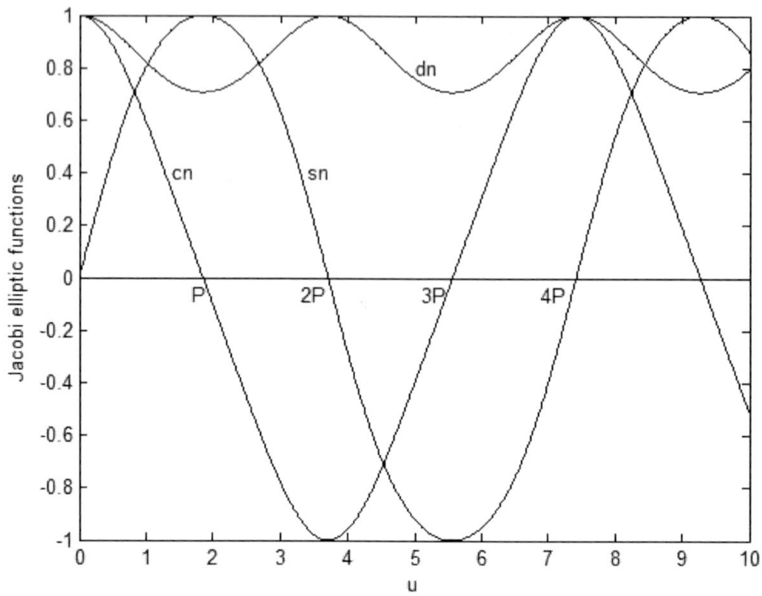

Figure 3.8. Elliptic functions with $k = 0.5, 4P(0.5) = 7.43$.

$$y(t) = A\,\mathrm{sn}\left(\sqrt{a - \frac{1}{2}bA^2}\ t\right);\ k^2 = \frac{bA^2}{2a - bA^2} \tag{3.71}$$

Similarly, the equation $y'' + ay + by^3 = 0;\ a, b > 0$ has a solution given by

$$y(t) = A\,\mathrm{cn}\left(\sqrt{a + bA^2}\ t\right)\ ;\ k^2 = \frac{bA^2}{2(a + bA^2)} \tag{3.72}$$

On the other hand, the equation $y'' - ay + by^3 = 0;\ a, b > 0$, has a solution given by

$$y(t) = A\,\mathrm{dn}\left(\sqrt{\frac{b}{2}}\ At\right)\ ;\ k^2 = 2\left(1 - \frac{a}{bA^2}\right) \tag{3.73}$$

Concerning the amplitude A and the nature of solutions, it is important to state the following notes:

1. The functions $\mathrm{sn}(u)$, and $A\mathrm{sn}(u)$ are not solutions of the same nonlinear differential equation.
2. It is not possible to obtain two linearly independent solutions with two arbitrary constants.

3. For the solutions (3.71), (3.72), and (3.73), the value A^2 is less than the ratio (a/b), arbitrary or less than the ratio $(2\,a/b)$ respectively.
4. Because the amplitude A can be evaluated from the initial condition, a relation between these conditions and the system parameters $a,$ and b exists.
5. For the solutions (3.72) and (3.73), it is clear that at $t = 0$, the amplitude A is equal to $y(0)$ because of $\mathrm{cn}(0) = \mathrm{dn}(0) = 1$. Therefore, for the solution (3.72), the initial condition can be arbitrary. For solution (3.73), the initial condition should satisfy $y(0) < \sqrt{(2\,a/b)}$ and simultaneously $y(0) > \sqrt{(a/b)}$. These conditions have to be satisfied to satisfy in turn, the condition $k^2 < 1$ and that k is positive.
6. For solution (3.71), the relation between the amplitude A, the initial condition, and the system parameters is more complicated. Because at $t = 0$, we have $y(0) = 0$, then A can be chosen arbitrarily. However, this contradicts the condition of positive k; i.e., $A^2 < (a/b)$. Therefore, A can be any value that satisfies the condition $A^2 < (a/b)$.
7. Finally, for the case $t \neq 0$, but $t = t_1$, the initial condition is $y(t_1) = y_1$, then

i. For the solution (3.72)

$$y_1 = A\,\mathrm{sn}\left(\sqrt{a - \frac{1}{2}bA^2}\;t_1\right)$$

This transcendental nonlinear algebraic equation has to be solved for the amplitude A. It is a numerical problem.

ii. Similarly, we will need an iterative numerical procedure to solve the solution (3.72) and the solution (3.73).

Example 3.26.

Consider the equation

$$y'' + 3y + y^3 = 0;\; y(0) = 1$$

For this case (point 5), the amplitude $A = y(0) = 1$, and hence according to the expression (3.72), the solution is

$$y(t) = \mathrm{cn}\left(\sqrt{3+1}\,t\right) = \mathrm{cn}(2t);\; k^2 = \frac{1}{6+2} = \frac{1}{8}$$

Let us check the solution; using the derivative formulas and the identities to obtain

$$y' = -2\,\mathrm{sn}(2t)\,\mathrm{dn}(2t)$$

$$y'' = -2\mathrm{sn}(2t)\left(-\frac{2}{8}\mathrm{sn}(2t)\mathrm{cn}(2t)\right) - 4\mathrm{dn}(2t)\big(\mathrm{cn}(2t)\mathrm{dn}(2t)\big)$$

$$y'' = \frac{1}{2}\text{sn}^2(2t)\,\text{cn}(2t) - 4\text{dn}^2(2t)\text{cn}(2t)$$

Substituting in the given equation yields

$$\frac{1}{2}\text{sn}^2(2t)\,\text{cn}(2t) - 4\text{dn}^2(2t)\text{cn}(2t) + 3\text{cn}(2t) + \text{cn}^3(2t) = 0$$

Using the second relations of the set 3.63

$$\frac{1}{2}\text{sn}^2(2t)\,\text{cn}(2t) - 4\text{cn}(2t)\left(1 - \frac{1}{8}\text{sn}^2(2t)\right) + 3\text{cn}(2t) + \text{cn}^3(2t) = 0$$

$$\text{sn}^2(2t)\,\text{cn}(2t) - 4\text{cn}(2t) + 3\text{cn}(2t) + \text{cn}^3(2t) = 0$$

$$\text{sn}^2(2t)\,\text{cn}(2t) - \text{cn}(2t) + \text{cn}^3(2t) = 0$$

$$\left(1 - \text{cn}^2(2t)\right)\text{cn}(2t) - \text{cn}(2t) + \text{cn}^3(2t) = 0$$

Therefore,

$$\text{cn}(2t) - \text{cn}^3(2t) - \text{cn}(2t) + \text{cn}^3(2t) = 0$$

Thus, the solution is correct. The reader can take another suitable value of the initial condition to carry the validation of his solution. To determine the period $4P$ of the solution, we have to compute

$$4P = 4\,\text{cn}^{-1}\left(0, \sqrt{0.125}\right) = 4 \times u_0 \cong 4 \times 1.745 = 6.98$$

The value 1.625 can be obtained from the cn table or by numerically solving the polynomial equation (taking more terms from the infinite series increases the accuracy).

$$1 - \frac{1}{2!}{u_0}^2 + \frac{1}{4!}(1 + 4k^2){u_0}^4 - \frac{1}{6!}(1 + 44k^2 + 16k^4){u_0}^6 = 0$$

Example 3.27.
Consider the equation

$$y'' - y + 2y^3 = 0;\ y(0) = 1$$

For this case (point 5), the amplitude $A = y(0) = 1$; however, the first inequality condition to have a solution according to (3.73) is not satisfied i.e.,

$$A^2 < \frac{2a}{b} = \frac{2}{2} = 1 \Rightarrow 1 < 1$$

Also, if $A = y(0) = 0.5$, the inequality $A^2 < \frac{2a}{b}$ is satisfied, but the inequality $A^2 > \frac{a}{b}$ does not satisfied; because it gives a negative value of k. Therefore, a solution cannot be obtained. However, for example, if the initial condition $y(0) = 3/4$, both inequalities are satisfied, and a solution can be found as

$$y(t) = \frac{3}{4}\,\mathrm{dn}\left(\frac{3}{4}\,t\right)\ ;\ k^2 = 2\left(1 - \frac{1}{2\left(\frac{3}{4}\right)^2}\right) = \frac{2}{9}$$

As can be noted, the solution depends on the value of parameters a and b of the general equation (3.69). They are -1 and 2 instead of 3 and 1 in the previous example. Furthermore, it depends on the initial condition.

The nonlinear differential equations considered in previous examples are two cases of the Known Duffing problem. The equation in example 3.26 represents the case of strong cubic hardening nonlinearity. On the other hand, the equation of exercise 3.33 illustrates the softening nonlinearity case. Finally, we emphasize that the discussion in this section represents only a small introduction to this rich topic.

EXERCISES

Section 3.2

3.1. Solve the following Euler-Cauchy equations

i. $x^3y''' - 3x^2y'' + 6xy' - 6y = 20\,x$

ii. $x^2y'' - 6y = 0;\ y(1) = 2, y'(1) = 1$

iii. $x^3y''' + 4x^2y'' + xy' - y = 0$

iv. $x^2y'' + 7xy' + 9y = e^x$

3.2. For some equations and in an analog way to what we did for the Euler-Cauchy equation, we can use other transformations such as to let $z = x^{-1}$ to solve the equation

$$x^3y''' + 2xy' + \frac{4}{x^2}y = \frac{1}{x^2}$$

Find a general solution.

3.3. If the roots of a second-order homogenous Euler-Cauchy equation are complex as $r_{1,2} = \alpha \pm j\beta$, then show that the general solution is given by

$$y(x) = x^{\alpha}[c_1 \cos(\beta \ln x) + c_2 \sin(\beta \ln x)]$$

3.4. Find a complete solution of the equation

$$x^2y'' + 4xy' + 2y = 2\ln x\,;\ y(1) = -1, y'(1) = 1$$

3.5. Find the ordinary points, regular singular points, and irregular singular points of the following differential equation

$$\text{i.}\,(1 - x^2)y'' + y' + y = 0, \text{ii.}\, y'' + \frac{2}{x}y' + \frac{3}{x(x-1)^3}y = 0$$

3.6. Consider the equation

$$x^2y'' + (3x - x^3)y' - 3y = 0$$

Show that the two exponents are differed by 4, and the method of Frobenius fails to give two independent solutions.

3.7. For the equation considered in example 3.4, apply the method of Frobenius to determine the recurrent formula, which generates the two linearly independent solutions.

3.8. For the equation, $x^2y'' + xy' + (x^2 - 4)y = 0$, apply the method of Frobenius to show that

$$y_1(x) = \sum_{k=0}^{\infty} \frac{(-1)^k x^{2(k+1)}}{2^{2k-1}\, k!\,(k+2)!}$$

Then use the solution form in equation (3.16) to find $y_2(x)$.

3.9. Follow the seven steps that are used in example 3.3 to find a general solution around the origin of the equation

$$(2x^2 + 1)y'' + 2xy' - 18y = 0$$

3.10. For the equation, $x^2y'' + (x - x^3)y' - (x^2 + 1)y = 0$, apply equation (3.11) to find the two exponents.

3.11. Show that the general solution at $x = 0$ of $x^2y'' - x(4-x)y' + (6-2x)y = 0$ is given by elementary functions as

$$y(x) = c_1x^2 + c_2x^2(e^{-x} - 1)$$

3.12. For each of the following equations, check whether the method of Frobenius can be used to obtain the complete general solution. If so, then find this solution at $x = 0$.

i. $x^2y'' + (x - x^3)y' - (x^2 + 1)y = 0$

ii. $(x^4 - x^2)y'' + 2xy' - 2(x^2 + 1)y = 0$

iii. $9x^2y'' + 9x(x+1)y' + (12x - 1)y = 0$

3.13. Show that the equation $x^2y'' + (x^2 - x)y' + y = 0$ has two equal exponents, and the first solution is $y_1(x) = xe^{-x}$. Use the solution formula (3.16) to determine the second solution $y_2(x)$.

3.14. For the equation $x^2y'' + y' + xy = 0$, show that it is possible to find two linearly independent solutions at the ordinary point $x = 1$; see example 3.7.

Section 3.3

3.15. Use the method of Frobenius to find two linearly independent solutions of Bessel's equation of order 3. What are the numerical results that can be concluded?

$$x^2y'' + xy' + (x^2 - 9)y = 0$$

3.16. Use the method of Frobenius and the results of Table 3.1 to solve Legendre's equation about the origin.

$$(1 - x^2)y'' - 2xy' + 2y = 0$$

3.17. Use the method of Frobenius to solve Chebyshev's equation.

$$(1 - x^2)y'' - xy' + 4y = 0$$

3.18. Consider the hypergeometric equation.

$$x(x-1)y'' + \left(\frac{2}{3} - 3x\right)y' - y = 0$$

Use the method of Frobenius to find a complete solution around the point $x = 0$. Then, determine the hypergeometric functions around both singular points.

3.19. For the following equation, check whether they are reducible to Bessel equations. If so, then determine a complete solution in terms of Bessel functions.

i. $x\,y'' + 2y' - 4xy = 0$

ii. $x^2y'' + 3xy' + (1 + x)y = 0$

iii. $x^2y'' + (3x^3 - x)y' - 3xy = 0$

iv. $x^2y'' - 2x^2y' + (x^4 + x^2 - 6)y = 0$

v. $x^2y'' + (x^2 - 2x)y' + (2 - x)y = 0$

vi. $x^2y'' + 2xy' + (5x^4 + 2)y = 0$

vii. $x^2y'' - xy' + (1 - x^2)y = 0$

viii. $x^2y'' + xy + x^2y = 0$

ix. $x^2y^{iv} - 4y = 0$

x. $x^4y^{iv} + 8x^3y''' + 12x^2y'' - 9y = 0$

Section 3.4-3.6

3.20. Consider the nonlinear equation

$$y' + ay^2 + \frac{b}{x}y = \frac{c}{x^2}$$

Show that the solution is given by

$$y(x) = \frac{r_1x^{r_1-1} + c\,r_2x^{r_2-1}}{x^{r_1} + c\,x^{r_2}}$$

where r_1, and r_2 are the solution of the equation $r(r-1) + br + ac = 0$, and c is a constant.

3.21. Use the results of exercise 3.20 to solve the equation

$$y' + 5y^2 + \frac{5}{x}y = \frac{1}{x^2}, y(1) = -1$$

3.22. Solve the following first-order nonlinear differential equations.

i. $y' + 3y^2 + 2y = 3;\ y(0) = 4$

ii. $y' + y^2 + x^2y + 3x = 0$

iii. $(x^2 + y^2)y' + 2x(2x + y) = 0$

iv. $y' + y^2 + 3y = 4$

3.23. Consider the general Riccati equation of the form

$$y' - y^2 + axy + bx^2 + c = 0$$

If a possible solution is given by $y = K_1 + K_2 + u$ then show that

$$K_1 = \sqrt{\frac{a}{2} + c}, K_2 = \frac{a}{2}$$

Moreover, if the condition $b = K_2(K_2 - a)$ is satisfied, then show that the Riccati equation can be reduced to the Bernoulli equation, which has the form

$$u' - 2K_1u - u^2 = 0$$

3.24. Use the results of exercise 3.23 to determine the equivalent Bernoulli equation, then solve the Riccati equation

$$y' - y^2 + 6xy - 9x^2 + 1 = 0$$

3.25. Show that the nonlinear equation $yy'' - y'^2 = 0$ has the solution given by

$$y(x) = c_2e^{c_1x}$$

3.26. Solve the following second-order nonlinear equations.

3.27. Use the method of a parameter to solve the nonlinear equation

$$y'''^2 + y''^2 - 2y'' = 0$$

Hint: use the functions $\varphi(t) = y''' = \sin 2t,\ \psi(t) = y'' = 2\cos^2 t$

3.28. Use the method of a parameter to solve the nonlinear equation

$$8y'' + 9y'^4 = 0$$

3.29. For the equation

$$xy'^2 - 2yy' + 2 = 0$$

Show that the solution is given by the implicit or parametric form, respectively as

$$32x^3 - 12x^2y^2 - 24c\,xy + 8cy^3 + 4c = 0$$

$$y = \frac{1}{2}ct^2 + \frac{4}{3t}, x = ct + \frac{2}{3t^2}$$

3.30. Show that the equation $y'' + ay + by^3 = 0$ has the solution (3.72).

3.31. Consider the use of elliptic functions to solve the nonlinear equation

$$y'' + 6y + 3y^3 = 0;\; y(0) = 1$$

i. Determine the solution $y(t)$ and check the correctness of the solution.
ii. Determine the period of the solution.

3.32. For the general strong cubic hardening nonlinearity

$$y'' + ay + by^3 = 0;\; y(0) = y_0, y'(0) = y_0'$$

It has the solution 3.72, show that the amplitude A is given by

$$A = \sqrt{-\frac{a}{b} + \frac{a + by_0^2}{b}\sqrt{1 + \frac{2by_0'^2}{a}}}$$

3.33. Consider the softening nonlinearity Duffing problem

$$y'' + 3y - y^3 = 0;\; y(0) \neq 0; y'(0) = 1$$

Based on the solution formula (3.71), determine a solution of the equation.

3.34. For the general strong cubic hardening nonlinearity, let the solution be

$$y(t) = A\,\mathrm{cn}(wt + \theta)$$

Show that the phase shift θ is equal to zero when $y(0) \neq 0,$ and $y'(0) = 0.$

Chapter 4

LAPLACE TRANSFORMS

4.1. INTRODUCTION

The Laplace transform is named after the mathematician and astronomer Pierre-Simon Laplace. The Laplace transform of a function is a complex variable function. Laplace transform is usually restricted to time-functions for $t > 0$; throughout the text, this restriction is kept. The Laplace transformation is developed to solve a particular category of linear ordinary nonhomogeneous differential equations given their initial conditions (initial-value problem). In this category, there are two types. The first is the linear constant-coefficients of differential equations CCDE of arbitrary order, which is appeared most popular in the applications. The nth-order equations have the form

$$y^{(n)} + a_{n-1}y^{(n-1)} + a_{n-2}y^{(n-2)} + \cdots + a_1 y' + a_0 y = f(t) \tag{4.1}$$

The given initial conditions are

$$y(0) = y_0, y'(0) = y_{10}, y''(0) = y_{20} \ldots, y^{(n-1)}(0) = y_{(n-1)0} \tag{4.2}$$

where $y^{(n)}$ is the nth derivative, i.e., $y^{(n)} = \frac{d^n y}{dt^n}$, and a_i are real constants; some of these constants may be equal to zero. The function of time $f(t)$ is the force input, which is given for $t \geq 0$. The second type is a specific class of linear variable-coefficients differential equations VCDE, which has the form

$$x^n y^{(n)} + a_{n-1}x^{n-1}y^{(n-1)} + \cdots + a_1 x y' + a_0 y = f(t) \tag{4.3}$$

The advantages of Laplace transform for solving these equations are:

1. Solving an algebraic equation instead of a differential equation
2. Obtaining the solution directly, that is, without finding first the general solution and then a particular one

3. Moreover, the method procedure is also carried out in one stage for the nonhomogeneous case, unlike the classical method.

Besides the solution to these differential equations, the Laplace transform is utilized to solve other mathematical problems. Consequently, it also becomes an essential tool for signal analysis and the control of dynamic systems.

4.2. DEFINITIONS AND THEOREMS

It is necessary first to know the constraints on applying this transformation. The following definitions and theorems explain the conditions for utilizing the Laplace transform. We will see that not every time functions can be transformed to have a Laplace function. For the next discussion, the Laplace variable will be denoted by the variable s, a Laplace function by $F(s)$, and $\mathcal{L}$ as a Laplace operator.

Definition 4.1.

The function $f(t)$ is said to piecewise continuous on the interval $[0, \infty)$ if every interval of the form $0 \le t_1 \le t \le t_2$ can be divided into a finite number of subintervals such that $f(t)$ is continuous in each subinterval's interior. It approaches finite limits as the time t approaches either endpoint of the interval from the interior shown in Figure 4.1.

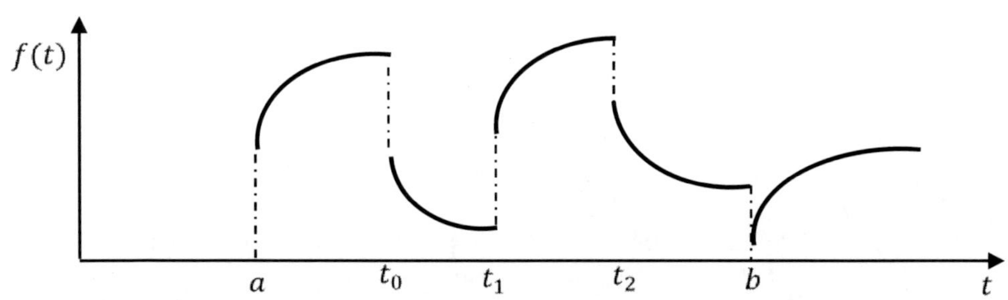

Figure 4.1. A piecewise time function.

To determine the Laplace transform for signals, essentially, we have to determine their mathematical definition. For example, the aperiodic time function shown in Figure 4.2 has the following mathematical definition.

$$f(t) = \begin{cases} 1, 0 \le t \le 1 \\ 0, 1 \le t \le 2 \\ 1, 2 \le t \le 3 \\ 0.5, t \ge 3 \end{cases}$$

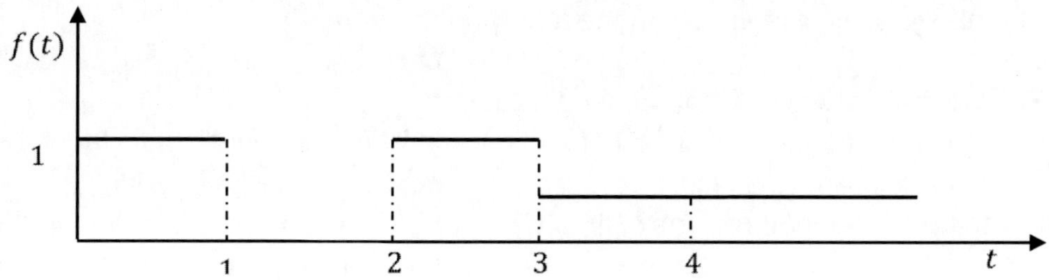

Figure 4.2. Aperiodic time-function.

Definition 4.2.

The function $f(t)$ is said to be of exponential order, if there exist scalars M, α, T such that

$$e^{-\alpha t}|f(t)| < M;\ \forall\, t > T \tag{4.4}$$

It means that as $t \to \infty$ the function $f(t)$ is bounded by the decreasing function $e^{-\alpha t}$. For instance:

- The function t^n is of an exponential order because $\lim_{t\to\infty} t^n\, e^{-\alpha t} = 0$.
- The function e^{t^2} is not of an exponential order because $\lim_{t\to\infty} e^{t^2}\, e^{-\alpha t} = \lim_{t\to\infty} e^{(t-\alpha)t} = \infty$ for every α.

Definition 4.3.

If $y = f(t)$ is a function whose domain is the positive real axis $0 \le t < \infty$, then the Laplace function $F(s)$ is given by the following integral.

$$\mathcal{L}\{y = f(t)\} = F(s) = \int_0^\infty f(t) e^{-st} dt \tag{4.5}$$

where $s = \sigma + jw$ is a complex variable. It is the Laplace transform of the function $y = f(t)$. When using the above definition, for some functions, we may meet the computational problem with infinite integration (improper integration); therefore, we should first answer the existence of this transform. Since by definition

$$\int_0^\infty f(t) e^{-st} dt = \lim_{k\to\infty} \{\int_0^k f(t) e^{-st} dt\} \tag{4.6}$$

Then we need to be sure that this limit exists before we can meaningfully discuss the function $F(s)$. The necessary condition but not sufficient that this limit exists is:

$$\lim_{t\to\infty} f(t)\, e^{-\alpha t} = 0 \tag{4.7}$$

The above discussion is put in the following theorem.

Theorem 4.1: Existence of Laplace transform

If $f(t)$ is a piecewise continuous function and of exponential order, then $f(t)$ has a Laplace transform.

For the elementary functions $\delta(t), u(t), t^n,\ e^{at},\ \sin wt, \cos wt$, definition 4.3 will be used to determine their Laplace transform as in the following examples.

Example 4.1.

Determine the Laplace transform of the delta function $\delta(t)$. It is an impulse located at the origin only, which is defined as

$$\delta(t) = \begin{cases} \frac{1}{\epsilon}, t = 0 \\ 0, \text{elsewhere} \end{cases}$$

Therefore,

$$F(s) = \mathcal{L}\{\delta(t)\} = \int_0^\infty \delta(t)e^{-st}dt = e^0 = 1$$

Example 4.2.

Determine the Laplace transform of the unit-step function $u(t)$, which is a unit pulse defined as

$$u(t) = \begin{cases} 0, t < 0 \\ 1, t \geq 0 \end{cases}$$

Therefore,

$$F(s) = \mathcal{L}\{u(t)\} = \int_0^\infty u(t)e^{-st}dt = \int_0^\infty 1. e^{-st}dt = -\frac{1}{s}e^{-st}|_0^\infty = -\frac{1}{s}(e^{-\infty} - e^0) = \frac{1}{s}, s > 0$$

Example 4.3.

Determine the Laplace transform of the function t^n where n is a positive integer.

$$F(s) = \mathcal{L}\{t^n\} = \int_0^\infty t^n e^{-st}dt = -\frac{1}{s}t^n e^{-st}|_0^\infty + \frac{n}{s}\int_0^\infty t^{n-1}e^{-st}dt$$

Using the rule of udv method of integration yields

$$F(s) = -\frac{1}{s}t^n e^{-st}|_0^\infty + \frac{n}{s}\int_0^\infty t^{n-1}e^{-st}dt = \frac{n}{s}\int_0^\infty t^{n-1}e^{-st}dt$$

$$F(s) = -\frac{1}{s^2} t^{n-1} e^{-st} |_0^\infty + \frac{n(n-1)}{s^2} \int_0^\infty t^{n-2} e^{-st} dt = \frac{n(n-1)}{s^2} \int_0^\infty t^{n-2} e^{-st} dt$$

By successive applying of the rule, we get

$$F(s) = \frac{n(n-1)(n-2)(n-3) \dots \dots}{s^n} \int_0^\infty t^{n-n} e^{-st} dt = \frac{n!}{s^n} \int_0^\infty e^{-st} dt$$

$$F(s) = \frac{n!}{s^n} \cdot \frac{1}{s} = \frac{n!}{s^{n+1}}, s > 0$$

Therefore, for example, the unit velocity function $f(t) = t$ has the Laplace transform

$$F(s) = \frac{1!}{s^{1+1}} = \frac{1}{s^2}, s > 0$$

Analogue, for noninteger *n*, we get

$$F(s) = \frac{\Gamma(n+1)}{s^{n+1}}, s > 0$$

where Γ is the generalized gamma function.

Example 4.4.

What is the Laplace transform of the exponential function e^{at} where a is a real or complex number?

$$F(s) = \mathcal{L}\{e^{at}\} = \int_0^\infty e^{at} e^{-st} dt = \int_0^\infty e^{(-s+a)t} dt = \frac{1}{-s+a} e^{(-s+a)t} |_0^\infty$$

$$F(s) = \frac{1}{s-a}, s > a$$

Let us assume that $a = jw$, then

$$F(s) = \mathcal{L}\{e^{jwt}\} = \int_0^\infty e^{jwt} e^{-st} dt = \frac{1}{s - jw} \cdot \frac{s + jw}{s + jw}$$

$$F(s) = \frac{s + jw}{s^2 + w^2} = \frac{s}{s^2 + w^2} + j \frac{w}{s^2 + w^2}$$

Since we know

$$e^{jwt} = \cos wt + j\sin wt$$

Then we can conclude the Laplace transform of the trigonometric functions $\sin wt$ and $\cos wt$ (see theorem 4.2 that states the linearity property of the Laplace transform) or

$$\mathcal{L}\{\cos wt\} = \frac{s}{s^2 + w^2}, \mathcal{L}\{\sin wt\} = \frac{w}{s^2 + w^2}$$

The above considered time functions are tabled in Table 4.1. The reader will remember the Laplace transform of these functions without regarding the process of integration. In literature, an extended table of transformation is available. For other time-domain functions that are usually found in the applications such as $te^{at}, t\sin wt, e^{at}\sin wt, \cosh wt, \sinh wt$, etc., the determination of their Laplace transform can be either found by direct computation of the integral or using Laplace transform properties. However, the reader will realize that using the Laplace properties is efficient and fast, as we will explain that in the next sections.

The Laplace transform properties will be given as a set of theorems; we will skip these theorems' proofs and direct the reader's attention to the theorems' usage in solving problems. However, the proofs of these theorems can be performed using formula (4.6) and simple mathematical manipulations. Besides, at the end of this chapter, the reader will find an extended Laplace transform table. The visible advantage of using these theorems is the fast determination of Laplace transforms and the extension of their applications.

Table 4.1. Laplace transform of essential elementary time functions

Time-domain function $f(t)$	**Laplace function $F(s)$**
$\delta(t)$	1
$u(t)$	$\frac{1}{s}$
t^n	$\frac{n!}{s^{n+1}}$
$e^{\pm at}$	$\frac{1}{s \mp a}$
$\cos wt$	$\frac{s}{s^2 + w^2}$
$\sin wt$	$\frac{w}{s^2 + w^2}$

Theorem 4.2.

The Laplace transform is a linear operator; therefore, if the Laplace transform of two time-functions $f(t)$ *and* $g(t)$ exists, then it holds

$$\mathcal{L}\{af(t) + bg(t)\} = a\mathcal{L}\{f(t) + b\mathcal{L}\{g(t)\} \tag{4.8}$$

The usefulness of this property is obvious to enlarge the set of time-domain functions that have Laplace transform.

Example 4.5.

$$\mathcal{L}\{3e^{-2t}+4\cos 4t\}=3\mathcal{L}e^{-2t}+4\mathcal{L}\cos 4t=\frac{3}{s+2}+\frac{4s}{s^2+16}=\frac{7s^2+8s+48}{s^3+2s^2+16s+32}$$

Example 4.6.

Determine the Laplace transform of the piecewise continuous function shown in Figure 4.2.

$$\mathcal{L}\{f(t)\}=\int_0^1 1.e^{-st}\,dt+\int_2^3 1.e^{-st}\,dt+\int_3^\infty 0.5\,e^{-st}\,dt$$

$$F(s)=\left(\frac{-e^{-s}+1}{s}\right)+\left(\frac{-e^{-3s}+e^{-2s}}{s}\right)+\frac{1}{2}\left(\frac{e^{-3s}}{s}\right)$$

$$F(s)=\frac{1}{s}\left(1-e^{-s}+e^{-2s}-\frac{1}{2}e^{-3s}\right)$$

In section 4.5, we will introduce another general approach for solving such problems.

Theorem 4.3.

If $f(t)$ has a Laplace transform denoted $\phi(s)$, then

$$\mathcal{L}\{t^n f(t)\}=(-1)^n\frac{d^n\phi(s)}{ds^n} \tag{4.9}$$

These multiplications by time function are found in many dynamic and signal problems.

Example 4.7.

What is the Laplace transform of the function $t\sin 5t$?

Since we know (from Table 1)

$$\mathcal{L}\{\sin 5t\}=\frac{5}{s^2+25}=\phi(s)$$

According to theorem 4.3; since $n = 1$, then

$$\mathcal{L}\{t\sin 5t\}=-\frac{d}{ds}\left[\frac{5}{s^2+25}\right]=-\frac{-5(2s)}{(s^2+25)^2}=\frac{10s}{(s^2+25)^2}$$

The reader can exercise the Laplace of $f(t) = t^2\cos 3\text{t}$ to be

$$F(s) = \frac{-54s + 2s^3}{(s^2+9)^3}$$

Theorem 4.4.

If $f(t)$ has a Laplace transform denoted $\phi(s)$, and if $f(t)/t$ has a limit as $t \to 0$ from the right, then

$$\mathcal{L}\left\{\frac{f(t)}{t}\right\} = \int_s^\infty \phi(s)ds \tag{4.10}$$

Example 4.8.

Determine the Laplace transform of time function

$$f(t) = \frac{\sin t}{t}$$

Since

$$\lim_{t\to 0}\left\{\frac{\sin t}{t}\right\} = 1$$

$$\mathcal{L}\{\sin t\} = \frac{1}{s^2+1}$$

Then according to theorem 4.4, we have

$$\mathcal{L}\frac{\sin t}{t} = \int_s^\infty \frac{1}{s^2+1}ds = \tan^{-1}(s)|_s^\infty = \frac{\pi}{2} - \tan^{-1}(s) = \cot^{-1}(s)$$

Theorem 4.5. First shifting theorem

If $f(t)$ has a Laplace transform denoted $F(s)$, then

$$\mathcal{L}\{e^{\pm at}f(t)\} = \mathcal{L}\{f(t)\}|_{s\to s\mp a} = F(s \mp a) \tag{4.11}$$

It means that multiplication in the time domain by exponential function causes a shift in the frequency domain. Signs in the power of the exponential function indicate a feature concerning the system's stability.

Example 4.9.

Find the Laplace transform of the following time-domain functions.

$$\mathcal{L}\{e^{-at}\cos bt\}, \mathcal{L}\{e^{-at}\sin bt\}, \text{ and } \mathcal{L}\{te^{at}\}$$

According to theorem 4.5, we have

$$\mathcal{L}\{e^{-at}\cos bt\} = \mathcal{L}\{\cos bt\}|_{s\to s+a} = \frac{s}{s^2+b^2}|_{s\to s+a} = \frac{s+a}{(s+a)^2+b^2}$$

Similarly,

$$\mathcal{L}\{e^{-at}\sin bt\} = \mathcal{L}\{\sin bt\}|_{s\to s+a} = \frac{b}{s^2+b^2}|_{s\to s+a} = \frac{b}{(s+a)^2+b^2}$$

$$\mathcal{L}\{te^{at}\} = \mathcal{L}\{t\}|_{s\to s-a} = \frac{1}{s^2}|_{s\to s-a} = \frac{1}{(s-a)^2}$$

Example 4.10.
Determine the Laplace transform of

$$f(t) = 3t\,e^{-3t}\cos\sqrt{2}\,t$$

The simultaneous use of theorems 4.3 and 4.5 yields

$$-3\frac{d}{ds}\left[\frac{s}{s^2+2}|_{s\to s+3}\right] = -3\frac{d}{ds}\left[\frac{s+3}{(s+3)^2+2}\right] = -3\frac{d}{ds}\left[\frac{s+3}{s^2+6s+11}\right]$$

$$\mathcal{L}\{3t\,e^{-3t}\cos\sqrt{2}\,t\} = \frac{3s^2+18s+21}{(s^2+6s+11)^2}$$

We will continue by introducing other theorems after we recognize the meaning of the Laplace inverse.

4.3. LAPLACE INVERSE

A meaningful question for a given Laplace function $F(s)$ is what is the corresponding $f(t)$, and how can we obtain the function $f(t)$? The following theorem answers these questions.

Theorem 4.6. Lerch's theorem
If $f(t)$ and $g(t)$ are both piecewise continuous on $[0,\infty)$ and of exponential order, and if $\mathcal{L}f(t) = \mathcal{L}g(t)$, then $f(t) = g(t)$. Henceforth, we will write $f(t) = \mathcal{L}^{-1}\{F(s)\}$ to indicate that $F(s) = \mathcal{L}f(t)$ and will call $\mathcal{L}^{-1}\{F(s)\}$ the inverse transforms of $F(s)$. The Laplace inverse operator is linear too, i.e.,

$$\mathcal{L}^{-1}\{aF_1 + bF_2\} = a\,\mathcal{L}^{-1}F_1 + b\,\mathcal{L}^{-1}F_2 \tag{4.12}$$

The existence question of the inverse is cumbersome to answer frequently. Usually, the available tables and some algebraic manipulations are used to obtain the inverse. The inverse has an essential role in solving the differential equations, as we will see later. The most straightforward technique to determine the inverse is the elementary partial fraction decompositions. Before introducing a general efficient way to carry out this technique, a simple example is given.

Example 4.11.

Determine $f(t)$ of the following Laplace function

$$F(s) = \frac{12s - 28}{s^3 - 3s^2 - 13s + 15}$$

$$f(t) = \mathcal{L}^{-1}\{F(s)\} = \mathcal{L}^{-1}\left\{\frac{12s - 28}{s^3 - 3s^2 - 13s + 15}\right\} = \mathcal{L}^{-1}\{\frac{12s - 28}{(s-5)(s+3)(s-1)}\}$$

Using the partial fraction decompositions and the linearity property, we can write

$$\mathcal{L}^{-1}\{F(s)\} = \mathcal{L}^{-1}\left\{\frac{1}{s-5}\right\} + \mathcal{L}^{-1}\left\{\frac{-2}{s+3}\right\} + \mathcal{L}^{-1}\{\frac{1}{s-1}\}$$

From Table 4.1, we obtain

$$f(t) = e^{5t} - 2e^{-3t} + e^{t}$$

Although the used technique in example 4.12 can be utilized for any rational ratio of $F(s)$, several theorems can be very useful in determining the Laplace transform more efficiently.

Theorem 4.7.

Let $f(t)$ be a piecewise continuous function on $[0, \infty)$ and of exponential order. If $\mathcal{L}f(t) = sG(s)$ and the inverse of the $G(s)$ is continuous for $t \geq 0$, and has a piecewise continuous derivative of exponential order on $[0, \infty)$, then

$$f(t) = \frac{d}{dt}[\,\mathcal{L}^{-1}\{G(s)\}] + g(0)\delta(t) \tag{4.13}$$

Theorem 4.8.

Let $f(t)$ be a piecewise continuous function on $[0, \infty)$ and of exponential order, and if $\mathcal{L}f(t) = \phi(s)/s$, and the inverse of the $\phi(s)$ exists, then

$$f(t) = \int_0^t [\,\mathcal{L}^{-1}\{\phi(s)\}]dt \tag{4.14}$$

Example 4.12.
Determine the following inverse.

$$\mathcal{L}^{-1}\{\frac{1}{s(s^2+1)}\}$$

A direct application of theorem 4.8 gives

$$f(t) = \int_0^t \left[\mathcal{L}^{-1}\left\{\frac{1}{(s^2+1)}\right\}\right] dt = \int_0^t \sin t\, dt = -\cos t\,|_0^t = 1 - \cos t$$

In this stage, the reader is encouraged to solve this problem using the technique of example 4.12, and he or she will need to work with complex variables, since

$$\frac{1}{s(s^2+1)} = \frac{k_1}{s} + \frac{k_2}{(s+j)} + \frac{k_3}{(s-j)};\; e^{\pm jt} = \cos t \pm j \sin t$$

The reader will recognize the efficacy of applying theorem 4.7.

Theorem 4.9. The inverse forms of theorems 4.3, 4.4, and 4.5.
If $f(t)$ has a Laplace transforms denoted $\phi(s)$, then

$$f(t) = \mathcal{L}^{-1}\{\phi(s)\} = -\frac{1}{t}\,\mathcal{L}^{-1}\left\{\frac{d\phi(s)}{ds}\right\} \tag{4.15}$$

$$f(t) = \mathcal{L}^{-1}\{\phi(s)\} = -t\,\mathcal{L}^{-1}\{\int_s^\infty \phi(s)\, ds\} \tag{4.16}$$

$$f(t) = \mathcal{L}^{-1}\{\phi(s)\} = e^{-at}\,\mathcal{L}^{-1}\{\phi(s-a)\} \tag{4.17}$$

In the following examples, the above theorems and the linearity concept of theorem 4.6 will be used to efficiently determine the Laplace inverse.

Example 4.13.
Find the inverse

$$\mathcal{L}^{-1}\left\{\ln\frac{s^2+1}{s(s+1)}\right\} = \mathcal{L}^{-1}\{\ln(s^2+1)\} - \mathcal{L}^{-1}\{\ln(s+1)\} - \mathcal{L}^{-1}\{\ln(s)\}$$

$$= -\frac{1}{t}\left[\mathcal{L}^{-1}\left\{\frac{d}{ds}\ln(s^2+1)\right\} - \mathcal{L}^{-1}\left\{\frac{d}{ds}\ln(s+1)\right\} - \mathcal{L}^{-1}\left\{\frac{d}{ds}\ln(s)\right\}\right]$$

$$= -\frac{1}{t}\left[\mathcal{L}^{-1}\left\{\frac{2s}{s^2+1}\right\} - \mathcal{L}^{-1}\left\{\frac{1}{s+1}\right\} - \mathcal{L}^{-1}\left\{\frac{1}{s}\right\}\right]$$

$$= -\frac{1}{t}(2\cos t - e^{-t} - u(t)) = \frac{e^{-t} - 2\cos t + u(t)}{t}$$

Example 4.14.
Find the inverse

$$\mathcal{L}^{-1}\left\{\frac{s+2}{(s^2+4s+5)^2}\right\} = -t\,\mathcal{L}^{-1}[\int_s^{\infty}\frac{s+2}{(s^2+4s+5)^2}\,ds]$$

$$-t\,\mathcal{L}^{-1}\left[\frac{1}{2}\int_s^{\infty}\frac{2s+4}{(s^2+4s+5)^2}\,ds\right] = -t\,\mathcal{L}^{-1}\{\frac{1}{2}\frac{1}{(s^2+4s+5)}|_s^{\infty}\}$$

$$\frac{1}{2}t\,\mathcal{L}^{-1}\{\frac{1}{(s+2)^2+1)} = \frac{1}{2}te^{-2t}\sin t$$

Example 4.15.
Find the inverse

$$\mathcal{L}^{-1}\left\{\frac{2s+7}{s^2+4s+13}\right\} = \mathcal{L}^{-1}\left\{\frac{2(s+2)+3}{(s+2)^2+9}\right\}$$

$$\mathcal{L}^{-1}\left\{\frac{2(s+2)}{(s+2)^2+9}\right\} + \mathcal{L}^{-1}\left\{\frac{3}{(s+2)^2+9}\right\}$$

$$= 2e^{-2t}\cos 3t + e^{-2t}\sin 3t = (2\cos 3t + \sin 3t)\,e^{-2t}$$

Example 4.16.
Find the inverse

$$\mathcal{L}^{-1}\left\{\frac{s}{(s+4)^4}\right\} = \frac{d}{dt}\left[\mathcal{L}^{-1}\left\{\frac{1}{(s+4)^4}\right\}\right] + g(0)\delta(t)) = \frac{d}{dt}\left[\frac{1}{3!}t^3e^{-4t}\right] + 0$$

$$= \frac{1}{6}(3t^2e^{-4t} - 4t^3e^{-4t}) = \left(\frac{1}{2} - \frac{2}{3}t\right)t^2e^{-4t}$$

4.4. The Solution of Differential Equations Using Laplace Transformation

As we introduced in section 4.1, the Laplace transform's essential use is to solve some widely used types of differential equations. First, we need to state a theorem that transforms the derivatives of the dependent variable.

Theorem 4.10.

If the Laplace transforms of the function $f(t)$ and $f'(t)$ exist, and if $f'(t)$ is continuous, then

$$\mathcal{L}\{f'(t)\} = s\mathcal{L}\{f(t)\} - f(0) = sF(s) - f(0) \tag{4.18}$$

Moreover, for positive integer *n*, if $f^{(n)}(t)$ is continuous, then we have

$$\mathcal{L}\{f^{(n)}(t)\} = s^n\mathcal{L}\{f(t)\} - s^{n-1}f(0) - s^{n-2}f'(0) - \cdots - sf^{(n-2)}(0) - f^{(n-1)}(0) \tag{4.19}$$

Note that equations (4.18) and (4.19) are equally applied for both sides of the differential equation.

Example 4.17.

Solve the following initial-value nonhomogeneous equation.

$$y'' + 4y' + 13 = \delta(t);\ y(0) = 2, y'(0) = 4$$

The three steps for solving are:

1) Take a Laplace transform of both sides of the given equation to obtain the equivalent algebraic equation (the Laplace of the delta function is taken from Table 1).

$$s^2Y(s) - sy(0) - y'(0) + 4sY(s) - y(0) + 13Y(s) = 1$$

$$s^2Y(s) - 2s - 4 + 4sY(s) - 2 + 13Y(s) = 1$$

$$(s^2 + 4s + 13)Y(s) = 2s + 7$$

2) Using simple algebraic manipulation, we can obtain an expression for the Laplace function $Y(s)$.

$$Y(s) = \frac{2s + 7}{s^2 + 4s + 13}$$

3) Finally, we determine the time-domain solution by implementing the Laplace inverse.

$$y(t) = \mathcal{L}^{-1}\left\{\frac{2s+7}{s^2+4s+13}\right\}$$

The reader can note that the required inverse was solved in example 4.15; therefore, the solution of the given equation is

$$y(t) = (2\cos 3t + \sin 3t)\, e^{-2t}$$

Note that the Laplace inverse can be obtained in various ways.

The reader may realize that the main difficulty in solving a differential equation is the inverse problem. However, for a complicated force function, the corresponding Laplace function's determination also requires a particular effort. Therefore, the efficient use of the previous theorems uses the Laplace transformation technique fruity tasks. On the other hand, the partial fraction decomposition to implement step 3 is sometimes preferable. In what follows, we discuss this decomposition.

As can be noted, manipulating step 2 always yields a ratio of two polynomials in terms of the variable s. The nominator polynomial $N(s)$ has an order m, which is less or at most equal to the polynomial order n of the denominator $D(s)$, i.e., $n \geq m$. The required decomposition depends on the order in the denominator. The following structures of the denominator polynomial and their decompositions have the forms.

1) The denominator can be factorized to n unrepeated terms of first-order polynomials, i.e.,

$$D(s) = s^n + c_{n-1}s^{n-1} + \cdots + c_1 s + c_0 = (s+p_1)(s+p_2)\ldots(s+p_n) \tag{4.20}$$

Then the required decomposition is

$$\frac{N(s)}{D(s)} = \frac{k_1}{s+p_1} + \frac{k_2}{s+p_2} + \cdots + \frac{k_n}{s+p_n} \tag{4.21}$$

2) The denominator can be factorized into several repeated first-order polynomials of power $q_1, q_2, \ldots q_r$, such that

$$\sum_{i=1}^{i=r} q_i = n;\; q_i > 1 \tag{4.22}$$

Or

$$D(s) = s^n + c_{n-1}s^{n-1} + \cdots + c_1 s + c_0 = (s+p_1)^{q_1}(s+p_2)^{q_2}\ldots(s+p_r)^{q_r} \tag{4.23}$$

Then the required decomposition is

$$\frac{N(s)}{D(s)}=\frac{k_{11}}{(s+p_1)^{q_1}}+\frac{k_{12}}{(s+p_1)^{q_1-1}}+\cdots+\frac{k_{1q_1}}{(s+p_1)}+\frac{k_{21}}{(s+p_2)^{q_2}}+\frac{k_{22}}{(s+p_2)^{q_2-1}}$$

$$+\cdots+\frac{k_{2q_2}}{(s+p_2)}+\cdots+\frac{k_{r1}}{(s+p_r)^{q_r}}+\frac{k_{r2}}{(s+p_r)^{q_r-1}}+\cdots+\frac{k_{rq_r}}{(s+p_r)} \qquad (2.24)$$

3) The denominator can be factorized into first-order brackets of a power one and brackets of a power greater than one. The required decomposition is mixed in the above two cases.

In some cases, depending on the denominator and the nominator, it is required to decompose in a different structure (unlike that used in equation (4.21)). For completeness, some preliminary examples will be given to learning the partial fraction decomposition technique. In other words, how the parameters of the decomposition, the roots $p_1, p_2, \ldots p_n$, and the coefficients $k_1, k_2 \ldots$, are determined.

Example 4.18.
Use partial fraction decomposition for

$$\frac{N(s)}{D(s)}=\frac{2}{s(s^2+5s+6)}$$

We can rewrite it as

$$\frac{N(s)}{D(s)}=\frac{2}{s(s+2)(s+3)}=\frac{k_1}{s}+\frac{k_2}{s+2}+\frac{k_3}{s+3}$$

Therefore, the roots are $p_1=0, p_2=-2, p_3=-3$. To compute the coefficients k_1, k_2, and k_3, we use the formula

$$k_i=\frac{N(s)}{D(s)}(s-p_i)|_{s=-p_i}$$

$$k_1=\frac{2}{s(s+2)(s+3)}(s+0)|_{s=0}=\frac{1}{3}$$

$$k_2=\frac{2}{s(s+2)(s+3)}(s+2)|_{s=-2}=-1$$

$$k_1=\frac{2}{s(s+2)(s+3)}(s+3)|_{s=-3}=\frac{2}{3}$$

The reader may check the correctness of the decomposition by ensuring that

$$k_1(s+2)(s+3)+k_2 s(s+3)+k_3 s(s+2)=2$$

$$\frac{s^2+5s+6}{3}-(s^2+3s)+\frac{2s^2+4s}{3}=2$$

Example 4.19.
Use partial fraction decomposition for

$$F(s)=\frac{N(s)}{D(s)}=\frac{s^2+5s+10}{(s+1)^3}$$

We can rewrite it as

$$\frac{N(s)}{\widetilde{D}(s)}=\frac{s^2+5s+10}{(s+1)^3}=\frac{k_{11}}{(s+1)^3}+\frac{k_{12}}{(s+1)^2}+\frac{k_{13}}{(s+1)}$$

Therefore, there are three repeated roots $p_{1,2,3}=p=-1$, and to compute the parameters k_1, k_2, k_3, we use the formula

$$k_{1r}=\frac{1}{(r-1)!}\frac{d^{r-1}}{ds^{r-1}}[F(s)(s-p)^n]|_{s=-p};\ r=1,2..n$$

For the considered example, we compute

$$k_{11}=\frac{1}{(1-1)!}\frac{d^0}{ds^0}\left[\frac{s^2+5s+10}{(s+1)^3}(s+1)^3\right]|_{s=-1}=1-5+10=6$$

$$k_{12}=\frac{1}{(2-1)!}\frac{d}{ds}\left[\frac{s^2+5s+10}{(s+1)^3}(s+1)^3\right]|_{s=-1}=(2s+5)|_{s=-1}=3$$

$$k_{13}=\frac{1}{(3-1)!}\frac{d^2}{ds^2}\left[\frac{s^2+5s+10}{(s+1)^3}(s+1)^3\right]|_{s=-1}=1$$

The partial fractional decomposition is

$$F(s)=\frac{s^2+5s+10}{(s+1)^3}=\frac{6}{(s+1)^3}+\frac{3}{(s+1)^2}+\frac{1}{(s+1)}$$

Example 4.20.
Use partial fraction decomposition for

$$F(s)=\frac{N(s)}{D(s)}=\frac{s^2+6s+6}{(s+1)^2(s+3)}$$

We can rewrite it as

$$F(s) = \frac{s^2 + 6s + 6}{(s+1)^2(s+3)} = \frac{k_{11}}{(s+1)^2} + \frac{k_{12}}{(s+1)} + \frac{k_3}{(s+3)}$$

Therefore, there are two repeated roots $p_{1,2} = p = -1$ and one single root $p_3 = -3$. To compute the coefficients, we compute first as in example 4.20 for the two repeated roots, then as in example 4.19 for the single root.

$$k_{11} = \frac{1}{(1-1)!}\frac{d^0}{ds^0}\left[\frac{s^2 + 6s + 6}{(s+1)^2(s+3)}(s+1)^2\right]|_{s=-1} = \frac{1}{2}$$

$$k_{12} = \frac{1}{(2-1)!}\frac{d}{ds}\left[\frac{s^2 + 6s + 6}{(s+1)^2(s+3)}(s+1)^2\right]|_{s=-1}$$

$$= \frac{(s+3)(2s+6) - (s^2 + 6s + 6)(1)}{(s+3)^2}|_{s=-1} = \frac{7}{4}$$

$$k_3 = \frac{s^2 + 6s + 6}{(s+1)^2(s+3)}(s+3)|_{s=-3} = -\frac{3}{4}$$

Then, the partial fractional decomposition is given as

$$F(s) = \frac{s^2 + 6s + 6}{(s+1)^2(s+3)} = \frac{1/2}{(s+1)^2} + \frac{7/4}{(s+1)} + \frac{-3/4}{(s+3)}$$

Note that in some practical cases, the denominator is factorized by the nature of the problem, i.e., it is unnecessary to have all first-order brackets as it will be done with example 4.23.

Example 4.21.
Use partial fraction decomposition for

$$F(s) = \frac{N(s)}{D(s)} = \frac{4s + 11}{s^2 + 6s + 13}$$

We can rewrite it as

$$F(s) = \frac{4s + 11}{(s+3)^2 + 4} = \frac{4s + 11}{(s + 3 + 2j)(s + 3 - 2j)}$$

$$F(s) = \frac{k_1}{(s + 3 + 2j)} + \frac{k_2}{(s + 3 - 2j)}$$

Therefore, there are two complex conjugate roots of $p_{1,2} = -3 \pm 2j$, and the coefficients k_1 and k_2 are computed as

$$k_1 = \frac{4s + 11}{(s + 3 + 2j)(s + 3 - 2j)}(s + 3 + 2j)|_{s=-3-2j}$$
$$k_1 = \frac{4(-3 - 2j) + 11}{(-3 - 2j + 3 - 2j)} = \frac{-1 - 8j}{-4j} \times \frac{j}{j} = 2 - \frac{1}{4}j$$

The value of the coefficient k_2 is the conjugate of k_1, thus it is $2 + \frac{1}{4}j$. The partial fractional decomposition is

$$F(s) = \frac{2 - \frac{1}{4}j}{(s + 3 + 2j)} + \frac{2 + \frac{1}{4}j}{(s + 3 - 2j)}$$

In this example, the computation with complex numbers cannot be avoided. However, if it is possible, it is preferable to prevent calculation with complex numbers.

Example 4.22.
Use partial fraction decomposition for

$$F(s) = \frac{N(s)}{D(s)} = \frac{s^2 + 3s + 5}{(s + 1)(s^2 + 6s + 13)}$$

As we see, in example 4.21, the decomposition will contain complex numbers because of the term $(s^2 + 6s + 13)$. It is possible to carry a different decomposition structure as

$$F(s) = \frac{k_1}{s + 1} + \frac{k_2 s + k_3}{(s^2 + 6s + 13)}$$

The coefficient k_1 can be found as we did previously, so

$$k_1 = \frac{s^2 + 3s + 5}{(s + 1)(s^2 + 6s + 13)}(s + 1)|_{s=-1} = \frac{3}{8}$$

Next, we carry out the usual algebraic manipulation

$$k_1(s^2 + 6s + 13) + (k_2 s + k_3)(s + 1) = s^2 + 3s + 5$$

Equating the coefficients of equal power yields

$$k_1 + k_2 = 1 \Rightarrow k_2 = 1 - \frac{3}{8} = \frac{5}{8}$$

$$13k_1 + k_3 = 5 \Rightarrow k_3 = 5 - \frac{13 \times 3}{8} = \frac{1}{8}$$

The third equating coefficient $6k_1 + k_2 + k_3 = 3$ can be used to check the obtained results.

$$6 \times \frac{3}{8} + \frac{5}{8} + \frac{1}{8} = 3 \Rightarrow \frac{24}{8} = 3$$

The obtained decomposition is

$$F(s) = \frac{\frac{3}{8}}{s+1} + \frac{\frac{5}{8}s + \frac{1}{8}}{(s^2 + 6s + 13)}$$

Let us return to the initial-value problem using partial fraction decomposition to determine the inverse of the obtained Laplace function. First, we will consider the problem, of example 4.17. The aim is to let the reader compare the two approaches.

Example 4.23.

The response of a dynamic system, which has specified initial conditions, is given as

$$y(t) = \mathcal{L}^{-1}\left\{\frac{2s+7}{s^2 + 4s + 13}\right\}$$

Let

$$F(s) = \frac{2s+7}{s^2 + 4s + 13} = \frac{k_1}{s + 2 - 3j} + \frac{k_2}{s + 2 + 3j}$$

The coefficients k_1 and k_2 are computed as follows

$$k_1 = \frac{2s+7}{(s+2-3j)(s+2+3j)}(s + 2 - 3j)|_{s=-2+3j}$$

$$k_1 = \frac{2(-2+3j)+7}{(-2+3j+2+3j)} = \frac{3+6j}{6j} = 1 - \frac{1}{2}j$$

Thus, $k_2 = 1 + \frac{1}{2}j$,

$$F(s) = \frac{1 - \frac{1}{2}j}{s + 2 - 3j} + \frac{1 + \frac{1}{2}j}{s + 2 + 3j}$$

The solution now is expressed by

$$y(t) = \mathcal{L}^{-1}\frac{1-\frac{1}{2}j}{s+2-3j} + \mathcal{L}^{-1}\frac{1+\frac{1}{2}j}{s+2+3j}$$

The complete solution is

$$y(t) = (1-\frac{1}{2}j)e^{-(-2-3j)t} + (1+\frac{1}{2}j)e^{-(-2+3j)t}$$

We should have a real solution; therefore, we use the known formulas of complex variable functions.

$$y(t) = (1-\frac{1}{2}j)e^{2t}(\cos 3t + j\sin 3t) + (1+\frac{1}{2}j)e^{2t}(\cos 3t - j\sin 3t)$$

$$y(t) = e^{2t}(\cos 3t + j\sin 3t) - \frac{1}{2}je^{2t}(\cos 3t + j\sin 3t) + e^{2t}(\cos 3t - j\sin 3t) + \frac{1}{2}je^{2t}(\cos 3t - j\sin 3t)$$

$$y(t) = e^{2t}\left(\cos 3t + \frac{1}{2}\sin 3t + \cos 3t + \frac{1}{2}\sin 3t\right) + je^{2t}\left(\sin 3t - \frac{1}{2}\cos 3t - \sin 3t + \frac{1}{2}\cos 3t\right)$$

Therefore, the complete solution of the given initial-value problem is

$$y(t) = e^{2t}(2\cos 3t + \sin 3t)$$

The reader can now notice the long computation with the complex roots and the efficiency of the approach that depends on applying the suitable theorems in solving the problem (see example 4.15).

Example 4.24.
Solve the initial-value problem

$$y'' + 5y' + 6y = 2u(t);\ y(0) = 1, y'(0) = -1$$

Taking the Laplace transform of both sides yields

$$s^2Y(s) - sy(0) - y'(0) + 5sY(s) - 5y(0) + 6Y(s) = \frac{2}{s}$$

$$(s^2+5s+6)Y(s)-s+1-5=\frac{2}{s}$$

$$(s^2+5s+6)Y(s)-s+1-5=\frac{2}{s}$$

$$Y(s)=\frac{s^2+4s+2}{s(s^2+5s+6)}=\frac{s^2+4s+2}{s(s+2)(s+3)}$$

$$Y(s)=\frac{k_1}{s}+\frac{k_2}{(s+2)}+\frac{k_3}{(s+3)}$$

where the decomposition coefficients are computed as

$$k_1=\frac{s^2+4s+2}{s(s+2)(s+3)}s|_{s=0}=\frac{1}{3}$$

$$k_2=\frac{s^2+4s+2}{s(s+2)(s+3)}(s+2)|_{s=-2}=1$$

$$k_3=\frac{s^2+4s+2}{s(s+2)(s+3)}(s+3)|_{s=-3}=-\frac{1}{3}$$

Therefore,

$$Y(s)=\frac{1/3}{s}+\frac{1}{(s+2)}-\frac{1/3}{(s+3)}$$

The complete solution is

$$y(t)=\mathcal{L}^{-1}Y(s)=\mathcal{L}^{-1}\frac{1/3}{s}+\mathcal{L}^{-1}\frac{1}{(s+2)}-\mathcal{L}^{-1}\frac{1/3}{(s+3)}$$

$$y(t)=\frac{1}{3}u(t)+e^{-2t}-\frac{1}{3}e^{-3t}$$

Example 4.25.

A fourth-order dynamic system at rest, which is exposure to exponential input, is described as

$$y^{iv}+2y'''+2y''+2y'+y=e^{-t};\ y(0)=y'(0)=y''(0)=y'''(0)=0$$

Taking Laplace transforms of both sides, and because all initial conditions are equal to zero, then we obtain

$$s^4Y(s) + 2s^3Y(s) + 2s^2Y(s) + 2sY(s) + Y(s) = \frac{1}{s+1}$$

$$(s^4 + 2s^3 + 2s^2 + 2s+)Y(s) = \frac{1}{s+1}$$

$$(s+1)^2(s^2+1)Y(s) = \frac{1}{s+1}$$

$$Y(s) = \frac{1}{(s+1)^3(s^2+1)}$$

Let us make a suitable decomposition of $Y(s)$.

$$Y(s) = \frac{k_{11}}{(s+1)^3} + \frac{k_{12}}{(s+1)^2} + \frac{k_{13}}{(s+1)} + \frac{k_2 s + k_3}{(s^2+1)}$$

The decomposition coefficients are computed from

$$k_{11} = \frac{1}{(1-1)!}\frac{d^0}{ds^0}\left[\frac{1}{(s+1)^3(s^2+1)}(s+1)^3\right]|_{s=-1} = \frac{1}{2}$$

$$k_{12} = \frac{1}{(2-1)!}\frac{d^1}{ds^1}\left[\frac{1}{(s+1)^3(s^2+1)}(s+1)^3\right]|_{s=-1} = \frac{-2s}{(s^2+1)^2}|_{s=-1} = \frac{1}{2}$$

$$k_{13} = \frac{1}{(3-1)!}\frac{d^2}{ds^2}\left[\frac{1}{(s+1)^3(s^2+1)}(s+1)^3\right]|_{s=-1} = \frac{1}{2}\frac{-2(s^2+1) + 8s^2(s^2+1)}{(s^2+1)^4}|_{s=-1}$$
$$= \frac{1}{4}$$

Rewriting $Y(s)$ in the form

$$Y(s) = \frac{\frac{1}{4}s^2 + s + \frac{5}{4}}{(s+1)^3} + \frac{k_2 s + k_3}{(s^2+1)}$$

Thus, we write

$$\left(\frac{1}{4}s^2 + s + \frac{5}{4}\right)(s^2+1) + (k_2 s + k_3)(s+1)^3 = 1$$

Equating the coefficients of equal power and solving for the unknown coefficients k_2 and k_3 gives

$$k_2 = -\frac{1}{4}, k_3 = \frac{1}{4}$$

The decomposed function becomes

$$Y(s) = \frac{1/2}{(s+1)^3} + \frac{1/2}{(s+1)^2} + \frac{1/4}{(s+1)} + \frac{-1/4s + 1/4}{(s^2+1)}$$

The complete solution is

$$y(t) = \mathcal{L}^{-1}Y(s)$$

$$y(t) = \mathcal{L}^{-1}\frac{1/2}{(s+1)^3} + \mathcal{L}^{-1}\frac{1/2}{(s+1)^2} + \mathcal{L}^{-1}\frac{1/4}{(s+1)} + \mathcal{L}^{-1}\frac{-1/4s}{(s^2+1)} + \mathcal{L}^{-1}\frac{1/4}{(s^2+1)}$$

$$y(t) = \frac{1}{2}.\frac{1}{2}t^2e^{-t} + \frac{1}{2}te^{-t} + \frac{1}{4}e^{-t} - \frac{1}{4}\cos t + \frac{1}{4}\sin t$$

$$y(t) = \frac{t^2 + 2t + 1}{4}e^{-t} - \frac{1}{4}(\cos t - \sin t)$$

For the next example, several previous theorems will be applied to determine the complete solution for solving the initial-value problem.

Example 4.26

Determine the response of the following displaced harmonic system forced under a sinusoidal input.

$$y'' + 4y = t\sin 2t;\ y(0) = 1, y'(0) = 0$$

Taking the Laplace transform of both sides yields

$$s^2Y(s) - sy(0) + 4Y(s) = -\frac{d}{ds}[\mathcal{L}\{\sin 2t\}]$$

$$(s^2+4)Y(s) - s = -\frac{d}{ds}[\frac{2}{s^2+4}]$$

$$(s^2+4)Y(s) = \frac{4s}{(s^2+4)^2} + s$$

$$Y(s) = \frac{4s}{(s^2+4)^3} + \frac{s}{(s^2+4)}$$

$$y(t) = \mathcal{L}^{-1}Y(s) = \mathcal{L}^{-1}\frac{4s}{(s^2+4)^3} + \mathcal{L}^{-1}\frac{s}{(s^2+4)}$$

The second Laplace inverse gives $\cos 2t$, while the first is determined using different theorems as follows.

$$\mathcal{L}^{-1}\frac{4s}{(s^2+4)^3} = -t\,\mathcal{L}^{-1}\left\{\int_s^\infty \frac{4s}{(s^2+4)^3}ds\right\} = -t\,\mathcal{L}^{-1}\left\{\left[-\frac{1}{(s^2+4)^2}\right]\Big|_s^\infty\right\}$$

$$= -t\,\mathcal{L}^{-1}\left\{-0 + \frac{1}{(s^2+4)^2}\right\} = -t\,\mathcal{L}^{-1}\left\{\frac{s}{s(s^2+4)^2}\right\}$$

$$= -t\int_0^t [\mathcal{L}^{-1}\{\frac{s}{(s^2+4)^2}\}]dt = t\int_0^t t\left[\mathcal{L}^{-1}\left\{\int_s^\infty \frac{s}{(s^2+4)^2}ds\right\}\right]dt$$

$$= t\left[\mathcal{L}^{-1}\left\{\frac{-\frac{1}{2}}{(s^2+4)}\Big|_s^\infty\right\}\right]dt = t\int_0^t t\left[\mathcal{L}^{-1}\left\{-0 + \frac{\frac{1}{2}}{(s^2+4)}\right\}\right]dt$$

$$= t\int_0^t t\left[\frac{1}{4}\sin 2t\right]dt = \frac{1}{4}t\left[-\frac{1}{2}t\cos 2t\Big|_0^t + \int_0^t \frac{1}{2}\cos 2t\,dt\right] = \frac{1}{4}t\left[-\frac{1}{2}t\cos 2t + \frac{1}{4}\sin 2t\right]$$

Thus, the response is given by

$$y(t) = \cos 2t + \frac{1}{4}t\left[-\frac{1}{2}t\cos 2t + \frac{1}{4}\sin 2t\right]$$

$$y(t) = \cos 2t + \frac{1}{16}t\sin 2t - \frac{1}{8}t^2\cos 2t$$

4.5. Laplace Transform of Especial Functions

The word especial in the title includes delayed time functions, a piece of time functions, and periodic time functions. These time functions are used mostly in signal theory analysis. When these signals represent the dynamic systems' input, then the output's nature will have some features that may not be included with other ordinary inputs. First, we will give examples of these time functions and their mathematical descriptions suitable for using the Laplace transform.

Example 4.27.

1) Consider the piece of the straight line shown in Figure 4.3.
The function is usually defined as

$$f(t) = \begin{cases} 0, t < 1 \\ t + 2, 1 \le t \le 3 \\ 0, t > 3 \end{cases}$$

Also, the function can be written in the form

$$g(t) = (t + 2)[u(t - 1) - u(t - 3)]$$

where the steps $u(t - 1)$ and $u(t - 3)$ are unit pulses delayed by 1 and 3 units of time, respectively. It is a suitable mathematical description for applying the Laplace transform. In general, any piece of function $f(t)$ between two time-points a and $b, a < b$ can be defined as

$$g(t) = f(t)[u(t - a) - u(t - b)]$$

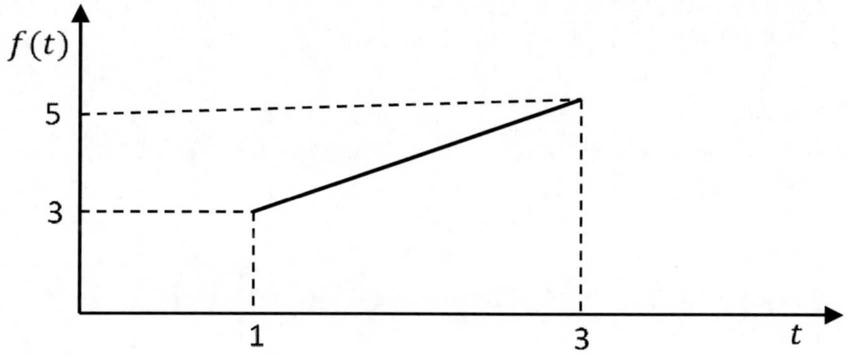

Figure 4.3. A piece of a straight line (Example 4.27).

2) Consider the time function piece that is defined as

$$f(t) = \begin{cases} 0, t < \frac{\pi}{2} \\ \sin t, \frac{\pi}{2} \le t \le \pi \\ 0, t > \pi \end{cases}$$

In a suitable expression, it is

$$g(t) = \sin t \left[u\left(t - \frac{\pi}{2}\right) - u(t - \pi)\right]$$

Theorem 4.11. Second shifting theorem
If for a time function, a Laplace function exists, then it holds that

$$\mathcal{L}\{f(t - a)u(t - a)\} = e^{-as}\,\mathcal{L}\{f(t)\} \qquad (2.25)$$

The theorem gives the transform of functions that have been translated (delayed) and cut off. The following two corollaries can almost use the application of the theorems:

Corollary 4.1.
For a time function that stats at a specific moment, this corollary states

$$\mathcal{L}\{f(t)u(t-a)\} = e^{-as}f(t+a) \tag{2.26}$$

Corollary 4.2.
It states, if

$$\mathcal{L}^{-1}\{\phi(s)\} = f(t)$$

Then,

$$\mathcal{L}^{-1}\{e^{-as}\phi(s)\} = f(t-a)u(t-a) \tag{2.27}$$

The following three examples illustrate the use of corollary 4.1.

Example 4.28.
What is the Laplace transform of the function example 4.27?

$$g(t) = (t+2)[u(t-1) - u(t-3)]$$

Taking the Laplace transform of both sides

$$\mathcal{L}\{g(t)\} = \mathcal{L}\{(t+2)u(t-1)\} - \mathcal{L}\{(t+2)u(t-3)\}$$

Using corollary 4.1 yields

$$\mathcal{L}\{g(t)\} = e^{-s}\mathcal{L}(t+1+2) - e^{-3s}\mathcal{L}(t+3+2)$$

$$\mathcal{L}\{g(t)\} = e^{-s}\mathcal{L}(t+3) - e^{-3s}\mathcal{L}(t+5)$$

$$\mathcal{L}\{g(t)\} = e^{-s}\left(\frac{1}{s^2} + \frac{3}{s}\right) - e^{-3s}\left(\frac{1}{s^2} + \frac{5}{s}\right)$$

$$= \frac{(1+3s)e^{-s} - (1-5s)e^{-3s}}{s^2}$$

Example 4.29.
Find the Laplace transform of the time function defined as

$$f(t) = \begin{cases} 0, t < 0.5 \\ \cos t, t \geq 0.5 \end{cases}$$

We rewrite the function as

$$g(t) = \cos t\, u(t - 0.5)$$

$$\mathcal{L}\{g(t)\} = e^{-s}\mathcal{L}\{\cos(t + 0.5)$$

$$\mathcal{L}\{g(t)\} = e^{-s}\mathcal{L}\{\cos t \cos 0.5 - \sin 0.5 \sin t\}$$

$$\mathcal{L}\{g(t)\} = e^{-s}\,[\cos 0.5\, \mathcal{L} \cos t - \sin 0.5 \mathcal{L} \sin t]$$

$$\mathcal{L}\{g(t)\} = e^{-s}\left[\cos 0.5 \frac{s}{s^2 + 1} - \sin 0.5 \frac{1}{s^2 + 1}\right]$$

$$\mathcal{L}\{g(t)\} = e^{-s}\frac{[s \cos 0.5 - \sin 0.5]}{s^2 + 1} = e^{-s}\frac{0.8775\, s - 0.4794}{s^2 + 1}$$

Example 4.30.
Find the Laplace transform of the time function defined as

$$f(t) = \begin{cases} \sin t\,, 0 \le t \le \pi \\ 0, t > \pi \end{cases}$$

We rewrite the function as

$$g(t) = \sin t[u(t - 0) - u(t - \pi)]$$

$$\mathcal{L}\{g(t)\} = e^{0}\mathcal{L}\{\sin t\} - e^{-\pi s}\,\mathcal{L}\{\sin(t + \pi)\}$$

$$\mathcal{L}\{g(t)\} = \frac{1}{s^2 + 1} + e^{-\pi s}\frac{1}{s^2 + 1} = \frac{1 + e^{-\pi s}}{s^2 + 1}$$

Theorem 4.12. Laplace Transform of periodic time functions

If a function $f(t)$ is periodic with a period T on $[0, \infty)$ and piecewise continuous on $0 \le t \le T$, then its Laplace transform is given by

$$\mathcal{L}\{f(t)\} = \frac{\int_0^T f(t)\, e^{-st} dt}{1 - e^{-Ts}} \tag{2.28}$$

Example 4.31.
Determine the Laplace transform of the periodic function shown in Figure 4.4.

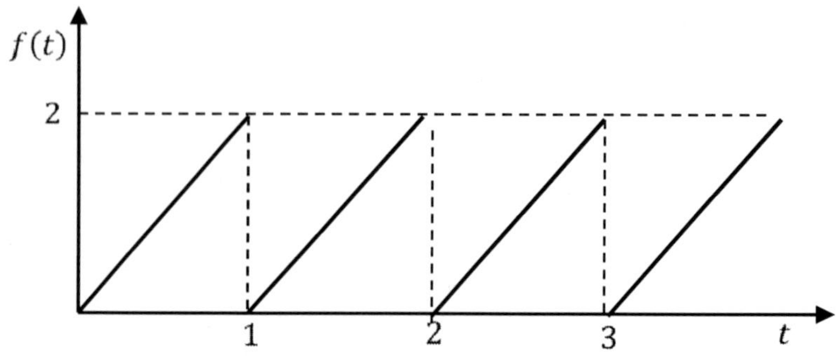

Figure 4.4. A periodic function (Example 4.31).

The time function has a period of 1.0, and it is defined in one period as $f(t) = 2t$ Therefore, using the results of theorem 4.10 yields

$$\mathcal{L}\{f(t)\} = \frac{\int_0^1 2t\, e^{-st} dt}{1 - e^{-s}} = \frac{2}{1 - e^{-s}}\left[-\frac{1}{s} te^{-st}|_0^1 + \frac{1}{s}\int_0^1 e^{-st}\, dt\right]$$

$$\mathcal{L}\{f(t)\} = \frac{2}{1 - e^{-s}}\left[-\frac{e^{-s}}{s} - \frac{e^{-s} - 1}{s^2}\right] = \frac{2[1 - (s+1)e^{-s}]}{s^2(1 - e^{-s})}$$

These delayed or periodic time functions can be a force or an input of a dynamic system modeled by a linear constant-coefficients differential equation. To solve an initial-value problem where its nonhomogeneous term is one of these types of time functions, we need to use theorems 4.9 and 4.10 as well as the corollaries 4.1 and 4.2.

Example 4.32.

A harmonic system at reset is described by

$$y'' + 4y = f(t);\ y(0) = y'(0) = 0$$

If the input force $f(t)$ is the cosine time function from $t = 0$ to $t = 2\pi$, determine the system response.

We rewrite the problem as

$$y'' + 4y = \cos t\,[1 - u(t - 2\pi)]\,;\ y(0) = y'(0) = 0$$

Taking Laplace transforms of both sides (the zero initial conditions are considered) yields

$$s^2Y(s) + 4Y(s) = \frac{s}{s^2 + 1} - e^{-2\pi s}\mathcal{L}\{\cos(t + 2\pi)\}$$

$$(s^2+4)Y(s)=\frac{s}{s^2+1}-e^{-2\pi s}\mathcal{L}\{\cos t\}$$

Solving for $Y(s)$ gives

$$Y(s)=\frac{1}{s^2+4}\left[\frac{s(1-e^{-2\pi s})}{s^2+1}\right]$$

Rearranging by separating exponential terms from the other gives

$$Y(s)=\frac{s}{(s^2+4)(s^2+1)}-\frac{s}{(s^2+4)(s^2+1)}e^{-2\pi s}$$

First, we find the Laplace inverse of the first term of $Y(s)$. Using a suitable decomposition (the reader can check it) gives

$$\mathcal{L}^{-1}\left\{\frac{s}{(s^2+4)(s^2+1)}\right\}=\mathcal{L}^{-1}\left\{\frac{\frac{1}{3}s}{(s^2+1)}\right\}-\mathcal{L}^{-1}\left\{\frac{\frac{1}{3}s}{(s^2+4)}\right\}$$

$$=\frac{1}{3}\cos t-\frac{1}{3}\cos 2t$$

For the second term of $Y(s)$, we apply Corollary 4.2 to obtain the inverse

$$\mathcal{L}^{-1}\frac{s}{(s^2+4)(s^2+1)}e^{-2\pi s}=\frac{1}{3}[\cos(t-2\pi)-\cos(2t-4\pi)]u(t-2\pi)]$$

According to the linearity property of the Laplace transform, all terms are added to obtain the complete solution becomes

$$y(t)=\frac{1}{3}\cos t-\frac{1}{3}\cos 2t-\frac{1}{3}[\cos(t-2\pi)-\cos(2t-4\pi)]u(t-2\pi)$$

Or simply

$$y(t)=\frac{1}{3}[\cos t-\cos 2t][1-u(t-2\pi)]$$

In Figure 4.5, the plot of the force input $f(t)$ and the output $y(t)$ (response of the harmonic system to the force input) is illustrated. As can be noted, the output ends at $t=2\pi$, which is the same time value as the input function is finished.

Example 4.33.

For the harmonic system that exposure to a periodic signal, determine the output.

$$y'' + y = f(t);\ y(0) = y'(0) = 0$$

where $f(t)$ is the periodic time function of example 4.31. Taking the Laplace transform of both sides

$$s^2Y(s) + Y(s) = F(s)$$

The complete solution is given as

$$y(t) = \mathcal{L}^{-1}\left\{\frac{2[1-(s+1)e^{-s}]}{s^2(s^2+1)(1-e^{-s})}\right\}$$

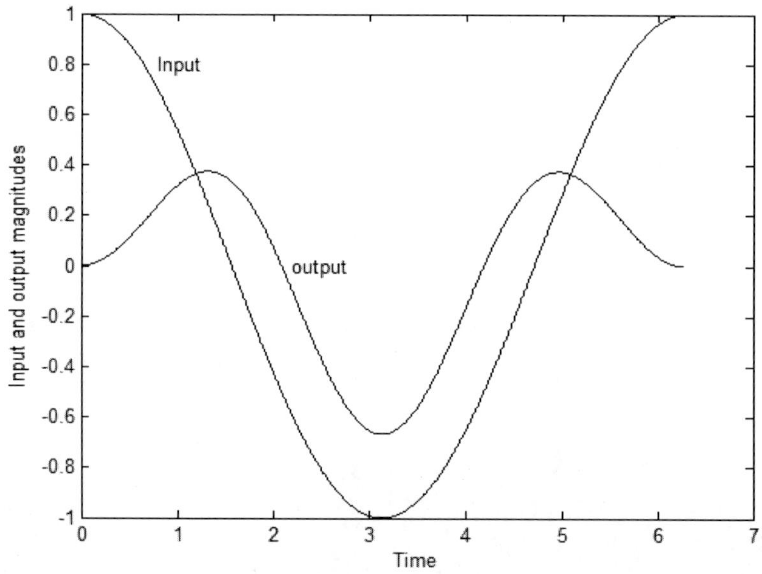

Figure 4.5. Input and output magnitudes (Example 4.32).

Rewrite as

$$y(t) = \mathcal{L}^{-1}\left\{\frac{2[1-(s+1)e^{-s}]}{s^2(s^2+1)}\sum_{m=0}^{\infty} e^{-ms}\right\}$$

$$y(t) = \mathcal{L}^{-1}\left\{\frac{2}{s^2(s^2+1)}\sum_{m=0}^{\infty} e^{-ms}\right\} - \mathcal{L}^{-1}\left\{\frac{2(s+1)}{s^2(s^2+1)}\sum_{m=0}^{\infty} e^{-(m+1)s}\right\}$$

$$y(t) = \mathcal{L}^{-1}\left\{\frac{2}{s^2(s^2+1)}\right\} + \mathcal{L}^{-1}\left\{\frac{2}{s^2(s^2+1)}\sum_{m=1}^{\infty} e^{-ms}\right\} - \mathcal{L}^{-1}\left\{\frac{2(s+1)}{s^2(s^2+1)}\sum_{m=1}^{\infty} e^{-ms}\right\}$$

$$y(t) = \mathcal{L}^{-1}\left\{\frac{2}{s^2(s^2+1)}\right\} + \mathcal{L}^{-1}\left\{\frac{2-2(s+1)}{s^2(s^2+1)}\sum_{m=1}^{\infty} e^{-ms}\right\}$$

Let us solve the first part of $y(t)$.

$$\mathcal{L}^{-1}\left\{\frac{2}{s^2(s^2+1)}\right\} = \mathcal{L}^{-1}\left\{\frac{2}{s^2}\right\} - \mathcal{L}^{-1}\left\{\frac{2}{(s^2+1)}\right\} = 2t - 2\sin t$$

$$\mathcal{L}^{-1}\left\{\frac{-2s}{s^2(s^2+1)}\sum_{m=1}^{\infty} e^{-ms}\right\} = \mathcal{L}^{-1}\left\{\varphi(s)\sum_{m=1}^{\infty} e^{-ms}\right\}$$

Using Corollary 4.2 of the second shifting theorem, we can write

$$\mathcal{L}^{-1}\left\{\frac{-2s}{s^2(s^2+1)}\sum_{m=1}^{\infty} e^{-ms}\right\} = \sum_{m=1}^{\infty} g(t-m).u(t-m)$$

where,

$$g(t) = \mathcal{L}^{-1}\left\{\frac{-2s}{s^2(s^2+1)}\right\} = \mathcal{L}^{-1}\left\{\frac{-2}{s\,(s^2+1)}\right\} = \int_0^t \left[\mathcal{L}^{-1}\left\{\frac{-2}{(s^2+1)}\right\}\right] dt$$

$$g(t) = -2\int_0^t \sin t\, dt = -2\cos t|_0^t = -2 + 2\cos t$$

Therefore,

$$\mathcal{L}^{-1}\left\{\frac{-2s}{s^2(s^2+1)}\sum_{m=1}^{\infty} e^{-ms}\right\} = \sum_{m=1}^{\infty}[-2 + 2\cos(t-m)]u(t-m)$$

The complete solution is

$$y(t) = 2t - 2\sin t + \sum_{m=1}^{\infty}[-2 + 2\cos(t-m)]u(t-m)$$

In Figure 4.6, for three periods, the complete solution is plotted. The complete solution in these periods is given by

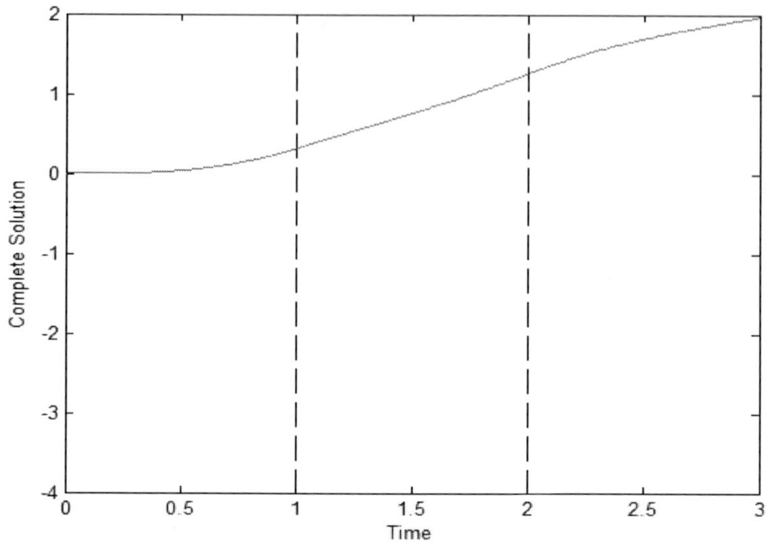

Figure 4.6. Dynamic outputs for three periods (Example 4.33).

$$y(t) = 2t - 2\sin t;\ 0 \le t \le 1$$

$$y(t) = 2t - 2\sin t + [-2 + 2\cos(t-1)];\ 1 \le t \le 2$$

$$y(t) = 2t - 2\sin t + [-2 + 2\cos(t-1)] - [2 + 2\cos(t-2)]\ ;\ 2 \le t \le 3$$

4.6. Convolution Theorem

One of the most considerable theoretical properties of the Laplace transform is the convolution theorem, which states the relation between two multiplied Laplace functions.

Theorem 4.13.

If the Laplace transform exists for the two-time functions $f(t)$and $g(t)$ then the following results hold

1. $$\mathcal{L}\{f(t)\}.\mathcal{L}\{g(t)\} = F(s)G(s) = \mathcal{L}\{\int_0^t f(t-\lambda)g(\lambda)d\lambda\} = \mathcal{L}\{\int_0^t f(\lambda)g(t-\lambda)d\lambda\} \quad (4.29)$$

2. $$\mathcal{L}^{-1}\{F(s)G(s)\} = \int_0^t f(t-\lambda)g(\lambda)d\lambda = \int_0^t f(\lambda)g(t-\lambda)d\lambda \quad (2.30)$$

The convolution is usually denoted simply as $f(t) * g(t)$, and the theorem is then stated as

$$\mathcal{L}\{f(t)\}.\mathcal{L}\{g(t)\} = \mathcal{L}\{f(t) * g(t)\} \tag{4.31}$$

Corollary 4.3.
If it is held that $\mathcal{L}g(t) = G(s)$, Then,

$$\mathcal{L}^{-1}\left\{\frac{1}{s}G(s)\right\} = \int_0^t g(t)d\lambda \tag{4.32}$$

The convolution theorem has been used in many fields; one crucial area is in the control theory. The classical control theory is based on the use of the Laplace transform. The next discussion deals with the transfer function concept, which is the essential topic in control theory.

Definition 4.4. Transfer function
For a single input-output system, the transfer function $G(s)$ of a linear constant-coefficients system is a Laplace ratio of the output $Y(s)$ to the input $F(s)$ provided that the initial conditions are all equal to zero, thus

$$G(s) = \frac{\mathcal{L}\{y(t)]}{\mathcal{L}\{f(t)]\}} = \frac{Y(s)}{F(s)} \Rightarrow Y(s) = G(s)F(s) \tag{4.33}$$

$$y(t) = \mathcal{L}^{-1}\{G(s)F(s)\} = \int_0^t f(t-\lambda)g(\lambda)d\lambda \tag{4.34}$$

This last expression carries the meaning of how the system convolutes the input signals to give a specific output.

Definition 4.5. Indicial admittance
The indicial admittance $A(s)$ is the response (output) of the system when the input function $f(t)$ is a unit pulse, or

$$Y(s) = \frac{G(s)}{s} = \frac{1}{s\,z(s)} \tag{4.35}$$

where $z(s)$ is a function of s whose coefficients depend solely on the system's parameters

Accordingly, for the linear systems, the convolution theorem states a very usable result, which says that:

If the indicial admittance $A(s)$ is known, then applying the convolution theorem, we can determine the response of the system for any other forcing (input) function, say $w(t)$. Then, it can be proved that for any input $w(t)$, the response $y(t)$ is given by

$$y(t) = \int_0^t A'(\tau)w(t-\tau)d\tau \tag{4.36}$$

$$y(t) = \int_0^t A(t-\tau)w'(\tau)d\tau + A(t)w(0) \tag{4.37}$$

Furthermore, if a unit impulse is applied to a linear time-invariant system, then the response of the system $h(t)$ will be

$$\mathcal{L}h(t) = \frac{\mathcal{L}f(t)}{z(s)} = \frac{\mathcal{L}\delta(t)}{z(s)} = \frac{1}{z(s)} \times \frac{s}{s} = s\mathcal{L}A(t) \tag{4.38}$$

$$h(t) = \frac{d}{dt}A(t) = A'(t) \tag{4.39}$$

Therefore, the system response to a unit impulse is the system's derivative response to a unit step function.

$$y(t) = \int_0^t h(t)w(t-\tau)d\tau \tag{4.40}$$

Example 4.34.
Determine the Laplace inverse of

$$\frac{1}{(s+a)(s+b)}$$

Let us take

$$F(s) = \frac{1}{(s+a)} \Rightarrow f(t) = \mathcal{L}^{-1}\left\{\frac{1}{(s+a)}\right\} = e^{-at}$$

$$G(s) = \frac{1}{(s+b)} \Rightarrow g(t) = \mathcal{L}^{-1}\left\{\frac{1}{(s+b)}\right\} = e^{-bt}$$

The convolution theorem leads to

$$\mathcal{L}^{-1}\{F(s).G(s)\} = \int_0^t f(t-\lambda)g(\lambda)d\lambda = \int_0^t e^{-a(t-\lambda)}e^{-b\lambda}d\lambda$$

$$= \int_0^t e^{-at+(a+b)\lambda} d\lambda = \frac{1}{a-b} e^{-at+(a+b)\lambda} |_0^t = \frac{e^{-bt} - e^{-at}}{(a-b)}$$

The reader can check the correctness of this answer by the method of partial fraction decomposition.

Example 4.35.
Solve the initial-value problem

$$y'' + 4y = \frac{1}{2}\sin 2t\,;\; y(0) = y'(0) = 0$$

Taking Laplace transforms of both sides

$$s^2Y(s) + 4Y(s) = \frac{1}{s^2+4}$$

$$y(t) = \mathcal{L}^{-1}\{\frac{1}{(s^2+4)^2}\}$$

$$F(s) = \frac{1}{(s^2+4)} \Rightarrow f(t) = \mathcal{L}^{-1}\left\{\frac{1}{(s^2+4)}\right\} = \frac{1}{2}\sin 2t$$

$$G(s) = \frac{1}{(s^2+4)} \Rightarrow g(t) = \mathcal{L}^{-1}\left\{\frac{1}{(s^2+4)}\right\} = \frac{1}{2}\sin 2t$$

Using the convolution theorem yields

$$y(t) = \mathcal{L}^{-1}\{F(s).G(s)\} = \frac{1}{4}\int_0^t \sin 2(t-\lambda)\sin 2\lambda\, d\lambda$$

$$y(t) = \frac{1}{4}\int_0^t \frac{\cos(4\lambda - 2t) - \cos 2t}{2} d\lambda = \frac{1}{8}\left[\frac{\sin(4\lambda - 2t)}{4} - \lambda\cos 2t]\right] |_0^t$$

$$y(t) = \frac{1}{8}\left[\frac{\sin(4t-2t)}{4} - t\cos 2t - \frac{\sin(0-2t)}{4} - 0\right]$$

$$y(t) = \frac{1}{8}\left[\frac{\sin 2t}{4} - t\cos 2t + \frac{\sin 2t}{4}\right] = \frac{1}{8}\left[\frac{\sin 2t}{2} - t\cos 2t\right]$$

Example 4.36.
Consider the dynamic system described by

$$y''' - y'' - 4y' - 4y = w(t);\; y(0) = y'(0) = y''(0) = 0$$

Determine an expression for the output for an arbitrary input $w(t)$.

Taking Laplace transforms on sides, collecting terms and solving for the function $Y(s)$ yields

$$Y(s) = \frac{W(s)}{s^3 - s^2 - 4s - 4} = \frac{W(s)}{(s^2-4)(s-1)} = \frac{W(s)}{(s+2)(s-2)(s-1)}$$

Let us write

$$Y(s) = F(s)W(s)$$

By the convolution theorem

$$y(t) = \int_0^t f(\lambda)W(t-\lambda)\, d\lambda$$

Therefore, we determine $f(t) \equiv f(\lambda)$,

$$f(t) = \mathcal{L}^{-1}\left\{\frac{1}{(s+2)(s-2)(s-1)}\right\} = \mathcal{L}^{-1}\left\{\frac{k_1}{s+2}\right\} + \mathcal{L}^{-1}\left\{\frac{k_1}{s-2}\right\} + \mathcal{L}^{-1}\left\{\frac{k_1}{s-1}\right\}$$

$$k_1 = \frac{1}{(s+2)(s-2)(s-1)}(s+2)|_{s=-2} = \frac{1}{12}$$

$$k_2 = \frac{1}{(s+2)(s-2)(s-1)}(s-2)|_{s=2} = \frac{1}{4}$$

$$k_3 = \frac{1}{(s+2)(s-2)(s-1)}(s-1)|_{s=1} = -\frac{1}{3}$$

$$f(t) = \frac{1}{12}e^{-2t} + \frac{1}{4}e^{2t} - \frac{1}{3}e^{t}$$

The required expression is

$$y(t) = \int_0^t [\frac{1}{12}e^{-2\lambda} + \frac{1}{4}e^{2\lambda} - \frac{1}{3}e^{\lambda}]W(t-\lambda)\, d\lambda$$

Example 4.37.

Consider The RLC electric circuit shown in Figure 4.7 and it is given that $R = 2\sqrt{L/C}$. The capacitor is not charged, and the coil is not energized. The switch is being closed at $t = 0$.

The loop equation of the circuit is, the input voltage is drooped across the three elements, thus

$$e_i(t) = Ri(t) + L\frac{di(t)}{dt} + \frac{1}{C}\int_{0+}^{t} i(t)dt$$

The initial conditions are zero since the capacitor is not charged and the coil is not energized. By taking the Laplace transform, we obtain

$$E_i(s) = \left(R + Ls + \frac{1}{Cs}\right)I(s)$$

The Laplace function of the output voltage $E_0(s)$ is the drop across the capacitance, thus

$$E_0(s) = \left(\frac{1}{Cs}\right)I(s)$$

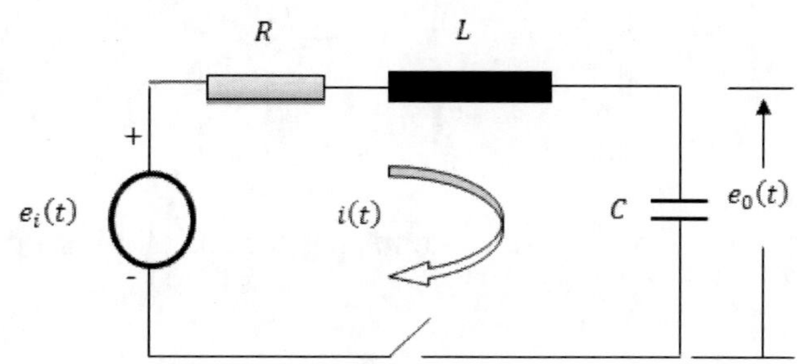

Figure 4.7. An RLC electric circuit (Example 4.38).

Then the transfer function of the electric circuit is

$$G(s) = \frac{E_0(s)}{E_i(s)} = \frac{1}{LCs^2 + RCs + 1}$$

Then, the function $z(s)$ is the denominator of the transfer function, which is and the initial admittance $A(s)$ are given, respectively, by

$$z(s) = LCs^2 + RCs + 1$$

$$A(s) = \frac{1}{s(LCs^2 + RCs + 1)} = \frac{1/LC}{s(s + p_1)(s + p_2)}$$

where

$$p_1 = \frac{R}{2L}\left(-1 + \sqrt{1 - \frac{4L}{CR^2}}\right), p_2 = \frac{R}{2L}\left(-1 - \sqrt{1 - \frac{4L}{CR^2}}\right)$$

Therefore, depending on the values of $R, L,$ and C, the coefficients p_1, and p_2 could be a real distinct, real repeated, or complex. However, for the given relation $R = 2\sqrt{L/C}$, we have

$$p_1 = p_1 = -\frac{R}{2L} = -\frac{1}{\sqrt{LC}}$$

Thus, the indicial admittance becomes

$$A(s) = \frac{1/LC}{s(s - \frac{1}{\sqrt{LC}})^2}$$

$$A(t) = \mathcal{L}^{-1}\left\{\frac{\frac{1}{LC}}{s\left(s - \frac{1}{\sqrt{LC}}\right)^2}\right]$$

Using the theorems previously introduced, we can find the unit pulse output.

$$A(t) = 1 + \frac{1}{\sqrt{LC}} te^{\sqrt{LC}t} - \frac{1}{L^2C^2} e^{\sqrt{LC}t}$$

The unit impulse output is the derivative.

$$A'(t) = \frac{1}{\sqrt{LC}} e^{\sqrt{LC}t} + te^{\sqrt{LC}t} - \frac{1}{\sqrt{L^3C^3}} e^{\sqrt{LC}t} = \left(\frac{1}{\sqrt{LC}} + t - \frac{1}{\sqrt{L^3C^3}}\right) e^{\sqrt{LC}t}$$

Another class of systems that employ the Laplace transform is the electromechanical systems, which include electric and mechanical elements to accomplish a given task. In practice, on one side, motors are the cardinal elements in any electromechanical automatic control system. On the other side, the linear control theories utilize the Laplace transform to analyze and design automated control systems. For all these types of motors, it is necessary to derive the related transfer function. Furthermore, the motor response for specific input-controlled signals is determined by using the Laplace transform theorems. The next example explores the use of Laplace transforms in deriving the transfer function of the armature-controlled DC motors.

Example 4.38.

The known electromechanical diagram of a DC armature-controlled motor is shown in Figure 4.8. In this figure, the applied voltage e_a represents the input to control the motor behavior, and the armature current, the resistance, and the inductance are i_a, R_a, L_a, respectively. The excitation constant field current is i_f and the back electromechanical voltage is denoted by e_b. The mechanical part is modeled by the torque delivered by the motor T, the equivalent moment of inertia plus the load referred to the motor shaft J and similarly, the equivalent viscous friction of the motor bearings f. The motor output is either the angular position θ or the angular velocity ω of the motor shaft. All units of the considered quantities are in the meter-Kilogram-second system. The time functions are the applied voltage, the armature current, the back electromechanical voltage, and the angular position and angular velocity. The initial conditions are assumed equal to zero; before applying the input voltage, the motor is in a halt mode.

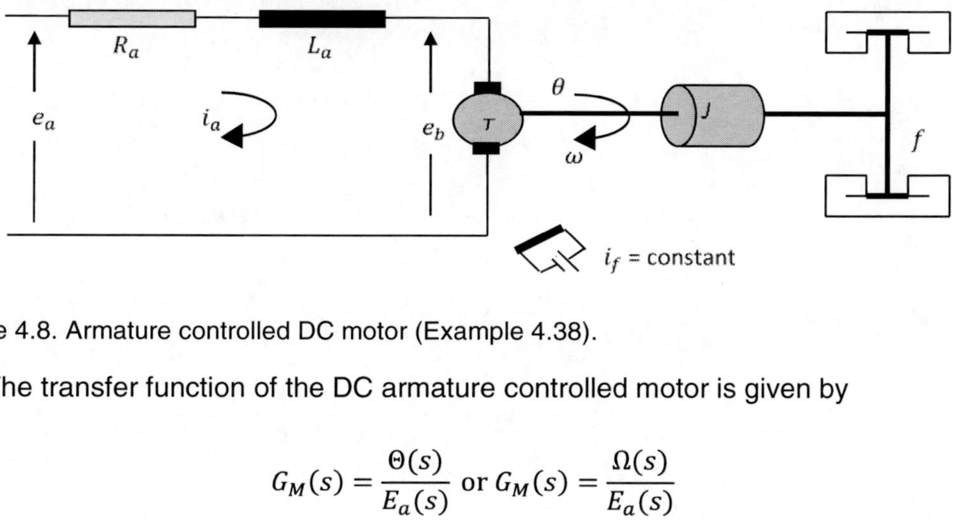

Figure 4.8. Armature controlled DC motor (Example 4.38).

The transfer function of the DC armature controlled motor is given by

$$G_M(s) = \frac{\Theta(s)}{E_a(s)} \text{ or } G_M(s) = \frac{\Omega(s)}{E_a(s)}$$

Then, it is desired to find the transfer function in terms of the motor parameters. The following physical relationships govern the dynamics of the motor.

- The differential equation of the armature electric circuit is

$$R_a i_a(t) + L_a \frac{di_a(t)}{dt} + e_b(t) = e_a(t)$$

- The differential equation of the rotating element is

$$J\frac{d^2\theta(t)}{dt^2} + f\frac{d\theta(t)}{dt} = T$$

- The back electromechanical voltage is proportional to the angular velocity by

$$e_b(t) = k_b\omega(t) = k_b\frac{d\theta(t)}{dt}$$

- The motor torque is proportional to both armature and field currents by

$$T = Ki_a(t)$$

Substituting for the torque and taking Laplace transforms of the three differential equations, we get

$$R_aI_a(s) + L_asI_a(s) + E_b(s) = E_a(s)$$

$$E_b(s) = k_bs\Theta(s)$$

$$Js^2\Theta(s) + fs\Theta(s) = KI_a(s)$$

Manipulating these algebraic equations, we can get

$$(R_a + L_as).\frac{(Js^2\Theta(s) + fs\Theta(s))}{K} + k_bs\Theta(s) = E_a(s)$$

Therefore,

$$\frac{\Theta(s)}{E_a(s)} = \frac{K}{s(L_aJ\,s^2 + (L_af + R_aJ)s + R_af + Kk_b)}$$

For armature controlled DC motors, the winding inductance is very small and can be neglected, and then a simplified transfer function is

$$\frac{\Theta(s)}{E_a(s)} = \frac{K_m}{s(T_ms + 1)}$$

where the gain and time constants of the motor are given by

$$K_m = \frac{K}{R_af + Kk_b}, T_m = \frac{R_aJ}{R_af + Kk_b}$$

Besides, we can obtain the transfer function relating the angular velocity to the input voltage simply as

$$\frac{\Omega(s)}{E_a(s)}=\frac{s\Theta(s)}{E_a(s)}=\frac{K_m}{(T_m s+1)}$$

Now, let us assume that the input voltage is equal to V constant voltage, and then the angular position response will be given by

$$\theta(t)=\mathcal{L}^{-1}\left\{\frac{K_m}{s(T_m s+1)}.\frac{V}{s}\right\}=K_m V(T_m e^{-t/T_m}+t-T_m)$$

The angular velocity response is

$$\omega(t)=\frac{d\theta}{dt}=K_m V\left(1-e^{-\frac{t}{T_m}}\right)$$

Therefore, the steady-state angular velocity (as $t\to\infty$) is equal to $K_m V$.

4.7. Advanced Applications

In this last section, we introduce some advanced applications of Laplace transform. These theorems have a wide range of applications in different fields; however, we will focus only on numerical computation.

Theorem 4.14. Final value theorem

If $\mathcal{L}\{f(t)\}, \mathcal{L}\{f'(t)\}, \text{and } \lim_{t\to\infty} f(t)$ exist, and if $sF(s)$ is analytic in the right-half of the s –plane, including the $jw-$ axis , except for a single pole at the origin, then

$$\lim_{t\to\infty} f(t)=\lim_{s\to 0} sF(s) \tag{4.41}$$

Therefore, replacing the approaching to infinity with approaching zero calculation has considerable advantages.

Theorem 4.15. Initial value theorem

If $\mathcal{L}\{f(t)\}, \mathcal{L}\{f'(t)\}, \text{and } \lim_{s\to\infty} sF(s)$ exist, then

$$\lim_{t\to 0} f(t)=\lim_{s\to\infty} sF(s) \tag{4.42}$$

Theorem 4.16. Integration formula

If $\mathcal{L}\{f(t)\}, \text{and } \lim_{s\to 0} F(s)$ exist, then the following integration formula is hold

$$\int_0^\infty f(t)dt = \lim_{s\to 0} F(s) \tag{4.43}$$

The formula (4.43) can be used to determine several significant improper integrations.

Theorem 4.17. Laplace of integration

If $\mathcal{L}\{f(t)\}$ is piecewise regular on $[0,\infty)$ and of exponential order, then the following formula is applied.

$$\mathcal{L}\left\{\int_a^t f(x)\,dx\right\} = \frac{1}{s}\mathcal{L}\{f(t)\} - \frac{1}{s}\int_0^a f(t)\,dt;\ a \geq 0 \tag{4.44}$$

Example 4.39.

Determine

$$\mathcal{L}\{e^{-3t}\int_0^t t\sin 2t\,dt\}$$

The determination of this Laplace function can be carried out by simultaneous use of theorems 4.16, 4.3, and 4.5, as follows:

$$\mathcal{L}\left\{\int_0^t t\sin 2t\,dt\right\} = \frac{1}{s}\mathcal{L}\{t\ \sin 2t\} + \frac{1}{s}\int_0^0 t\sin 2t\,dt = \frac{1}{s}\mathcal{L}\{t\ \sin 2t\}$$

$$= -\frac{1}{s}\frac{d}{ds}\left(\frac{2}{s^2+4}\right) = -\frac{1}{s}\cdot\frac{-4s}{(s^2+4)^2} = \frac{4}{(s^2+4)^2}$$

Then the final answer is obtained.

$$\mathcal{L}\left\{e^{-3t}\int_0^t t\sin 2t\,dt\right\} = \frac{4}{(s^2+4)^2}|_{s=s+3} = \frac{4}{[(s^2+3)^2+4]^2}$$

To check that this procedure is valid, we determine first the integral

$$\int_0^t t\sin 2t\,dt = \frac{1}{4}\sin 2t - \frac{1}{2}t\cos 2t$$

Therefore,

$$\mathcal{L}\left\{e^{-3t}\int_0^t t\sin 2t\,dt\right\} = \mathcal{L}\left\{\frac{1}{4}e^{-3t}\sin 2t\right\} - \mathcal{L}\left\{\frac{1}{2}e^{-3t}t\cos 2t\right\}$$

$$=\frac{1}{4}\frac{2}{s^2+4}|_{s=s+3}-\frac{1}{2}\left(-\frac{d}{ds}\left(\frac{s}{s^2+4}\right)|_{s=s+3}\right)$$

$$=\frac{1/2}{(s^2+3)^2+4}-\frac{1}{2}\left(\frac{s^2-4}{(s^2+4)^2}\right)|_{s=s+3}=\frac{1}{2(s^2+3)^2+4}-\frac{(s+3)^2-4}{2[(s^2+3)^2+4]^2}$$

$$=\frac{4}{[(s^2+3)^2+4]^2}$$

Example 4.40.

Compute the value of the improper integration

$$I=\int_0^\infty \frac{e^{-t}\cos t-e^{-2t}\cos 2t}{t}dt$$

Following theorem 4.15 yields

$$I=\lim_{s\to 0}\left[\mathcal{L}\left\{\frac{e^{-t}\cos t}{t}\right\}-\mathcal{L}\left\{\frac{e^{-2t}\cos 2t}{t}\right\}\right]$$

Using theorem 4.4 gives

$$I=\lim_{s\to 0}[\int_s^\infty [\mathcal{L}\{e^{-t}\cos t\}]ds-\int_s^\infty [\mathcal{L}\{e^{-2t}\cos 2t\}]ds]$$

Using theorem 4.5 gives

$$I=\left[\left[\int_s^\infty \frac{s+1}{(s+1)^2+1}ds-\int_s^\infty \frac{s+2}{(s+2)^2+4}ds\right]\right.$$

Evaluating the integrations yields

$$I=\lim_{s\to 0}[\frac{1}{2}\ln(s^2+2s+2)|_s^\infty-\frac{1}{2}\ln(s^2+4s+8)|_s^\infty]$$

$$I=\lim_{s\to 0}\left[-\frac{1}{2}(\ln(s^2+2s+2)+\frac{1}{2}(\ln(s^2+4s+8)\right]$$

Carrying out the limit gives

$$-\frac{1}{2}\ln 2+\frac{1}{2}\ln 8=-\frac{1}{2}\ln 2+\frac{3}{2}\ln 2=\ln 2$$

Example 4.41.

Compute the value of the following improper integral

$$I = \int_0^\infty \frac{\sin at}{bt} dt$$

$$I = \lim_{s\to\infty} \left[\mathcal{L}\left\{\frac{\sin at}{bt}\right\}\right]$$

$$I = \lim_{s\to\infty} [\frac{1}{b}\int_s^\infty [\mathcal{L}\{\sin at\}]ds] = \frac{1}{b}\lim_{s\to\infty}[\int_s^\infty \frac{a}{s^2+a^2}ds] = \frac{1}{b}\lim_{s\to\infty}\left[\tan^{-1}\left(\frac{s}{a}\right)|_s^\infty\right]$$

$$I = \frac{1}{b}\lim_{s\to\infty}\left[\frac{\pi}{2} - \tan^{-1}\left(\frac{s}{a}\right)\right] = \frac{\pi}{2b}$$

Thus the value of the integration does not depend on the value of a. Moreover, it can be shown that for any real values of a and b,

$$\int_0^\infty \frac{\cos at}{bt} dt = \infty$$

Another result of using Laplace's operator is the application of solving the Euler-Cauchy differential equation (see section 3.2). This particular type of linear variable-coefficients differential equation can be solved based on using theorem 4.3. To illustrate, we introduce the solution of a general third-order homogenous Euler-Cauchy equation. A general form of such an equation is

$$t^3\,y''' + at^2\,y'' + bt\,y' + c\,y = 0;\ \text{and given values of } y(0), y'(0), y''(0) \quad (4.45)$$

Applying Theorem 4.3 gives

$$-\frac{d^3}{ds^3}\mathcal{L}y''' + a\frac{d^2}{ds^2}\mathcal{L}y'' - b\frac{d}{ds}\mathcal{L}y' + c\mathcal{L}y' = 0$$

With the aid of theorem 4.10, it can be easily shown that

$$\mathcal{L}\{ty'''\} = -\frac{d}{ds}\mathcal{L}y''' = -s^3\frac{dY(s)}{ds} - 3s^2Y(s) + 2y(0)s + y'(0) \quad (4.46a)$$

Consequently, we can obtain

$$\mathcal{L}\{t^2y'''\} = \frac{d^2}{ds^2}\mathcal{L}y''' = s^3\frac{d^2Y(s)}{ds^2} + 6s^2\frac{dY(s)}{ds} + 6Y(s)s - 2y(0) \quad (4.46b)$$

$$\mathcal{L}\{t^3y'''\} = \frac{d^3}{ds^3}\mathcal{L}y''' = -s^3\frac{d^3Y(s)}{ds^3} - 9s^2\frac{d^2Y(s)}{ds^2} - 18s\frac{dY(s)}{ds} - 6Y(s) \qquad (4.46c)$$

Similarly, we can find relations for the terms $\mathcal{L}\{t^2y''\}$ and $\mathcal{L}\{t\,y'\}$ as

$$\mathcal{L}\{t^2y''\} = s^2\frac{d^2Y(s)}{ds^2} + 4s\frac{dY(s)}{ds} - 2Y(s) \qquad (4.46d)$$

$$\mathcal{L}\{t\,y'\} = -s\frac{dY(s)}{ds} - Y(s) \qquad (4.46e)$$

Substituting relations 4.46 c-e in equation (4.45) and arranging terms to obtain

$$s^3\frac{d^3Y(s)}{ds^3} + (9-a)s^2\frac{d^2Y(s)}{ds^2} + (18-4a+b)s\frac{dY(s)}{ds} + (6-2a+b-c)Y(s) = 0 \qquad (4.47)$$

Next, if we let $Y(s) = s^m$ as a possible solution, then

$$\frac{dY(s)}{ds} = ms^{m-1}, \frac{d^2Y(s)}{ds^2} = m(m-1)s^{m-2}, \frac{d^3Y(s)}{ds^3} = m(m-1)(m-2)s^{m-3}$$

Substituting these derivative terms in equation (4.47), dropping the common factor s^m and arranging, we get the following algebraic equation.

$$m^3 + (6-a)m^2 + (11-3a+b)m + (6-2a+b-c) = 0 \qquad (4.48)$$

The roots of equation (4.48) can be of different cases; one possible case is when all roots are negative distinct integer values r_1, r_2, and r_3. For such a case, the general solution of the third-order Euler-Cauchy equation is given by

$$y(t) = c_1\mathcal{L}^{-1}s^{r_1} + c_2\mathcal{L}^{-1}s^{r_2} + c_3\mathcal{L}^{-1}s^{r_3} \qquad (4.49)$$

Therefore,

$$y(t) = c_1t^{|r_1|-1} + c_2t^{|r_2|-1} + c_3t^{|r_3|-1} \qquad (4.50)$$

Clearly, in the solution, the numerical value $(1/(r-1)!)$ is contained in the integration constant. Another case is the multiplicity of two or three negative values of the roots. The linear combination of the solution components is secured by multiplying the components by $\ln t$ or $(\ln t)^2$, respectively. Besides, it is possible to have noninteger negative values of the roots. If these values are less than (-1), the general solution can be obtained by the Laplace transform in the sixth line of Table 4.2. For example, assume $r = -1.5$, then the corresponding time-domain term is $\sqrt{t}$.

Similarly, for other equation orders, one can derive to obtain the corresponding algebraic equation.

Example 4.42.

Determine a general solution of the Euler-Cauchy ODE

$$t^3y''' - 2t^2y'' + 2ty' = 0$$

By comparing to equation (4.45), we obtain $a = -2, b = 2, c = 0$. Therefore, the equivalent algebraic equation becomes

$$m^3 + 8m^2 + 19m + 12 = 0 \Longrightarrow (m+1)(m+3)(m+4) = 0$$

Therefore, the roots are $r_1 = -1, r_2 = -3$, $r_3 = -4$, and according to formula (4.50), the general solution is

$$y(t) = c_1t^{1-1} + c_2t^{3-1} + c_3t^{4-1} = c_1 + c_2t^2 + c_3t^3$$

Table 4.2. Extended table of Laplace transforms

	$f(t)$; $t \in [0, \infty)$	$F(s)$; $s > 0$
1	$\delta(t)$	1
2	$u(t)$	$1/s$
3	t	$1/\ s^2$
4	t^n; n is a positive real integer	$n!/\ s^{n+1}$
5	t^a; a is a real noninteger	$\Gamma(a+1)/\ s^{a+1}$
6	$\sqrt[n]{t}$	$\Gamma(\frac{1}{n}+1)/s^{\frac{1}{n}+1}$
7	$e^{\pm at}$	$1/(s \mp a)$
8	$\sin wt$	$w/(s^2 + w^2)$
9	$\cos wt$	$s/(s^2 + w^2)$
10	$\sinh wt$	$w/(s^2 - w^2)$
11	$\cosh wt$	$s/(s^2 - w^2)$
12	$t \sin wt$	$2ws/(s^2 + w^2)^2$
13	$t \cos wt$	$(s^2 - w^2)/(s^2 + w^2)^2$
14	$t^n e^{\pm at}$	$n!/\ (s \mp a)^{n+1}$
15	$e^{\pm at} \sin wt$	$w/[(s \mp a)^2 + w^2]$
16	$e^{\pm at} \cos wt$	$(s \mp a)/[(s \mp a)^2 + w^2]$
17	$e^{\pm at}\, t \sin wt$	$2w(s \mp a)/((s \mp a)^2 + w^2)^2$
18	$e^{\pm at} t \cos wt$	$((s \mp a)^2 - w^2)/((s \mp a)^2 + w^2)^2$
19	$f(t)u(t-a)$	$e^{-as}\ \mathcal{L}\{f(t+a)\}$
20	$\int_0^\infty f(t)dt$	$\lim_{s\to 0}[\mathcal{L}\{f(t)\}$
21	$\int_a^t f(x)dx$	$\frac{1}{s}\left[\mathcal{L}\{f(t)\} - \int_0^a f(t)dt\right]$

Example 4.43.
Determine a general solution of the Euler-Cauchy ODE

$$t^3y''' - 2t^2\,y'' + 4t\,y' - 4y = 0$$

By comparing to equation (4.45), we obtain $a = -2, b = 2, c = 0$. Therefore, the equivalent algebraic equation becomes

$$m^3 + 8m^2 + 21m + 18 = 0 \Longrightarrow (m+2)(m+3)^2$$

Therefore, the roots are $r_1 = -2, r_2 = -3,\ r_3 = -3$; therefore, the general solution is

$$y(t) = c_1 t\ + c_2 t^2 + c_3 t^2 \ln t$$

Example 4.44.
Determine a general solution of the Euler-Cauchy ODE

$$t^3y''' + 1.5t^2\,y'' = 0$$

Following the steps of finding the roots, we can find $r_1 = -1,\ r_2 = -2, r_3 = -1.5$; therefore, the general solution becomes

$$y(t) = c_1 + c_2 t + c_3\sqrt{t}$$

Note that the numerical value $\Gamma(0.5 + 1)$ is contained within the constant of integration c_3.

EXERCISES

Section 4.2

4.1. Use definition 4.3 to prove theorem 4.3.

4.2. Use definition 4.3 to determine the Laplace transform of the piecewise continuous function shown in Figure 4.9.

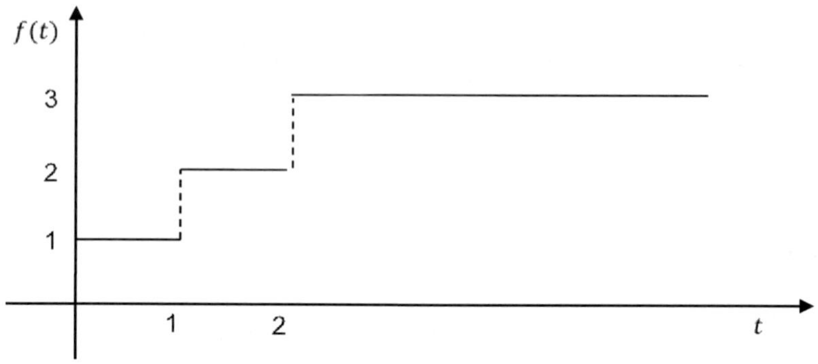

Figure 4.9. Piecewise function (exercise 4.2).

4.3. Determine the Laplace transform of time-function defined by

$$f(t) = \begin{cases} t, 0 \le t \le 1 \\ t - 1, 1 < t < \infty \end{cases}$$

4.4. Use definition 4.3 to determine the Laplace transform of

$$f(t) = t \cosh 2t$$

4.5. Use direct calculation to show that

$$\mathcal{L}\{t \sin t\} \neq (\mathcal{L}\{t\}).(\mathcal{L}\{\sin t\})$$

4.6. Use the linearity property of the Laplace transform to determine

$$\mathcal{L}\{\cos(\omega t + \varphi) + \sin(\omega t - \varphi)\}; \quad \omega, \varphi \in \Re$$

4.7. Prove items 4 and 5 in Table 4.2, where the Gamma function has the following identities.

$$\Gamma(x+1) = x\Gamma(x)$$

$$x(x+1)(x+2)\ldots.(x+n-1) = \frac{\Gamma(x+n)}{\Gamma(x)}$$

4.8. Prove items 17 and 18 in Table 4.2.

4.9. Show that

$$\mathcal{L}\left\{\frac{1}{2w^3}[\sin(wt) - wt\cos(wt)]\right\} = \frac{1}{(s^2+w^2)^2}$$

4.10. Apply the suitable theorems to determine the Laplace transform (use tables to check your answer) of the following time functions.

i. $f(t) = t\cos(5t + 4) + t^2 \sin 2t$

ii. $f(t) = t^2 \cos^2(wt)$

iii. $f(t) = e^{-2t}t^2 \sin 5t$

iv. $f(t) = t^3 e^{-2t}$

v. $f(t) = 3t^3 + 6t^{0.25}$

vi. $f(t) = (e^{-3t} \sin 2t - e^{-t} \sin 4t)/t$

vii. $f(t) = e^{-3t} \cosh 2t$

Section 4.3-4.4

4.11. Use the elementary partial fraction decomposition to determine $f(t)$ of the following Laplace function.

$$F(s) = \frac{s + 4}{s^3 + 6s + 11s + 6}$$

4.12. Find the decomposition coefficients k_1, k_2, k_3, and k_4, so that

$$\frac{1}{(s^2 + 4s + 9)(s^2 - 2s + 5)} = \frac{k_1 s + k_2}{s^2 + 4s + 9} + \frac{k_3 s + k_4}{s^2 - 2s + 5}$$

4.13. Show that

$$\mathcal{L}^{-1}\left(\frac{1}{s(s + a)(s + b)}\right) = \frac{1}{ab}\left(1 - \frac{b}{b - a}e^{-at} + \frac{a}{b - a}e^{-bt}\right)$$

4.14. Show that the unit step response $c(t)$ of a standard closed-loop unity negative feedback second-order control system, which is described by the natural frequency w_n, and the damping ratio ξ, is given by

$$c(t) = \mathcal{L}^{-1}\left(\frac{w_n^2}{s(s^2 + 2\xi w_n\, s + w_n^2)}\right) = 1 - \frac{1}{\sqrt{1-\xi^2}} e^{-\xi w_n t} \sin(w_n\sqrt{1-\xi^2}\; t + \varphi);$$

$$\varphi = \cos^{-1}(\xi)$$

4.15. A control system has the following transfer function

$$\frac{C(s)}{R(s)} = \frac{s+3}{s^2 + 6\,s + 25}$$

Determine the response $c(t)$ if a unit step input is applied.

4.16. Prove theorem 4.7 using the following Laplace functions.

a) $F(s) = s/(s+2)$

b) $F(s) = s^2/(s^2+4)$

4.17. Apply the suitable theorems to determine the following Laplace inverses (use partial fraction method as well).

i. $F(s) = 5/(s+4)(s-1)(s^2+1)$

ii. $F(s) = (s+5)/s(s^2+5s+6)$

iii. $F(s) = (3s^2+8s+9)/(s+3)(s+1)^2$

iv. $F(s) = (s^3+4s^2+4s-5)/(s+3)(s+1)^3$

v. $F(s) = 1/(s+1)(s^3+3s^2+7s+5)$

vi. $F(s) = \ln((s^2-1)/(s^2-4))$

vii. $F(s) = (s+5)/(s+1)^3$

viii. $F(s) = (s-3)/(s^2-2s+10)$

ix. $F(s) = s/(s^2+3)(s^2+4)$

x. $F(s) = (2s+7)/(s^2+10s+50)$

4.18. Use Laplace transforms to solve the following initial-value problems:

i. $y'' + 6y' + 8y = 3e^{-2t};\ y(0) = 1, y'(0) = -1$

ii. $y'' + 4y = 3\cos 2t;\ y(0) = 1;\ y'(0) = 2$

iii. $y''' + y = t^2 + 1;\ y(0) = y'(0) = y''(0) = 0$

iv. $y''' - y'' + 9y' - 9y = 0, y(0) = 0, y'(0) = 3, y''(0) = 0$

v. $y'' - 2y' + y = t\,e^{-2t};\ y(0) = 1, y'(0) = 0$

vi. $y'' + 4y' + 7y = t\,e^{-t}\cos t\ ;\ y(0) = \ y'(0) = 0$

vii. $y'' + 5y' + 6y = 1 + 3e^{-2t};\ y(0 = 1, y'(0) = -1$

viii. $y^{iv} + 6y'' + 8y = u(t);\ y(0) = y'(0) = y''(0) = y'''(0) = 0$

4.19. For a second-order Euler-Cauchy equation

$$t^2y'' + a\,t\,y' + b\,y = 0$$

Show that the equivalent algebraic equation is given by

$$m^2 + (3 - a)m + (2 - a + b) = 0$$

4.20. Determine the general solution of the following Euler-Cauchy equation (also, consider the results of exercise 4.19).

i. $t^2y'' - t\,y' + y = 0$

ii. $t^2y'' + 1.5\,ty' - 0.5\,y = 0$

iii. $t^3y''' - 3t^2y'' + 6\,t\,y' - 6y = 0$

iv. $t^3y''' - 6y = 0$

v. $t^3y''' + 3t^2y'' + ty' = 0$

4.21. A first-order differential equations system can also describe a dynamic system; it is called the state-space representation. The representation is

$$x'(t) = Ax(t) + B\,u(t), x(0) = X_0$$

$$y(t) = C\,x(t)$$

where A and B are the coefficient and input matrices, respectively, the matrices C is the output matrix, x is the state space vector, and y is the output (response). Use Laplace transforms to derive the output expression.

Section 4.5

4.22. Determine the mathematical expressions of the signals shown in Figure 4.10 that are suitable to apply the second shift theorem to determine their Laplace transform.

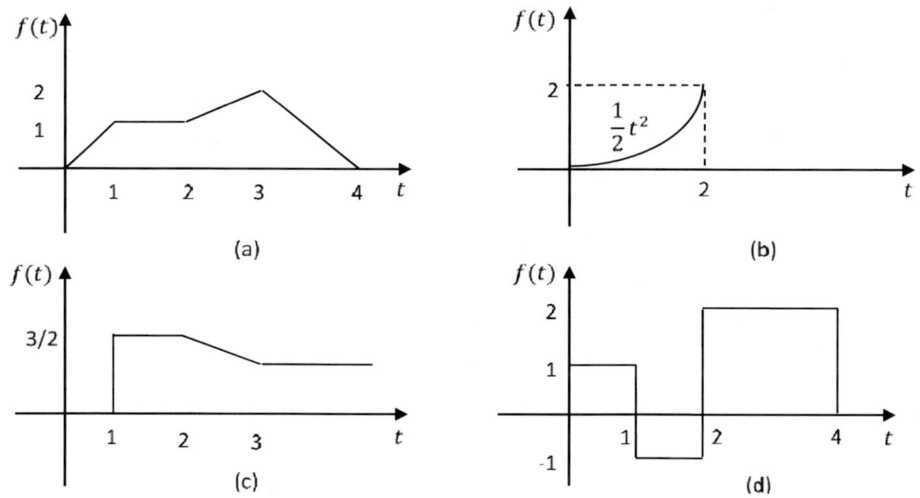

Figure 4.10. A set of different signals (exercise 4.22).

4.23. Determine the Laplace transforms of the following signals:

i. A signal defined as

$$f(t) = (t^2 + 3t + 2)[u(t-2) - u(t-3)]$$

ii. A signal defined as

$$f(t) = \begin{cases} \sin t; \ 0 \le t \le \pi/2 \\ 1; \ t \ge \pi/2 \end{cases}$$

iii. A signal defined as

$$f(t) = \begin{cases} \cos t; \ 0 \le t \le 2\pi \\ 0; \ t \ge 2\pi \end{cases}$$

iv. A full-wave rectified signal, which is a periodic time function defined in one period as

$$f(t) = |\sin t|,\ 0 \le t \le \pi$$

v. A periodic signal defined in one period as

$$f(t) = \begin{cases} t;\ 0 \le t \le 1 \\ 1;\ 1 \le t \le 2 \end{cases}$$

vi. A periodic sawtooth signal defined in one period as

$$f(t) = \begin{cases} t;\ 0 \le t \le 1 \\ -\frac{1}{2}t + \frac{3}{2};\ 1 \le t \le 3 \end{cases}$$

4.24. Use Laplace transforms to solve the following initial-value problems.

i. $y' + 5y = f(t), y(0) = 0$, where $f(t)$ is the signal in exercise 4.23-i.

ii. $y'' + 4y' + 4y = f(t), y(0) = y'(0) = 0$, where $f(t)$ is the signal in exercise 4.23-iii.

4.25. For the full-wave rectified signal $f(t)$ given in exercise 4.23-iv, determine the response of the following harmonic system.

$$y'' + 9y = f(t), y(0) = y'(0) = 0$$

4.26. Solve

$$y'' + 5y' + 6y = |\cos 2t|, y(0) = 1, y'(0) = 0$$

Section 4.6-4.7

4.27. Use Laplace transforms to compute the following improper integrations.

i. $\int_0^\infty ((e^{-2t} - e^{-5t})/t)dt$

ii. $\int_0^\infty ((\sin t + \sin 3t)/5t)dt$

iii. $\int_0^\infty (e^{-3t} \sin 2t/t)\, dt$

iv. $\int_0^\infty t\, e^{-3t} \cos \sqrt{2}\, t\, dt$

4.28. Determine the Laplace function of

$$f(t) = e^{-2t} \int_0^t (1 + t\ \cos 2t)\ dt$$

4.29. Show that

$$\int_0^\infty \frac{e^{-at}\cos bt - e^{-pt}\cos qt}{t} dt = \frac{1}{2}[\ln(p^2 + q^2) - \ln(a^2 + b^2)]$$

4.30. An open-loop control system is shown in Figure 4.11. Determine the steady-state output, i.e., $\lim_{t \to \infty} c(t)$.

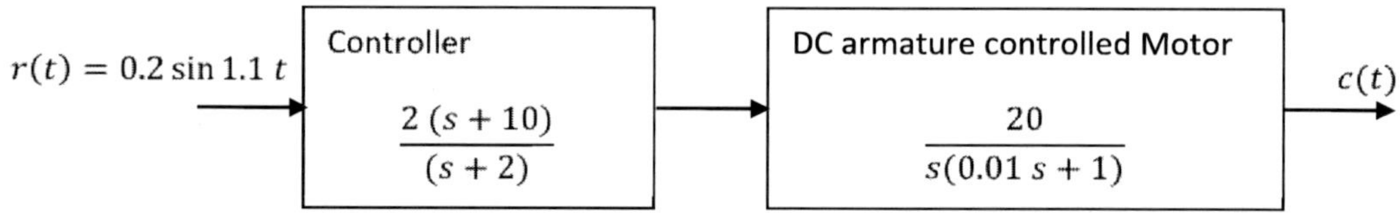

Figure 4.11. A control system (exercise 4.30).

4.31. For a field controlled DC motor, the transfer function is given by

$$\frac{\Theta(s)}{E_f(s)} = \frac{K_m}{s(T_e s + 1)(T_m s + 1)}$$

If the field voltage E_f of V volt is applied to the motor, use the final value theorem to determine the steady-state revolution per minute (rpm) of the motor.

4.32. Use Laplace transforms to analyze the electric circuit shown in Figure 4.12; the transfer function $V_0(s)/i(s)$ is to be considered.

4.33. Determine the differential equation that governs the control circuit shown in Figure 4.13, and then find the transfer function $U(s)/E(s)$.

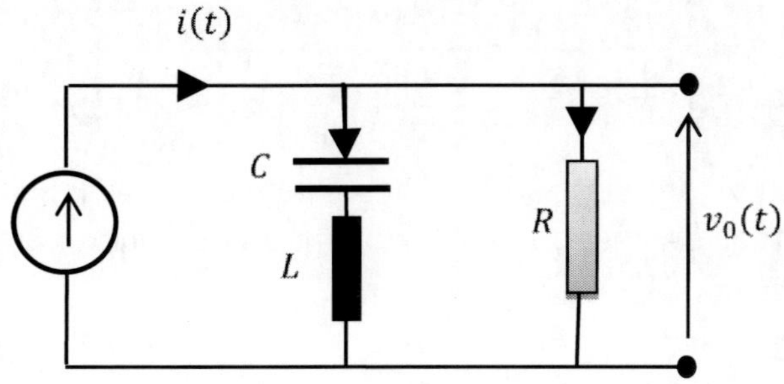

Figure 4.12. An Electric circuit (exercise 4.32).

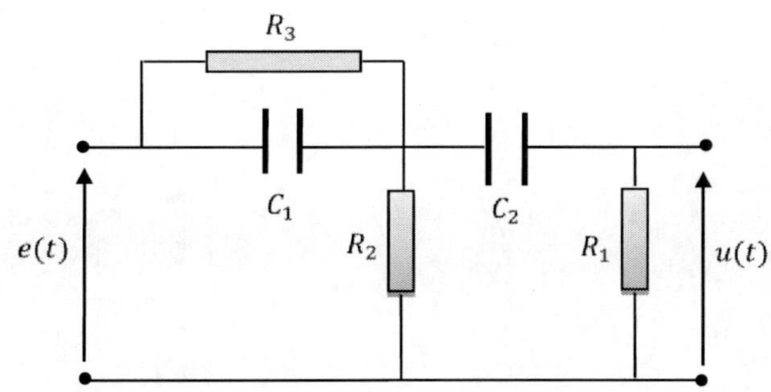

Figure 4.13. A Controller circuit (exercise 4.33).

4.34. Consider the thermal system shown in Figure 4.14. A heater coil supplying heat energy $u(t)$ is embedded into the outer chamber of temperature T_2. The temperature in the inner chamber is T_1 and the environment temperature is constant T_0. It is given that the product of mass m and specific heat c of the outer and inner chambers are 0.5 and 2, respectively. The interface thermal conductance k_{ij} between chambers is equal to 1, and between the outer chamber and outside is 0.5. It is required to determine the Laplace function of the temperature inside the inner chamber $T_1(s)$.

4.35. The control system for a ship yaw heading can be represented by a transfer function relating the heading $\Theta_h(s)$ to the commanded heading $\Theta_{com}(s)$ as follows

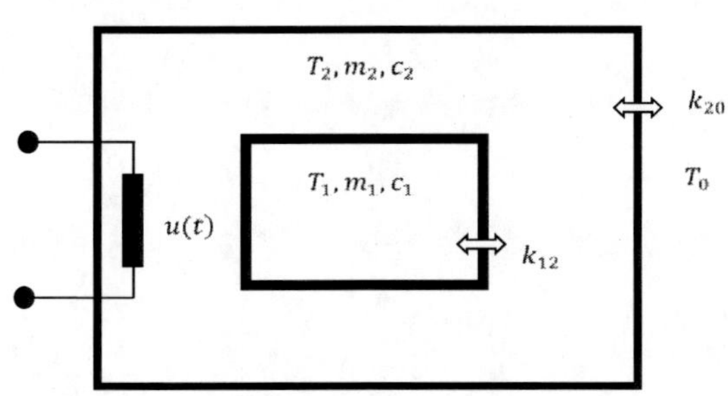

Figure 4.14. A Thermal system (exercise 4.34).

$$\frac{\Theta_h(s)}{\Theta_{com}(s)} = \frac{48.5(s + 0.85)}{s(0.25s + 1)(10s + 1)}$$

i. Write the differential equation describing the yaw heading dynamic system.
ii. Obtain the yaw response to a command heading of 4 degrees.

Chapter 5

NUMERICAL METHODS

5.1. INTRODUCTION

Readers already know that many problems do not have an exact solution in terms of elementary functions for one reason or another. As it is known, some types of definite integrations do not have implicit forms of solutions to obtain the numeric integration value. Another known numerical problem is the roots' finding of a nonlinear algebraic equation, including the polynomial equation of order greater than four. In this context, as we mentioned previously in chapter three, several differential equations do not have the exact form of a solution, and thus a mathematical problem is aroused. Although many numerical problems are already have been developed, there are still unsolved problems exactly. Since many engineering applications have in their solution frameworks one or more of such numerical or mathematical problems, then there will be a need to have an alternative solution other than the exact one.

The numerical solution is the alternative way for numerical problems whenever there is no exact formula for calculation. Similarly, for a mathematical problem that does not have an exact solution, there are two possible forms of solutions. These are the approximated and numerical forms. The difference between these two types of solution forms is essential to be defined. The approximated form is another form of exact form, usually in the form of a series of polynomials or elementary functions. In other words, the approximated form contains the independent variable explicitly. On the other hand, the numerical form is a step-by-step numerical calculation (usually in an iterative way). Numerical values substitute the independent variable during the iterative procedure.

Numerical methods are many for each type of problem. They have been developed over the years, and still, until now, new methods appear from time to time. In general, for a particular type of problem, the methods are competitive for the need of a suitable initial estimate, fast convergence, less memory requirement, and higher accuracy. The rapid progress in microelectronics and computer science encourages scientists, mathematicians, and engineers to develop new methods that are more accurate, fast, and require fewer memory locations. This progress enables us to use old methods, where at the time of their development, it was impractical to use them due to the longtime of computation they needed.

Studying the subject of numerical methods should contain both the algorithm to obtain the numerical solution and the method features analysis to determine the applicability of its use. The algorithm procedures are often not so difficult to follow, while the analysis requires a sufficient, if not a profound, knowledge of mathematical propositions and theories. Therefore, dealing with all known methods for each problem in one reasonable length chapter is impossible. However, to be familiar with different developing methods, we will give more than one method for each treated problem. More can be found in the specialist books to increase the knowledge about the numerical methods.

All the topics that we will consider in this chapter are easily solved using MATLAB software. We will also explain how the reader can use the MATLAB tool to solve some of these topics. Other software tools also exist to solve numerical problems; however, MATLAB is the more powerful and easy-to-use software. Therefore, one may argue why to study these details while there is such a software tool. Engineering readers should know these principles; thus, they will be able to deal with complicated problems, which may not be included in these software packages. On the other hand, engineers may require a practical implementation to solve only a specific numerical problem or a small set of problems without the aid of these software tools, which need a large amount of memory for setup.

5.2. Definitions and Concepts

Before starting the numerical solution of some considerable numerical and mathematical problems, we prefer to give some basic definitions and concepts common to all methods for all types of problems.

Definition 5.1. Numerical problem and mathematical problem

We mean a problem given by a clearly defined relationship between a finite number of input and output data by a numerical problem. A numerical problem's data must be represented by real numbers that can be handled for computation in finite time. For instance, finding the roots of a nonlinear algebraic equation is a type of numerical problem. On the other hand, mathematical problems cannot be described by a finite number of real numbers but are defined by the basic mathematical relations that can analyze the behavior output for a given input. The differential equation is a clear example of a mathematical problem.

Definition 5.2. Numerical iteration

Most numerical methods, if not all, iteratively finish the task, i.e., starting from a specific initial value(s), computing formula, using the results to compute again, etc. The computation formula is usually called a recurrent formula. The iteration is stopped when an insignificant difference is noticed between two successive computations when a prescribed stopping criterion is satisfied.

Definition 5.3. Convergence, divergence, and numerical instability

After a sufficient number of iterations, if it is given three successive numerical solutions, which are represented by the vectors, $X_i, X_{i+1},$ and X_{i+2} and that for small values $\epsilon_1,$ and ϵ_2 the following norms satisfy the inequalities.

$$|X_{i+1} - X_i| \leq \epsilon_1 \tag{5.1}$$

$$|X_{i+2} - X_{i+1}| \leq \epsilon_2 \tag{5.2}$$

Then, the numerical solution is said to be converged if

$$\epsilon_1 > \epsilon_2 \leq \epsilon \tag{5.3}$$

where ϵ defines the accuracy requirement of the numerical solution. Otherwise, the numerical solution diverges. If the norm inequalities 5.1 and 5.2 are satisfied and then violated during iterations, numerical instability occurs during iterations. It is not always possible to predict these features before running the computation algorithm. However, for some problems, rigorous mathematic is developed to explain that a specific numerical method is converged and convergence conditions. The numerical method may have a slow convergence feature, i.e., only after a large finite number of iterations, the norm inequalities are satisfied to fulfill the accuracy requirements. Therefore, the time for completion of a specific numerical task is directly proportional to the convergence feature. This could be enough reason to ignore a considered method for a prescribed numerical problem.

Definition 5.4. Initial vector

By an initial vector (eventually, an initial value), we mean the vector of values assigned to the independent variables before starting the method algorithm procedure. It is denoted usually, as X_0 (eventually, x_0). It is worth mentioning that some methods may not require having an initial vector to start; this is why methods of a considered type of problem are classified according to their requirement of the initial vector. The initial vector values have to be assigned suitably to ensure the convergence of the numerical solution. The initial vector must consider initial conditions for problems having such conditions, for example, the initial value problem of differential equations. However, in other cases, a specific initial vector's choice may represent an obstacle. An unsuitable vector may cause either a very slow convergence or numerical instability or even a divergence of the method.

Definition 5.5. Accuracy

For a convergent numerical method, the accuracy is terminology associated with the numerical solution and not with the numerical method because it depends on the number of iterations. The performance of the numerical solution is accordingly evaluated. Therefore, two numerical methods that solve the same problem may reach the same results after a different number of iterations. The achievement of the desired accuracy is measured by the absolute difference (or the norm) between two successive iterations. For instance, if we require accuracy to be better than 10^{-4}, this means to have an absolute value, which is less than 10^{-4}. The accuracy of calculation depends on the computer system itself; in some sensitive numerical problems, the used system does affect the obtained results.

Definition 5.6. Memory requirement

During iterations, it is required to memorize intermediate values that will be used in the next calculation; thus, a memory demand has appeared. Besides the demands of memory to memorize the input and output vectors (matrices), one recurrent formula in the algorithm requires one or more locations of memory to memorize the previous values. The memory requirement is increased as the number of variables increases; for example, the memory requirement of the numerical solution of a system of algebraic linear or nonlinear equations is directly proportional to the number of these equations. Fortunately, computer technology's high progress makes the memory requirement not a crucial issue in most cases.

Definition 5.7. Numerical algorithm

Generally, by an algorithm, we mean several sequence computation operations and some logical relations that perform a single task using a computer. The numerical algorithm is arithmetic operations (without logical relations) that the computer can perform on a set of inputs to obtain definite results. Examples of numerical algorithms are so many such as the long division of a polynomial by another of a lower order, Horner's algorithm of computing the resultant value of assigning a number to an nth-order polynomial, finding the square root of a number, and so on. For a considered mathematical or numerical problem, possibly more than one numerical algorithm performs the task.

5.3. The Solution of Nonlinear Algebraic Equations

As the reader realizes that, a second-order algebraic equation is the simple nonlinear equation, which has the form

$$Ax^2 + Bx + C = 0 \tag{5.4}$$

The roots are found easily by the square law in closed-form as

$$x_{1,2} = \frac{-B \mp \sqrt{B^2 - 4AC}}{2A} \tag{5.5}$$

Despite this, we (the computers) need to use a numerical procedure to perform the square root. For instance, the $\sqrt{m}$ can be determined by the recurrent formula

$$r_n = \frac{1}{2}\left(r_{n-1} + \frac{m}{r_{n-1}}\right), n = 1,2,3 \tag{5.6}$$

where r_0 is the lowest integer value, which approximates the square root $\sqrt{m}$. For example, for $\sqrt{3}$, if we set $r_0 = 1$, then we get sequentially the values 2, 1.75, 1.732142875, 1.73205081, 1.732050808. After these five iterations, the obtained result is accurate enough (better than 10^{-8}) value of $\sqrt{3}$.

Nonlinear algebraic equations are of two types. The nth-order polynomial and of what is called transcendental equations. The standard forms of these two types are, respectively

$$x^n + a_{n-1}x^{n-1} + a_{n-2}x^{n-2} + \cdots + a_1 x + a_0 = 0, a_0 \neq 0 \tag{5.7}$$

$$h(x) - g(x) = 0 \tag{5.8}$$

where the coefficients a_i are real numbers, and the functions h and g are continuous nonlinear functions. It is sufficient for a transcendental equation to have at least one of these two functions as a nonlinear one. Two examples of these two types are:

$$x^4 + 3x^3 + 11x^2 + 5x + 7 = 0$$

$$\sin x - x^3 + 2 = 0$$

Due to the subject's large content and a better explanation, we will divide this section into some subsections to consider several topics.

5.3.1. The Solution of Transcendental Equations

Consider a transcendental equation of the form

$$f(x) = 0 \tag{5.9}$$

where $f(x)$ is a continuous nonlinear function of a real variable x.

The numerical problem is to find the variable x that satisfies equation (5.9) in some interval $[a, b]$, when there is no exact analytic solution, i.e., to find numerically only one real root. The interval $[a, b]$ may be a given known, estimated, or unknown. The necessary condition of the solution existence is

$$f(a)f(b) < 0 \tag{5.10}$$

When the condition (5.10) is satisfied, we do not know how many roots are in the given interval. However, plotting the equation in the plane $x - f(x)$ could determine precisely the number of roots in the prescribed or assumed interval. There are several numerical methods for finding one real root of a transcendental equation; obviously, the graphical method is not concerned. We will consider only four methods, which have different features. The reader can find more numerical methods in the specialist books of numerical analysis.

1. The Bisection Method:

This method may be the oldest one that is used to solve nonlinear algebraic equations. It is based on the idea of dividing the given or assumed interval into two sections and checking the necessary condition (5.10) to allocate in which section the root is there. Then, a new smaller interval is obtained and repeated until the accuracy requirement is satisfied. With the aid of Figure 5.1, the method procedure is explained.

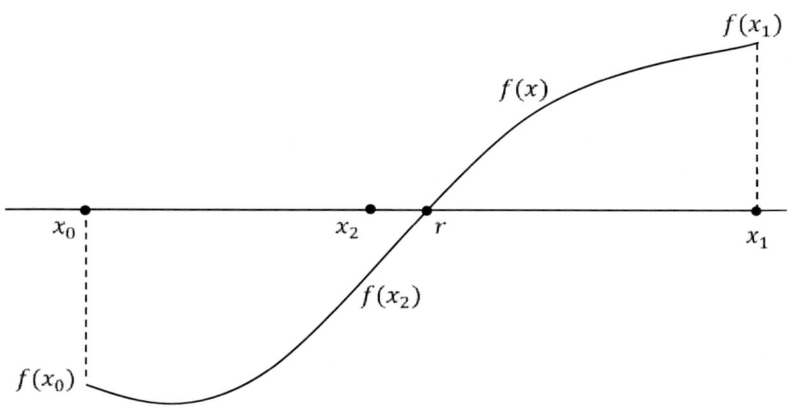

Figure 5.1. Bisection method concepts.

The method procedure is written in the following points:

- Assume some accuracy requirement given as a small value δ
- Let the given interval $I =< x_0, x_1 >$ and $x_2 = \frac{1}{2}(x_0 + x_1)$
- Because (as shown in Figure 5.1) it holds that $f(x_0)f(x_2) > 0, f(x_1)f(x_2) < 0$, then the new smaller interval is $I_1 =< x_2, x_1 >$
- if $f(x_2) = 0 \pm \delta$, then $r = x_2$, otherwise
- Let $x_3 = \frac{1}{2}(x_2 + x_1)$ and continue noting that the intervals $I_2 > I_3 > I_4 .... > I_n$, and the desired root $r = \frac{1}{2}(x_n + x_{n-1})$.

Whenever there is a root in the given interval, the method always converges. However, the convergence speed depends entirely on the equation itself; it could be the case that the convergence will be very slow. Besides, the method does not need an initial value; this advantage makes the method suitable as a starting method to find the initial value to implement another method. Finally, the method requires two memory locations to store two previous values.

Example 5.1.

In an *RLC* electric circuit, for given values of the capacitance *C* and inductance *L*, it is required to determine the proper resistance *R* to dissipate energy at a specified rate. The corresponding mathematical model for the problem is found to be

$$e^{-0.005\,R}\cos\left(0.05\sqrt{2000-0.01R^2}\right)=0.01$$

Determine the resistance value with an accuracy of 10^{-3}.

To solve the problem, we set the problem as a transcendental equation as

$$f(R)=e^{-0.005\,R}\cos\left(0.05\sqrt{2000-0.01R^2}\right)-0.01=0$$

Next, we have to determine a suitable interval of searching, considering that the value of *R* is always a positive real value. The maximum positive value of *R* is when

$$0.05\sqrt{2000-0.01R^2}=0 \Rightarrow R=447.2\ \Omega$$

The value of *R* could be less than this value. Thus, let us test the following assumed interval $I =< 100,400 >$. Thus,

$$f(100)=e^{-0.5}\cos\left(0.05\sqrt{2000-100}\right)-0.01=-0.35679$$

$$f(400)=e^{-2}\cos\left(0.05\sqrt{2000-1600}\right)-0.01=0.06312$$

Therefore, a positive real root exists. A first approximation, say R_1 is

$$R_1=\frac{400-100}{2}+100=250 \Rightarrow f(250)=-0.0901$$

Since the function value is negative, then the new smaller interval, say I_1 becomes $< 250,400 >$. The second and third approximations and their corresponding function values will be

$$R_2=\frac{400-250}{2}+250=325 \Rightarrow f(325)=-0.0032$$

$$R_3=\frac{400-325}{2}+325=\ 362.5 \Rightarrow f(325)=0.0322$$

Since the function value has changed its sign, the desired resistance value becomes within the new interval $< 325, 262.5 >$. Table 5.1 is arranged to show the principle of applying the bisection method for this problem. Then the desired resistance value is 327.9296 Ohms. The accuracy of calculation after nine iterations is within one Ohm; in practice, integer values are recommended or $R = 328\ \Omega$.

Table 5.1. Iterations of the bisection method in example 5.1

Approximation	Root R_i	$f(R_i)$	I_i
1	250	-0.091	$< 250, 400 >$
2	325	- 0.0032	$< 250, 325 >$
3	362.5	0.0322	$< 325, 362.5 >$
4	343.75	0.0151	$< 325, 343.75 >$
5	334.375	0.0061	$< 325, 334.375 >$
6	329.6875	0.0015	$< 325, 329.6875 >$
7	327.34375	-0.000805	$< 329.6875, 327.34375 >$
8	328.51562	0.00036	$< 327.34375, 328.51562 >$
9	327.9296		
	$\|R_9 - R_8\|$ =|327.9296 -328.51562| = 0.586		

2. The Simple Iteration Method

The method is based on the idea of rewriting equation (5.9) in the form

$$x = \varphi(x) \tag{5.11}$$

Further, it is assumed that there exists an interval $I =< a, b >$, where there is at least one real root $r \subset I$, which satisfies the following condition

$$\varphi(I) \subset I;\ \forall x \in I :\ \varphi(x) \in I \tag{5.12}$$

The function $\varphi(x)$ is not unique, i.e., for a considered equation, one can have more than one type of the function $\varphi(x)$. The method algorithm is in the following three steps:

i. Select a number $\delta > 0$ and initial or estimated value $x_0 \in I$.
ii. Compute the next approximated values from the recurrent formula $x_i = \varphi(x_{i-1});\ i = 0, 1, 2, ...$
iii. If $|x_{i+1} - x_i| \leq \delta$, then stop and set $r = x_i$, otherwise, continue in finding a new approximated value.

Example 5.2.

Find one real root in the interval $I =< 1.5, 2 >$ of the transcendental equation

$$f(x) = \frac{1}{4}x^2 - \sin x = 0$$

Let us first check the necessary condition 5.10 for the existence of a real root in the given interval.

$I =< 1.5, 2 >,$

$$f(1.5)f(2) = (-0.4349)(0.0907) < 0$$

Therefore, there is a real root in the given interval. Next, we will consider two forms of equation (5.11) and compute simultaneously for both. We can rewrite the recurrent formulas, as in the following two forms.

$$x_i = 2\sqrt{\sin x_{i-1}}$$

$$x_i = \sin^{-1}\left(\frac{x_{i-1}^2}{4}\right)$$

Let us assume that the initial value is one of the ends of the given interval, say $x_0 = 2$; thus, Table 5.2 depicts six iterations for both recurrent formulas. After these iterations, the root is 1.93392, and the obtained accuracy is better than 10-3 ($|x_6 - x_5| = 6.4 \times 10^{-4}$). We can continue to obtain, after eight iterations, a value of 1.93377 with an accuracy better than 10^{-4}.

Since only one previous value is required to compute the new one, the method demands only one memory location to store the old value. The convergence of the simple iteration method depends on the initial value. For instance, if we take $x_0 = 1.5$, then to reach the same accuracy, the number of iterations will be increased. However, the method's main drawback is the selection of a suitable recurrent formula that ensures convergence.

Table 5.2. Simple iteration method of example 5.2

i	$x_i = 2\sqrt{\sin x_{i-1}}$	$x_i = \sin^{-1}\left(\frac{x_{i-1}^2}{4}\right)$
1	1.90714	1.57079
2	1.94316	0.66472 $\notin I$
3	1,93025	0.11008 $\notin I$
4	1.93503	0.00306 $\notin I$
5	1.93328	$< 10^{-5}$
6	1.93392	
	$\lvert x_6 - x_5\rvert = 6.4 \times 10^{-4}$	Does not give a root in I

There will be several recurrent formulas in a complicated transcendental equation, and one has to search for the suitable one. Moreover, when the interval is not specified, the convergence problem will be challenging to analyze. For these reasons, the simple iteration method is of interest only for simple problems.

3. Secant Method

The method is in the category of what is called the interpolation method. The method is based on finding the root of the linear equation that is defined by the straight line joining the endpoints of the given interval, and we continue in an iterative way using the recurrent formula

$$x_{i+1} = x_i - \frac{(x_i - x_{i-1})f(x_i)}{f(x_i) - f(x_{i-1})}, i = 1,2, \ldots \quad (5.13)$$

Therefore, during iterations, $f(x)$ is computed once; however, three memory locations are required to memorize three previous computations x_i, x_{i-1}, $f(x_{i-1})$, which are required for a new computing value. This feature represents fast computation and excessive memory requirements. In the first iteration, the two ends of the given interval are used to compute the root's first approximation. One advantage of using this method is the function $f(x)$ needs not be differentiable but only continuous. Finally, the convergence property of this method is better than the previous two methods.

Example 5.3.

The uniform flow condition of water in a rectangular channel is modeled by the Manning equation as

$$Q = \frac{\sqrt{s}}{n}(Wh)^{\frac{5}{3}}(W + 2h)^{-\frac{2}{3}}$$

where Q is the flow rate, s is the inclination (slope) of the channel, n is the Manning roughness coefficient, W is the channel width, and h is the depth of water in the channel. When it is desired to find the water depth in the channel for a constant flow rate, we come with a transcendental equation. For the following numerical values $Q = 15\ m^3/s$, $W = 4.6\ m$, $s = 0.002$, and $n = 0.017$, find the water's depth in the channel.

Substituting the given values, we get the following transcendental equation.

$$f(h) = h^{\frac{5}{3}}(4.6 + 2h)^{-\frac{2}{3}} - 0.4482 = 0$$

Let us assume that the suitable interval of the solution is $< 1, 2.5 >$ meters, i.e., $h_0 = 1, h_1 = 2.5$, thus using the recurrent formula (5.13), we get the first approximation as

$$h_2 = 2.5 - \frac{1.5\left(2.5^{\frac{5}{3}}(4.6 + 2 \times 2.5)^{-\frac{2}{3}} - 0.4482\right)}{\left(2.5^{\frac{5}{3}}(4.6 + 2 \times 2.5)^{-\frac{2}{3}} - 0.4482\right) - \left(1^{\frac{5}{3}}(4.6 + 2 \times 1)^{-\frac{2}{3}} - 0.4482\right)}$$
$$= 1.3345$$

Continuing to obtain the second approximation

$$h_3 = 1.3345 - \frac{(1.3345 - 2.5)\left(1.3345^{\frac{5}{3}}(4.6 + 2 \times 1.3345)^{-\frac{2}{3}} - 0.4482\right)}{\left(1.3345^{\frac{5}{3}}(4.6 + 2h)^{-\frac{2}{3}} - 0.4482\right) - \left(2.5^{\frac{5}{3}}(4.6 + 2 \times 2.5)^{-\frac{2}{3}} - 0.4482\right)}$$
$$= 1.3684$$

$$h_4 = 1.3716$$

Hence, the water depth is 1.3716 m; the accuracy is better than one centimeter.

4. Newton's Method

Newton's method belongs to the category of fast convergence methods. It is based on assuming the existence of an approximated root x_r in a given interval $I = < a, b >$, and then the function $f(x)$ is expanded by Taylor series around the root x_r. It is not essential to know the accuracy of this approximation, but only the condition's satisfaction in equation (5.10). Therefore, we can write

$$f(x_r) = f(x) + (x_r - x)f'(x) + \frac{1}{2}(x_r - x)^2 f''(x) + \cdots = 0 \qquad (5.14)$$

Upon neglecting the second and higher terms yields

$$x_r \cong x - \frac{f(x)}{f'(x)} \qquad (5.15)$$

To improve the accuracy of the first approximation (usually called the initial value x_0), a recurrent formula is accordingly set, as

$$x_{i+1} = x_i - \frac{f(x_i)}{f'(x_i)}, i = 0,1,2 \ldots \qquad (5.16)$$

The recurrent formula requires only one stored previous value. The method drawback is twofold; the first is that not every initial value ensures the convergence of the method, and the second is the computation of the derivatives in iterations. However, for arbitrary $x_0 \in I$, the necessary conditions of the method convergence are:

- Over the whole interval $I = < a, b >$, the second derivative $f''(x)$ does not change its sign.
- Provided that condition 5.10 is satisfied and that $f'(x) \neq 0$, then the following inequalities hold.

$$\left|\frac{f(a)}{f'(a)}\right| < (b-a), and \left|\frac{f(b)}{f'(b)}\right| < (b-a) \tag{5.17}$$

Therefore, additional computation is imposed on iterations to test the convergence of the method. Working the method on computers requires numerical evaluation of the first and second derivatives; for complicated equations, computation's burden is increased significantly. Furthermore, two different suitable initial values (lead to convergence) may result in different speeds of convergence, and there is no general rule that indicates which initial condition gives faster convergence. For instance, in practice, the transcendental equation arises when required to determine the moment at which the output of a dynamic system possesses a specified value. The next example illustrates Newton's method.

Example 5.4.

It is given that the output of a dynamic system is

$$c(t) = \sin t - t\cos t$$

The problem is to determine the moment that the output reaches the value $(\pi/2)$. First, we write the corresponding transcendental equation. It is easily read as

$$f(t) = t\cos t - \sin t + \frac{\pi}{2} = 0$$

Let us consider an interval $1 \le t \le 2$; we have to check for the validity of this interval for finding the root of the equation. Thus, the first and second derivatives are

$$f'(t) = -t\sin t, f''(t) = -\sin t - t\cos t$$

We can find that $f''(x)$ does not change its negative sign over the interval $I =< 1, 2 >$, and the computation of formulas 5.17 gives

$$\left|\frac{f(1)}{f'(1)}\right| = \left|\frac{\cos(1) - \sin(1) + \frac{\pi}{2}}{-\sin(1)}\right| = 1.599 > 1,$$

$$\left|\frac{f(2)}{f'(2)}\right| = \left|\frac{2\cos(2) - \sin(2) + \frac{\pi}{2}}{-2\sin(2)}\right| = 0.09 < 1$$

Therefore, not for every initial value, the method will converge. However, let us consider $t_0 = 1$, and use the recurrent formula

$$t_{i+1} = t_i - \frac{t_i\cos t_i - \sin t_i + \pi/2}{-t_i\sin t_i}, i = 0,1,2 \dots$$

Table 5.3 shows the results when the initial value is $t_0 = 1$.

Therefore, the output reaches the value $(\pi/2)$ at $t = 1.90569589$, with an accuracy better than 10^{-6}, which is obtained only after five iterations. Newton's method is a fast convergent method compared to previously discussed methods, but it needs a suitable initial value. The combination of Newton's method and, for instance, the bisection method constitutes a very efficient way of solving complicated transcendental nonlinear equations. Starting with the bisection method for a few approximations, then the last approximated result from the bisection method is utilized as an initial value for Newton's method. Furthermore, with a slight modification, Newton's method can find one complex conjugate pair of roots.

Table 5.3. Newton method solution for example 5.4

i	t_i	t_{i+1}	$\lvert t_{i+1} - t_i \rvert$
0	1.0000	2.5088	1.5088
1	2.5088	1.8054	0.7034
2	1.8054	1.9071	0.1017
3	1.9071	1.9057	0.0014
4	1.90569589	1.90569572	1.6×10^{-7}

5.3.2. The Solution of Polynomial Equations

Consider the nth-order real polynomial

$$p(x) = a_0x^n + a_1x^{n-1} + a_2x^{n-2} + \cdots + a_{n-1}x + a_n \tag{5.18}$$

Unlike the transcendental equation, the polynomial equation has a defined set of roots always. A polynomial equation of nth–order has precisely n roots, either all real (positive and negative), all complex conjugate, or some real and others are complex conjugate roots. Based on Descartes' rule, the possible configuration of the roots can be determined. This rule states:

1. The maximum number of positive roots is equal to the number of sign reversals of the polynomial $p(x)$ and the difference between the number of sign reversals and positive roots is an even number.
2. The maximum number of negative roots is equal to the number of sign reversals of the polynomial $p(-x)$ and the difference between the number of sign reversals and negative roots is an even number.

All roots of the polynomial equation $p(x) = 0$, lie in the interval $r < |x| < R$, where

$$R = 1 + \frac{1}{|a_0|} \max_{k=1,2,\dots n} \{|a_k|\} \tag{5.19}$$

$$\frac{1}{r} = 1 + \frac{1}{|a_n|} \max_{k=0,1,2,\ldots n-1} \{|a_k|\} \tag{5.20}$$

This simple information guides the determination of the roots of a polynomial equation efficiently. To illustrate how the above information can be established, we consider a simple example.

Example 5.5.

Consider the fourth-order polynomial equation

$$2x^4 + 7x^3 + 8x^2 - 5x - 12 = 0$$

Then $p(-x) = 0$ is

$$2x^4 - 7x^3 + 8x^2 + 5x - 12 = 0$$

We conclude:

1. There are four roots: They are either all real or two real, and one pair of a complex conjugate, or two pairs of the complex conjugate.
2. There is only one sign reversal of $p(x)$, so there is at a maximum one positive root (i.e., either one or no one), and the difference between the number of sign reversals and the number is either $1 - 0 = 1$, which is odd or $1 - 1 = 0$, which is even. Therefore, there is certainly one positive root.
3. There are three sign reversals of $p(-x)$; thus, there are at maximum three negative roots. The difference between the number of sign reversals and the number is either $3 - 1 = 2$, which is even; $3 - 2 = 1$, which is odd; or $3 - 3 = 0$, which is even. Therefore, there are certainly either two negative or three negative roots.
4. Therefore, the roots are either one positive and three negative or one positive, one negative, and one pair of the complex conjugate.
5. The equation coefficients are: $a_0 = 2, a_1 = 7, a_2 = 8, a_3 = -5, a_4 = -12$, then we calculate

$$R = 1 + \frac{1}{2}\max\{7, 8, 5, 12\} = 7$$

$$\frac{1}{r} = 1 + \frac{1}{12}\max\{2, 7, 8, 5\} \Rightarrow r = 0.6$$

Then the interval of the roots is $0.6 < |x| < 7$. It means that the positive root is greater than 0.6 and less than 7. The magnitude of the complex root, if any, also lies between these two values. The absolute values of the negative roots are also within this interval.

If it is not possible to factorize the polynomial equation to a product of several brackets, then numerical methods have to be used to solve these equations. However, based on Brown's method, the third-order polynomial equations can be solved in a closed-form to find the three roots.

The following cases have to be considered to determine the roots of a polynomial equation of an order greater than three.

Case I: When all roots are real, then

- Find one real root using any of the methods of the previous subsection; let that root be named r_1.
- Perform the long division $p(x)/(x - r_1)$ to eliminate the obtained root, and set the results

$$q_{n-1} = \frac{p_n(x)}{(x - r_1)}$$

Horner schema can be used to perform the division.

- Repeat the above two steps for q_{n-1} to find a second root r_2 and q_{n-2} and so on.

Case II: When there are real roots and one complex conjugate pair, then

- Using the procedure of case I to find all real roots one by one until a quadratic equation is obtained due to the long divisions.
- Use the square law to find the complex conjugate pair of roots.

Case III: When there are real roots and more than one complex conjugate pair, then

- Using the procedure of case I to find all real roots one by one until an even-order polynomial is obtained due to the long divisions, the last equation's order is twice the number of pairs of roots left.
- Use the modified Newton method to find the complex conjugate pair of roots one by one.

It is worth mentioning that when the real roots are integers, then rounding the results of previous single root-finding methods will make the remainder of the long division equal zero. This method permits a precise procedure; however, it will impose some difficulties in determining accurate results when the real roots are not integers. Several methods, such as Bairstow's method, use an efficient procedure for division in solving the problem. The subject is very considerable, and one may refer to the numerical books to ascertain these methods.

For completeness, we explain one procedure of division by a quadratic polynomial. Let us consider an nth–order polynomial that is given in 5.18, and two values r_1 and r_2 are given, such to set a quadratic polynomial as

$$d(x) = (x - r_1)(x - r_2) = x^2 - px + q;\ p = r_1 + r_2, q = r_1 r_2$$

The task is to perform the long division $p(x)/d(x)$, which has the form

$$w(x) = \frac{p(x)}{d(x)} = v_0 x^{n-2} + v_1 x^{n-3} + \cdots + v_{n-2}$$

Then we can determine the coefficients v_i as follows:

$$v_{-1} = 0$$

$$v_0 = a_0$$

$$v_k = a_k + pv_{k-1} + qv_{k-2}, k = 1,2, \ldots n-1$$

$$v_n = a_n + qv_{n-2}$$

Two typical engineering examples are the characteristic equation's solution to determine the eigenvalues and to explore a dynamic system's stability.

Example 5.6.

A linear dynamic system's stability problem is explored by finding the root of the system's characteristic equation. For a particular linear dynamic system, it is given the following characteristic equation.

$$p(s) = 2s^6 + 8s^5 + 23s^4 + 48s^3 + 61s^2 + 70s + 40 = 0$$

Show that the dynamic system is critically stable if it is given that the system has one complex conjugate pair $r_{1,2} = \pm j\sqrt{2.5}$.

First, we have to find the other 3 roots. Since $p(s)$ has no sign reversals, then there are no real positive roots. On the other hand, the polynomial $p(-s)$ has six sign reversals. Therefore, there is a maximum of 6 - 4 = 2 real negative roots. Accordingly, the remaining four roots are either all negative real or two real negative roots and another complex conjugate pair. The quadratic equation of the given roots is

$$d(s) = s^2 + 2.5$$

If we carry out the long division $p(s)/d(s)$, we obtain the equation

$$w(s) = 2s^4 + 8s^3 + 18s^2 + 28s + 16 = 0$$

For the last equation, we use the Newton method with $s_0 = 0$, to obtain the third root.

$$s_1 = 0 - \frac{16}{28} = -0.5714$$

$$s_2 = -0.5714 - \frac{4.5981}{13.7726} = -0.9295$$

$$s_3 = -0.9295 - \frac{0.5936}{8.8485} = -0.9966 \Rightarrow r_3 = -1$$

Again, we perform the long division $w(s)/(s+1)$; thus, we get a third-order equation from which we can find the fourth root using Newton's method $r_4 = -2$. Finally, another long division by $(s+2)$ gives

$$2s^2 + 2s + 8 = 0 \Rightarrow r_{5,6} = -\frac{1}{2} \pm j\frac{\sqrt{15}}{2}$$

Therefore, the fifth and sixth roots are the complex conjugate pair with negative real values. According to Lyapunov's theorem, since all three roots have negative real values with one pair of a pure complex conjugate, then the system is critically stable.

5.3.3. The Solution of Nonlinear Systems

An nth-order system of nonlinear algebraic equations has a form

$$\begin{aligned} f_1(x_1, x_2 \dots x_n) &= 0 \\ &\vdots \\ f_n(x_1, x_2 \dots x_n) &= 0 \end{aligned} \tag{5.21}$$

where, $f_i, i = 1,2 \dots n$ are nonlinear functions of the variables $x_1, x_2 \dots x_n$. In matrix form, we write

$$F(X) = 0$$

Our task is to find only one real vector $X = [x_1, x_2 \dots x_n]$ that satisfies the system (5.21), whereas a given initial vector X_0 is given. We will not explore all existing methods for solving nonlinear systems but affirm that a real solution may not exist for some system parameters. The selection of a particular initial vector is a crucial problem. The most popular method for solving nonlinear systems is the modification of Newton's method. The following recurrent formula gives it.

$$X_{i+1} = X_i - J^{-1}(X_i)F(X_i), i = 0,1,2 \dots \tag{5.22}$$

where J is the Jacobian matrix given by

$$J = \begin{bmatrix} \frac{\partial f_1}{\partial x_1} & \frac{\partial f_1}{\partial x_2} & \cdots & \frac{\partial f_1}{\partial x_n} \\ \frac{\partial f_2}{\partial x_1} & \frac{\partial f_2}{\partial x_2} & \cdots & \frac{\partial f_2}{\partial x_n} \\ \cdots & \cdots & & \cdots \\ \frac{\partial f_n}{\partial x_1} & \frac{\partial f_n}{\partial x_2} & \cdots & \frac{\partial f_n}{\partial x_n} \end{bmatrix} \tag{5.23}$$

The nonlinear function F and the inverse of the Jacobin matrix are computed for every approximated value X_i. The convergence conditions are very complicated, and they will not be included here; however, if we have a suitable initial vector and a regular Jacobian matrix, the method converges. The iteration-stopping criterion is that the Euclidian norm is less than a definite accuracy requirement.

$$||X_i - X_{i-1}|| \leq \epsilon \tag{5.24}$$

Example 5.7.
Find a real solution to the nonlinear system

$$x^2 + xy + y^2 = 3$$
$$\sin x - y^2 = 0$$

Assuming the following initial values $x_0 = 1,\ y_0 = 1$. The function $F(x, y)$ is

$$F(x,y) = \begin{bmatrix} f_1 \\ f_2 \end{bmatrix} = \begin{bmatrix} x^2 + xy + y^2 - 3 \\ \sin x - y^2 \end{bmatrix}$$

We prepare first the inverse of the Jacobin matrix; according to 5.23, we have

$$J = \begin{bmatrix} \frac{\partial f_1}{\partial x} & \frac{\partial f_1}{\partial y} \\ \frac{\partial f_2}{\partial x} & \frac{\partial f_2}{\partial y} \end{bmatrix} = \begin{bmatrix} 2x + y & x + 2y \\ \cos x & -2y \end{bmatrix}$$

$$J^{-1} = \frac{1}{(-4xy - 2y^2 - x\cos x - 2y\cos x)} \begin{bmatrix} -2y & -x - 2y \\ -\cos x & 2x + y \end{bmatrix}$$

Substituting the initial values in the $F(x_0, y_0)$ and $J^{-1}(x_0, y_0)$, we obtain

$$F(X_0) = \begin{bmatrix} 0 \\ -0.1585 \end{bmatrix}, J^{-1}(X_0) = \begin{bmatrix} 0.2624 & 0.3937 \\ 0.0709 & -0.3937 \end{bmatrix}$$

The first approximation is

$$\begin{bmatrix} x_1 \\ y_1 \end{bmatrix} = \begin{bmatrix} 1 \\ 1 \end{bmatrix} - \begin{bmatrix} 0.2624 & 0.3937 \\ 0.0709 & -0.3937 \end{bmatrix} \begin{bmatrix} 0 \\ -0.1585 \end{bmatrix} = \begin{bmatrix} 1.0624 \\ 0.9376 \end{bmatrix}$$

Then the followings are the next approximations

$$\begin{bmatrix} x_1 \\ y_1 \end{bmatrix} = \begin{bmatrix} 1.06366224 \\ 0.93495865 \end{bmatrix}, \begin{bmatrix} x_2 \\ y_2 \end{bmatrix} = \begin{bmatrix} 1.06366402 \\ 0.93495500 \end{bmatrix}$$

With these values, we have accuracy better than 10^{-4}, and therefore, the required one real solution is $x = 1.06366, y = 0.934955$.

5.3.4. Solutions by MATLAB

The famous MATLAB software has a powerful numerical toolbox that can be used to find the roots of nonlinear transcendental equations and polynomial equations in only one MATLAB function. Furthermore, a MATLAB program can be written to solve many numerical problems that need different numerical techniques. These programs may also include intermediate computations, tests for convergence, control of accuracy, and plots of the given interval function. For a nonlinear transcendental equation, one real root can be found by the MATLAB function $\boldsymbol{fzero}$. For a given function $f(x)$ and an initial value x_0, it will have the form

$$\text{fzero}(@(x)\ f(x), x_0)$$

For instance, for the equation of example 5.2, we write the following one line-command in the command screen to obtain the result:

$$\text{fzero}(@(x) 0.25 * x^2 - \sin(x), 1.5, 2.5)$$

Thus, we obtain a root equal to 1.9339. The same result will be obtained with $x_0 = 2$. Also, This function gives a message erroneous when the initial value setting is not correct.

For polynomial equations, all roots can be found by the MATLAB function $\boldsymbol{roots}$. For a given polynomial $p(x)$, where its coefficients are put as a row vector P, the function has the form

$$\text{roots}([P])$$

For example, the roots of equation $6x^3 + 17x^2 + 27x + 20 = 0$ can be found from the MATLAB command

$$\text{roots}([6 \quad 17 \quad 27 \quad 20])$$

5.4. POLYNOMIAL APPROXIMATION AND INTERPOLATION

In some engineering applications, it may be desired to express approximately, in a specific interval $[a, b]$, a particular complicated continuous function by a polynomial of specified order, i.e.,

$$f(x) = p(x) + \text{error}(x), \forall x \in [a, b] \tag{5.25}$$

$$f(x) \cong p(x) = a_n x^n + a_{n-1} x^{n-1} + \cdots + a_1 x \ + + a_0 \tag{5.26}$$

The idea behind this approximation is to take the advantage of mathematically manipulating with polynomials rather than with complicated functions without scarifying the implementation accuracy. It is much easier and faster to compute function value, differentiation, integrations, etc., with polynomials. The validity of such approximation only occurs in the given interval, and it may be changed dramatically for other ranges. For a definite accuracy requirement, if a proper order of the polynomial is selected, a method to find the coefficients completes the desired approximation. For performing this approximation (finding the coefficients $a_i; i = 1,2..n$), there are many techniques and methods, including but not limited to the following:

1. Approximation based on Taylor series
2. Simple interpolation technique
3. Approximation based on least-squares error method
4. Linear regression and quadratic techniques
5. Least-squares approximation by orthogonal polynomials
6. Finite differences technique
7. Lagrange interpolation form

This section will consider only four techniques: the Lagrange interpolation form, approximation based on the least square error, the linear and quadratic regression, and the finite difference technique.

5.4.1. Approximation Based on Least-Squares Error Method

Consider a function $f(x)$, which has to be approximated in a specifically given interval $[a, b]$; by an nth-order polynomial $p_n(x)$. Let the error function be denoted $E(x)$, and it is defined as

$$E(x) = [p_n(x) - f(x)]^2 \tag{5.27}$$

The square of the difference is to account for negative as well as positive errors. The method is to determine the coefficients of the polynomial such that the function $E(x)$ is minimum, i.e.,

$$a_i = \min\{E(x)\}, \forall x \in [a,b], i = 0,1,2 \ldots . n \tag{5.28}$$

There are two cases to be considered as follows:

1. The continuous form: if $f(x)$ is a continuous function, then the minimization procedure results in the following $(n+1)$ algebraic system of linear equations

$$\begin{bmatrix} c_{11} & c_{12} & \cdots & c_{1,n+1} \\ c_{21} & c_{22} & \cdots & c_{2,n+1} \\ \vdots & \vdots & & \vdots \\ c_{n+1,1} & c_{n+1,2} & \cdots & c_{n+1,n+1} \end{bmatrix} \begin{bmatrix} a_0 \\ a_1 \\ \vdots \\ a_n \end{bmatrix} = \begin{bmatrix} b_0 \\ b_1 \\ \vdots \\ b_n \end{bmatrix} \tag{5.29}$$

where,

$$c_{ij} = \int_a^b x^{i+j-2}\, dx = \frac{1}{i+j-1}\left[b^{i+j-1} - a^{i+j-1}\right];\ i,j = 1,2 \ldots n+1 \tag{5.30}$$

$$b_i = \int_a^b x^i f(x)\, dx\ ;\ i = 0,1,2 \ldots n \tag{5.31}$$

Formula (5.30) indicates that matrix C is symmetric, i.e., we need to calculate only $\frac{1}{2}n(n+1)$ times.

2. The discrete form: when the function values are considered only at *m* discrete values where *m* is usually greater than $(n+1)$. In this case, the algebraic system of linear equations is given by

$$C_{m\times(n+1)} \begin{bmatrix} a_0 \\ a_1 \\ \vdots \\ a_n \end{bmatrix} = \begin{bmatrix} b_0 \\ b_1 \\ \vdots \\ b_n \end{bmatrix} \tag{5.32}$$

where,

$$c_{ij} = x_i^{j-1}\ ; i = 1,2 \ldots m, j = 0,1,2 \ldots n+1 \tag{5.33}$$

$$b_k = f(x_k);\ k = 0,1,2 \ldots n \tag{5.34}$$

$$x_0 = a, x_m = b \tag{5.35}$$

Since matrix C is not a square matrix, then we can only solve it by squaring the system. The squaring can be done by multiplying from left both sides of the system, in equation (5.32), by the transpose matrix of C, i.e., to have the following square $(n+1)$ system.

$$C^T C \begin{bmatrix} a_0 \\ a_1 \\ \vdots \\ a_n \end{bmatrix} = C^T \begin{bmatrix} b_0 \\ b_1 \\ \vdots \\ b_n \end{bmatrix} \tag{5.36}$$

When the integrals (5.31) cannot be found explicitly in an analytic form, the discrete form could be the second choice. This solution can be carried out by selecting a suitable number of equidistance points for solving the system (5.36). In practice, the desired accuracy is achieved only when the number of points is large enough. However, in most cases, due to a large amount of computation, computers are used to carry out this approximation method.

Example 5.8.

We want to approximate the function $f(x) = x \sin x$ in the interval $[0, \pi]$ using a fourth-order polynomial.

First, we calculate the 25 entries of the matrix C (due to symmetry of the matrix, they are 15 computations and here, only some of them are shown) and the five entries of the vector b, using formulas (5.30 and 5.31) respectively as follows

$$c_{11} = \frac{1}{1+1-1}[\pi^{1+1-1} - 0] = \pi$$

$$c_{12} = c_{21} = \frac{1}{1+2-1}[\pi^{1+2-1} - 0] = \frac{1}{2}\pi^2$$

$$c_{13} = c_{31} = \frac{1}{1+3-1}[\pi^{1+3-1} - 0] = \frac{1}{3}\pi^3$$

$$\vdots$$

$$\vdots$$

$$c_{55} = \frac{1}{5+5-1}[\pi^{5+5-1} - 0] = \frac{1}{9}\pi^9$$

$$b_0 = \int_0^{\pi} x^0 x \sin x \, dx = \pi,$$

$$b_1 = \int_0^{\pi} x^1 x \sin x \, dx = \pi^2 - 4$$

$$b_2 = \int_0^{\pi} x^2 x \sin x \, dx = \pi^2 - 6\pi,$$

$$b_3 = \int_0^{\pi} x^3 x \sin x \, dx = \pi^4 - 12\pi^2 + 48$$

$$b_4 = \int_0^{\pi} x^4 x \sin x \, dx = \pi^5 - 20\pi^3 + 120\pi$$

Next, we determine the coefficients vector by solving the norm system, i.e.,

$$\begin{bmatrix} a_0 \\ a_1 \\ a_2 \\ a_3 \\ a_4 \end{bmatrix} = \begin{bmatrix} \pi & \pi^2/2 & \pi^3/3 & \pi^4/4 & \pi^5/5 \\ \pi^2/2 & \pi^3/3 & \pi^4/4 & \pi^5/6 & \pi^6/6 \\ \pi^3/3 & \pi^4/4 & \pi^5/5 & \pi^6/6 & \pi^7/7 \\ \pi^4/4 & \pi^5/5 & \pi^6/6 & \pi^7/7 & \pi^8/8 \\ \pi^5/5 & \pi^6/6 & \pi^7/8 & \pi^8/8 & \pi^9/9 \end{bmatrix}^{-1} \begin{bmatrix} \pi \\ \pi^2 - 4 \\ \pi^2 - 6\pi \\ \pi^4 - 12\pi^2 + 48 \\ \pi^5 - 20\pi^3 + 120\pi \end{bmatrix}$$

The solution gives

$$a_0 = 0.0452,\ a_1 = -0.4302, a_2 = 1.944, a_3 = -0.7616, a_4 = 0.0584$$

Therefore, the fourth-order approximated polynomial is

$$p_4(x) = 0.0584x^4 - 0.7615x^3 + 1.944x^2 - 0.4302x + 0.0452$$

Let us divide the interval to 21 points by a step of $0.05\,\pi$; the function and polynomial values of these points are calculated to obtain the second norm (square root) of the sum squares of differences $E(x)$. It can be found that

$$\sqrt{\sum_{i=0}^{i=20} E(x_i)} = 0.0793$$

Furthermore, the calculations are used for plotting an illustrative plot. Figure 5.2 shows the plot. Therefore, despite the good locking of the plots, in terms of function and polynomial values at any point inside the interval, the approximation accuracy is better than only 10^{-1}. However, accuracy can be improved only by increasing the order of the polynomial. To solve the same problem using the discrete form, let us consider ten function values given in Table 5.4; this number is selected without any accuracy reference, but only to explain the discrete approach.

Table 5.4. The generated data of example 5.8

x_i	0	$\pi/9$	$2\pi/9$	$3\pi/9$	$4\pi/9$	$5\pi/9$	$6\pi/9$	$7\pi/9$	$8\pi/9$	π
$x_i \sin x_i$	0	0.1194	0.4488	0.9096	1.3751	1.7188	1.8138	1.5706	0.9551	0

If a fourth-order polynomial is assumed, then the matrix C will have dimensions of (10×5) as follows.

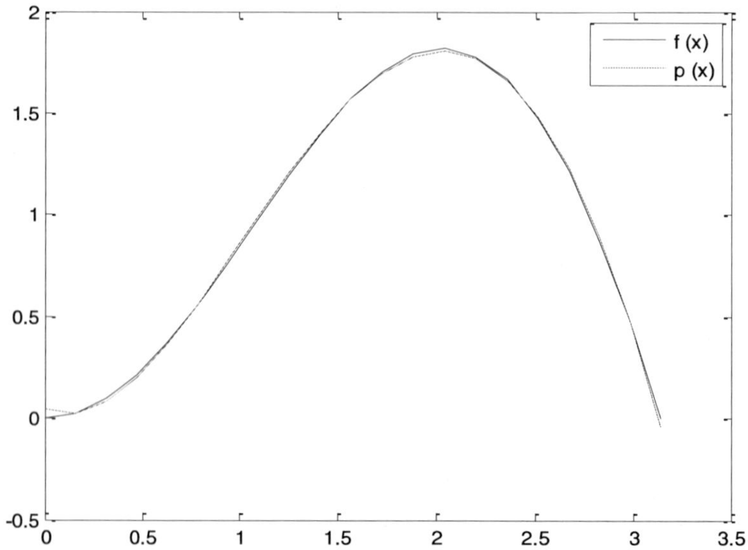

Figure 5.2. Least-square approximation (Example 5.8).

$$C = \begin{bmatrix} 1 & 0 & 0 & 0 & 0 \\ 1 & \pi/9 & (\pi/9)^2 & (\pi/9)^3 & (\pi/9)^4 \\ 1 & 2\pi/9 & (2\pi/9)^2 & (2\pi/9)^3 & (2\pi/9)^4 \\ \vdots & \vdots & \vdots & \vdots & \vdots \\ 1 & \pi & \pi^2 & \pi^3 & \pi^4 \end{bmatrix}$$

The function values in Table 5.5 are used to compute the coefficients $b_k, k = 1,2 \dots 10$. The norm system 5.36 becomes

$$\begin{bmatrix} 10 & 15.7079 & 34.7762 & 86.1285 & 227.6442 \\ 15.7079 & 34.7263 & 86.1285 & 227.6442 & 626.1719 \\ 34.7263 & 86.1285 & 227.6442 & 626.1719 & 1769.9575 \\ 86.1285 & 227.6442 & 626.1719 & 1769.9575 & 5102.5321 \\ 227.6442 & 626.1719 & 1769.9575 & 5102.5321 & 14929.6122 \end{bmatrix} \begin{bmatrix} a_0 \\ a_1 \\ a_2 \\ a_3 \\ a_4 \end{bmatrix} = \begin{bmatrix} 8.9084 \\ 16.5282 \\ 33.9620 \\ 74.4562 \\ 171.3437 \end{bmatrix}$$

The solution of the above linear system results in the following fourth-order polynomial.

$$p_4(x) = 0.0577x^4 - 0.7428x^3 + 1.8668x^2 - 0.3313x + 0.0114$$

As can be seen, the polynomial coefficients are almost similar to that of the continuous case. However, the accuracy has to be calculated to make a comparison between the two approaches. For sure, the reader notes that the computation burden can only be handled in a reasonable time by computers.

5.4.2. Linear and Quadratic Regressions

Experiments are usually carried out to describe a relationship between the dependent variable and one or more independent variables. For example, readings (data) of a current in a multi-elements electric circuit or a temperature in a furnace are recorded to establish the relationships as functions of time. For many reasons, repeating the experiments will not usually give the same readings, and hence there is a need to fit these data to the best possible suggested formula. This procedure means that we have to select a specific type of formula whose parameters are unknown. This selection of formula may be following what the theory of the considered relationship defines. In general, the technique of finding these parameters is called regression.

When the type of selected formula is simply a straight line, then a linear regression technique is the corresponding terminology. However, with some mathematical manipulation, the linear regression results can also be used for many nonlinear assumed formulas, as we will see throughout this section. Consider an experiment result that is given in a table, as

$$x_i : y_i = f(x_i);\ i = 1,2 \dots m \tag{5.37}$$

where, $(x_{i+1} - x_i)$ is not necessarily equal over the given approximation interval. Besides, during experiments, usually for the same value of the variable x different values of $f(x)$ are obtained. These differences could be significant to be considered in the data table.

We wish to fit these table data by a straight line, which passes through as much as possible. The straight line is simply defined, as

$$y(x) = ax + b \tag{5.38}$$

where a and b are unknown parameters.

One primary method to find these parameters is to apply the technique of the least square error, the error between the table data y_i and the corresponding (for the same x_i) values obtained from the straight-line equation. The results can be set in the following formulas.

$$a = \frac{m\sum_{i=1}^{m} x_i y_i - \sum_{i=1}^{m} x_i \sum_{i=1}^{m} y_i}{m\sum_{i=1}^{m} x_i^2 - \left(\sum_{i=1}^{m} x_i\right)^2} \tag{5.39}$$

And

$$b = \frac{\sum_{i=1}^{m} x_i^2 \sum_{i=1}^{m} y_i - \sum_{i=1}^{m} x_i y_i \sum_{i=1}^{m} x_i}{m\sum_{i=1}^{m} x_i^2 - \left(\sum_{i=1}^{m} x_i\right)^2} \tag{5.40}$$

The linear regression results can be used to fit data that may be governed by nonlinear relation, say of the form $w = f(t)$, such as

$$w = f(t) = k_1 + \frac{k_2}{t}, w = k_1 e^{k_2 t}, w = e^{k_1 + k_2 t}, w = \sqrt{k_1 + k_2 t}, \ldots \text{etc.}$$

For such nonlinear relations, the procedure is to convert first the given data table $[t_i, w_i]$ to another data table $[x_i, y_i]$ such that $x = g(t)$, and $y = h(w)$ where g and h functions are to be determined to obtain the linear relation of the straight line (5.38). Next, the linear regression parameters a and b are calculated using the formulas (5.39 and 5.40). Finally, the nonlinear parameters k_1 and k_2 are determined according to the used transformation.

For example, let us consider that the given data must fit the nonlinear relation $w = k_1 e^{k_2 t}$ and then we construct a new data table as $[x_i = t_i,\ y_i = \ln(w_i)]$, then after finding the linear parameters a, and b, the nonlinear parameters k_1, and k_2 are calculated as $k_1 = e^b$, and $k_2 = a$, since

$$\ln w = \ln k_1 + \ln(e^{k_2 t}) = \ln k_1 + k_2 t \Leftrightarrow y = ax + b$$

Example 5.9.

It is given the following experimental data Table 5.5.

Table 5.5. Input data for example 5.9

t_i	1.20	2.50	3.80	4.66	5.53	6.33	7.68	9.11
w_i	35.9375	6.9851	4.1641	3.4657	3.0505	2.7981	2.4880	2.3319

If it is known that this data fits an empirical nonlinear relation of the form

$$w = k_1 e^{k_2/t}$$

Use the linear regression technique to determine the nonlinear parameters k_1 and k_2.

The transformation to a linear formulation is done as follows:

$$\ln w = \ln k_1 + \ln\left(e^{k_2/t}\right) = \ln k_1 + \frac{k_2}{t} \Leftrightarrow y = ax + b$$

where,

$$y = \ln w, k_1 = e^b, k_2 = a$$

Therefore, we construct a new data table in terms of the two variables $[x_i = t^{-1}, y_i = \ln(w_i)]$ that is suitable for linear regression. Table 5.6 lists the data for the new variables.

Table 5.6. The transformed input data of example 5.9

$x_i = t^{-1}$	0.8333	0.400	0.2632	0.2146	0.1808	0.1580	0.1269	0.1098
$y_i = \ln(w_i)$	3.5818	1.9437	1.4265	1.2429	1.1153	1.0289	0.9115	0.8467

Now we calculate the following four summations:

$$\sum_{i=1}^{9} x_i = 2.2899, \sum_{i=1}^{9} x_i^2 = 1.0564, \sum_{i=1}^{9} x_i y_i = 4.9803, \sum_{i=1}^{9} y_i = 12.0974$$

Using formulas (5.39 and 5.40) gives $a = 3.7948,\ b = 0.4288$. Therefore,

$$k_1 = e^{0.4288} = 1.5354, k_2 = 3.7948$$

Finally, the nonlinear fitting formula has the form

$$w = 1.5354e^{3.7848/t}$$

The minimal value of the square root of the squared error sum is 0.0495. It is worth mentioning that if we wrongly select the fitting formula, then this minimal value will be considerable. For example, if we choose the relation $w = e^{k_1 + k_2 t}$, then the minimum value will be 22.339.

Similarly, a second-order polynomial can be assumed instead of the straight line to formulate a quadratic regression. The second-order polynomial has the form

$$y(x) = ax^2 + bx + c \tag{5.41}$$

For an m-data table, when the least square error method is used, the quadratic regression parameters $a, b,$ and c can be calculated from the following formulas.

$$a = \frac{(s_1s_3 - s_2^2)\sum_{i=1}^{m} y_i + (s_1s_2 - ms_3)\sum_{i=1}^{m} x_iy_i + (ms_2 - s_1^2)\sum_{i=1}^{m} x_i^2y_i}{\Delta} \tag{5.42}$$

$$b = \frac{(s_2s_3 - s_1s_4)\sum_{i=1}^{m} y_i + (ms_4 - s_2^2)\sum_{i=1}^{m} x_iy_i + (s_1s_2 - ms_3)\sum_{i=1}^{m} x_i^2y_i}{\Delta} \tag{5.43}$$

$$c = \frac{(s_2s_4 - s_3^2)\sum_{i=1}^{m} y_i + (s_2s_3 - s_1s_4)\sum_{i=1}^{m} x_iy_i + (s_1s_3 - s_2^2)\sum_{i=1}^{m} x_i^2y_i}{\Delta} \tag{5.44}$$

where,

$$\Delta = m(s_2s_4 - s_3^2) + s_1(s_2s_3 - s_1s_4) + s_2(s_1s_3 - s_2^2) \tag{5.45}$$

$$s_q = \sum_{i=1}^{m} x_i^q \,;\; q = 1,2,3,4 \tag{5.46}$$

Furthermore, several nonlinear relationships can be determined using a suitable transformation using the quadratic regression as with linear regression. For instance, for nonlinear function

$$w = f(t) = \sqrt{k_1t^2 + k_2t + k_3}$$

The given data table $[t_i, w_i]$ can be transformed into $[t_i, w_i^2]$, and for the new data table, we determine the quadratic parameters, which are equal to the nonlinear relation parameters.

Example 5.10.
It is given the following experimental data Table 5.7.

Table 5.7. Input data of example 5.10

t_i	1	1.15	1.25	1.45	1.6	1.75	1.95	2.25	2.4	2.75	3.15
w_i	1.589	1.4999	1.4492	1.3723	1.3386	1.3273	1.3481	1.451	1.5301	1.769	2.1064

If it is thought that the given data fits a theoretical nonlinear relation of the form $w = \sqrt{k_1t^2 + k_2t + k_3}$, use the quadratic regression technique to determine the parameters k_1, k_2, and k_3.

The transformation to quadratic formulation is done as follows:

$$w^2 = k_1t^2 + k_2t + k_3$$

Thus, the transformed table data is performed for the same t data and square of w, as in Table 5.8.

Table 5.8. The transformed input data of example 5.10

t_i	1	1.15	1.25	1.45	1.6	1.75	1.95	2.25	2.4	2.75	3.15
$y_i = w^2$	2.5249	2.2497	2.1003	1.8831	1.7918	1.7618	1.8174	2.1053	2.3412	3.1294	4.4369

Computing according to the formulas (5.42-5.46), we obtain

$$\Delta = 96.0546, s_1 = 20.7, s_2 = 43.72, s_3 = 101.6602, s_4 = 254.4563, s_5 = 26.142, s_6 = 52.9002$$

$$s_7 = 121.,4698, a = 1.3616, b = -4.7604, c = 5.9234$$

In this example, the parameters of the nonlinear relation are the same as the parameters of the quadratic regression, i.e.,

$$k_1 = a = 1.3616, k_2 = b = -4.7604, k_3 = c = 5.9234$$

Therefore, the nonlinear relation is

$$w = \sqrt{1.3616\,t^2 - 4.7604\,t + 5.9234}$$

5.4.3. Interpolation and Finite Differences

Another related issue of numerical approximation is the interpolation technique. For a given data table that represents an unknown function $f(x)$, when for a point x_p that is not included in the table, the function value is required, then we have to interpolate that value from the table. If the polynomial approximation techniques are used first, then the desired function value for an arbitrary point within the approximation interval can be obtained by direct computation.

The simplest way to find $f(x_p)$ is to assume a straight-line relationship between two points of the data table before and after the point x_p; say points x_b, and x_a and their corresponding function values of table $f(x_b)$ and $f(x_a)$. Then we compute

$$f(x_p) = \frac{f(x_a) - f(x_b)}{(x_a - x_b)} x_p + \frac{x_a f(x_b) - x_b f(x_a)}{(x_a - x_b)} \tag{5.47}$$

$$f(x_p) = \begin{bmatrix} x_p^2 & x_p & 1 \end{bmatrix} \begin{bmatrix} x_1^2 & x_1 & 1 \\ x_2^2 & x_2 & 1 \\ x_3^2 & x_3 & 1 \end{bmatrix}^{-1} \begin{bmatrix} f(x_1) \\ f(x_2) \\ f(x_3) \end{bmatrix} \tag{5.48}$$

The mathematical expression (5.47) is called the linear interpolation formula. If the data table does not represent a linear relation, then for an accurate evaluation of a particular interpolated point, three data points $[x_1, f(x_1)]$, $[x_2, f(x_2)]$, and $[x_3, f(x_3)]$, have to be taken, and hence a quadratic interpolation is considered. The computation formula

For m - point data table, if all data points are used, then the computation formula will be

$$f(x_p) = \begin{bmatrix} x_p^{m-1} & \cdots & x_p & 1 \end{bmatrix} \begin{bmatrix} x_1^{m-1} & \cdots & x_1 & 1 \\ x_2^{m-1} & \cdots & x_2 & 1 \\ \vdots & \vdots & \vdots & \vdots \\ x_m^{m-1} & \cdots & x_m & 1 \end{bmatrix}^{-1} \begin{bmatrix} f(x_1) \\ f(x_2) \\ \vdots \\ f(x_m) \end{bmatrix} \tag{5.49}$$

However, if the data table represents a polynomial of the order less than m, it will not be necessary to use the formula (5.49). It will be enough to take only several points equal to the polynomial order plus one. For instance, in formula (5.48), we use three points for the second-order polynomial. This arrangement will avoid the unnecessary computation of a larger dimension inverse matrix. On the other hand, if the data table does not represent a polynomial but a general function, the accuracy improves as the number of points increases. However, in this case, the right order, which gives the highest accuracy, is unknown. Therefore, some simulation runs will be necessary.

An alternative way of interpolation computation is to use what is called the differences interpolation formulas. To illustrate the concept of the differences, let us consider the given information in Table 5.9. The differences D_{ij} are the divided differences, and δ_j are the interpolation differences. In Table 5.9, the final divided differences D_{14} is itself the interpolation differences δ_4, which is given by

$$D_{14} = \frac{D_{23} - D_{13}}{x_4 - x_0} \tag{5.50}$$

If the table of these five data points $(m = 5)$ represents a fourth-order polynomial p_4, then the difference D_{14} should be equal to zero; otherwise, the data table represents a general unknown function $f(x)$.

Several different interpolation formulas utilize these divided differences (or the interpolation differences) in a way or another to compute an interpolated point. Some of these formulas are called forward, which use the differences $D_{1j}, j = 1,2 \ldots m-1$, others are named backward, which use the differences $D_{4j}, j = 1,2..m-1$. Besides, other formulas use differences in a zigzag way or what are called centered finite divided differences. We will introduce only one of the forward formula types, named Newton's divided differences interpolation formula. It is given as

Table 5.9. Concept of differences

x_i	$f(x_i)$	Differences δ_1	Differences δ_2	Differences δ_3	δ_4
x_0	$f(x_0)$				
		$D_{11} = [f(x_1) - f(x_0)]/(x_1 - x_0)$	$D_{12} = \dfrac{D_{21} - D_{11}}{x_2 - x_0}$		
x_1	$f(x_1)$			$D_{13} = \dfrac{D_{22} - D_{12}}{x_3 - x_0}$	
		$D_{21} = [f(x_2) - f(x_1)]/(x_2 - x_1)$	$D_{22} = \dfrac{D_{31} - D_{21}}{x_3 - x_1}$		
x_2	$f(x_2)$			$D_{23} = \dfrac{D_{32} - D_{22}}{x_4 - x_1}$	D_{14}
		$D_{31} = [f(x_3) - f(x_2)]/(x_3 - x_2)$	$D_{32} = \dfrac{D_{41} - D_{31}}{x_4 - x_2}$		
x_3	$f(x_3)$				
		$D_{41} = [f(x_4) - f(x_3)]/(x_4 - x_3)$			
x_4	$f(x_4)$				

$$f(x_p) = f(x_0) + (x_p - x_0)D_{11} + (x_p - x_0)(x_p - x_1)D_{12} + \cdots$$
$$+(x_p - x_0)(x_p - x_1)(x_p - x_2) \ldots (x_p - x_{m-1})D_{m-1} \tag{5.51}$$

Moreover, if we set in formula (5.51) the variable x instead of x_p, then we will get a polynomial of $(m-1)$ order that approximates the function $f(x)$ when the difference $D_{1,m-1}$ is not equal to zero. However, if $D_{1,k}$ is the number of columns (where $k \leq m-1$), in which all differences are equal, then the considered data table represents a polynomial of order $(k-1)$. Formula (5.51) is suitable for calculating the interpolation function values near the beginning of the table.

Example 5.11

Given five-point data in Table 5.10 and it is required to determine the interpolated function value $f(1.35)$.

Table 5.10. Numerical data for example 5.11

x_i	0	1	2	2.5	3
$f(x_i)$	−4	1	10	16	23

First, we use all five points and substituting in formula (5.49); we set $x_p = 1.35$ to get

$$f(1.35) = [1.35^4 \quad 1.35^3 \quad 1.35^2 \quad 1.35 \quad 1]\begin{bmatrix} 0 & 0 & 0 & 0 & 1 \\ 1.0^4 & 1.0^3 & 1.0^2 & 1 & 1 \\ 2^4 & 2^3 & 2^2 & 2 & 1 \\ 2.5^4 & 2.5^3 & 2.5^2 & 2.5 & 1 \\ 3^4 & 3^3 & 3^2 & 3 & 1 \end{bmatrix}^{-1}\begin{bmatrix} -4 \\ 1 \\ 10 \\ 16 \\ 23 \end{bmatrix} = 3.695$$

Next, we construct the differences as in Table 5.11.

Table 5.11. Differences of example 5.11

x_i	$f(x_i)$	Differences δ_1	Differences δ_2	Differences δ_3	δ_4
0	−4				
		$D_{11} = [1+4]/(1-0) = 5$			
1	1		$D_{12} = \frac{9-5}{2-0} = 2$	$D_{13} = 0$	
		$D_{21} = [10-1]/(2-1) = 9$			
2	10		$D_{22} = \frac{12-9}{2.5-1} = 2$	$D_{23} = 0$	$D_{14} = 0$
		$D_{31} = [16-10]/(2.5-2) = 12$			
2.5	16		$D_{32} = \frac{14-12}{3-2} = 2$		
		$D_{41} = [23-16]/(3-2.5) = 14$			
3	23				

Since $\delta_3 = 0$ then the given data table represents a polynomial of second-order; in fact, the fifth point is not necessary to determine the polynomial order and use all five points to apply formula 5.49. In this example, it is enough to consider only four points. Using Newton divided differences formula (5.51) gives

$$f(1.35) = -4 + (1.35-0)5 + (1.35-0)(1.35-1)2 = 3.695$$

It is the same answer obtained by formula (5.49). Furthermore, if we use formula (5.51) with the variable x, then we obtain the second-order polynomial as

$$p_2(x) = 2x^2 + 3x - 4 = (2x+3)x - 4$$

For further interpolation evaluation and by considering the speed computation, it is faster to compute by direct evaluation using the polynomial expression than to use Newton's formula. It is desired for the former to carry two multiplications plus two algebraic additions, while for the latter, it needs three multiplications and five algebraic additions.

Example 5.12.

Assume the function $f(x) = 8x^2 \sin x - 2$. It is required to approximate the function as a polynomial by considering equidistance 11 points and applying Newton's formula.

For this purpose, we select the interval [0, 1] and a step of 0.1 to calculate the function values. Next, we determine the differences in Table 5.12 (short format is used to show the table values; actual calculations were long formatted). The final difference is

$$\delta_{10} = \frac{-0.0012 - (-0.0013)}{1-0} = 0.0001$$

Table 5.12. Differences for example 5.12

x_i	$f(x_i)$	δ_1	δ_2	δ_3	δ_4	δ_5	δ_6	δ_7	δ_8	δ_9
0	−2									
		0.0799								
0.1	−1.992		2.38							
		0.556		7.6687						
0.2	-1.9364		4.6806		−1.3101					
		1.492		7.1446		−1.2411				
0.3	−1.7872		6.824		−1.9307		0.1358			
		2.857		6.3725		−1.1596		0.0595		
0.4	−1.5015		8.7537		−2.5105		0.1775		−0.0055	
		4.604		5.3682		−1.0531		0.551		−0.0013
0.5	−1.0411		10.3462		−3.037		0.216		0.0067	
		6.673		4.1534		−0.9235		0.0498		−0.0012
0.6	−0.3738		11.5922		−3.488		0.2509		−0.0078	
		8.992		2.7538		−0.7729		0.0436		
0.7	0.5253		12.4183		−3.8852		0.2814			
		11.475		1.1997		−0.6041				
0.8	1.6729		12.7783		−4.1873					
		14.031		−0.4752						
0.9	3.076		12.6357							
		16.558								
1	4.7318									

Since $\delta_{10} \neq 0$, the given function can be approximated by a polynomial of order 10. Applying the Newton formula to get

$$p_{10}(x) = -2 + 0.0799x + 2.38x(x-0.1) + 7.6687x(x-0.1)(x-0.2)$$

$$-1.3101x(x-0.1)(x-0.2)(x-0.3) + \cdots + 0.0001x(x-0.1)\ldots(x-0.9)$$

$$p_{10}(x) = 1.1353 \times 10^{-4}x^{10} - 1.8506 \times 10^{-3}x^9 + 3.5481 \times 10^{-4}x^8 + 6.6362 \times 10^{-2}x^7 +$$

$$1.7235 \times 10^{-4}x^6 - 1.3333x^5 + 1.588 \times 10^{-5}x^4 + 7.999x^3 + 1.994 \times 10^{-7}x^2 - 6.794 \times 10^{-9}x - 2$$

If, for example, we compute $p_{10}(0.5324) = -0.8574$, while for the given function, we obtain $f(0.5324) = -0.8490$, thus the absolute error is 0.0085. This 10th order polynomial approximates the given $f(x)$ with an accuracy of better than 10^{-2} over the interval $[0, 1]$. Of course, the accuracy decreases dramatically outside this interval.

5.4.4. Lagrange Interpolation Formula

Another known technique of a function approximation is named after the French scientist Joseph Lagrange. In the interval $[a, b]$, the function values of the function $f(x)$

in non-equal spacing $(n+1)$ points, say $x_0, x_1 \dots x_n$, are sufficient to obtain an approximated polynomial $p_n(x)$ using the following formula.

$$f(x) \cong p_n(x) = \sum_{i=0}^{n} l_i(x) f(x_i) \tag{5.52}$$

where l_i are the Lagrange coefficients given by

$$l_i(x) = \prod_{\substack{0 \le m \le n \\ m \ne i}} \frac{x - x_m}{x_i - x_m} \tag{5.53}$$

$$l_i(x) = \frac{(x - x_0)(x - x_1)(x - x_2) \dots (x - x_{i-1})(x - x_{i+1}) \dots (x - x_{n-1})(x - x_n)}{(x_i - x_0)(x_i - x_1)(x_i - x_2) \dots (x_i - x_{i-1})(x_i - x_{i+1}) \dots (x_i - x_{n-1})(x_i - x_n)}$$

The distances between the considered points are not necessarily equal; however, to simplify the calculation, equal distances h are preferable. Therefore, if $x_0 = a$ and $x_{n+1} = b$, then

$$x_i = x_{i-1} + h;\ i = 1,2 \dots n+1;\ h = (b-a)/n$$

Example 5.13.

In a specific electromagnetic configuration, an exerted force F on an electric charge as a function of distance d is given by

$$F(d) = \frac{6.5631\, d}{(d^2 + 0.64)^{\frac{3}{2}}};\ d \in [1,2]$$

Use the Lagrange form to determine a third-order polynomial that approximates the exerted force.

The following equidistance ($h = (2-1)/3 = 1/3$ points are $(1, 4/3, 5/3, 2)$, and the corresponding function values are $3.125, 2.3277, 1.7312, 1.3133$, respectively. Using the formula (5.53) gives

$$l_0(d) = \frac{\left(d - \frac{4}{3}\right)\left(d - \frac{5}{3}\right)(d-2)}{\left(1 - \frac{4}{3}\right)\left(1 - \frac{5}{3}\right)(1-2)} = -4.5d^3 + 22.5d^2 - 37d + 20$$

$$l_1(d) = \frac{(d-1)\left(d - \frac{5}{3}\right)(d-2)}{\left(\frac{4}{3} - 1\right)\left(\frac{4}{3} - \frac{5}{3}\right)\left(\frac{4}{3} - 2\right)} = 13.5d^3 - 63d^2 + 94.5d - 45$$

$$l_2(d) = \frac{(d-1)\left(d-\frac{4}{3}\right)(d-2)}{\left(\frac{5}{3}-1\right)\left(\frac{5}{3}-\frac{4}{3}\right)\left(\frac{5}{3}-2\right)} = -13.5d^3 + 58.5d^2 - 81d + 36$$

$$l_3(d) = \frac{(d-1)\left(d-\frac{4}{3}\right)(d-\frac{5}{3})}{(2-1)\left(2-\frac{4}{3}\right)(2-\frac{5}{3})} = 4.5d^3 - 18d^2 + 23.5d - 10$$

Applying formula (5.52) gives the desired third-order polynomial as

$$F(d) \cong p_3(d) = -0.0997d^3 + 1.3023d^2 - 5.0206\, d + 6.943$$

Before ending this section, we would like to mention that we can solve the problem of fitting experimental data or approximating a polynomial of a selected order by invoking the MATLAB function ***polyfit*** and evaluating by the function ***polyval***. Eventually, all the mentioned techniques and methods of approximation can be easily programmed by MATLAB code.

5.5. Numerical Differentiation

Because of the availability of differentiation rules, the exact closed-form solution of the differentiation is always possible. However, there are at least two reasons to use numerical differentiation. First, some functions are so complicated that applying the rules is tedious, especially when the computation is required at a particular point within a given interval. The second reason is when the derivatives are necessary for programmed problems. There are two ways of computing the derivatives numerically. The first is to approximate the function by a suitable order polynomial, as shown previously. The second way is to use a specific number of function values to derive formulas of computation. In the second way, when these function values are computed in correspondence to equidistance points, the computation formulas are utilized efficiently. However, these two ways are overlapped.

Consider a given function $f(x)$, and it is required to compute $f'(x_0)$ and let the approximated polynomial of the function be as given in equation (5.26), then

$$f'(x_0) = \sum_{i=1}^{n} (n-i+1)a_{n-i+1}x_0^{n-i} \tag{5.54}$$

Formula (5.54) can be programmed easily. In this case, the point x_0 should belong to the interval of approximation. It is recommended to select a relatively small interval, around the point x_0. However, when the derivatives are required at different points, the interval should contain all these points. Similarly, we can compute the second derivative from

$$f''(x_0) = \sum_{i=1}^{n} (n-i+1)(n-i)a_{n-i+1}x_0^{n-i-1} \tag{5.55}$$

Example 5.14.

The distance d of a moving object as a function of time is given by

$$d(t) = \frac{e^t}{1+\sin t}$$

Compute the velocity and acceleration after 0.5 seconds.

It is known that the velocity $v(t)$ and acceleration $a(t)$ are the first $d'(t)$ and second derivatives $d''(t)$, respectively. Instead of applying differentiation rules, let us approximate the distance function as a 4th – order polynomial (which can be determined by any previously mentioned techniques, say simple interpolation technique).

$$p_4(t) = 0.104t^4 - 0.1169t^3 + 0.4879t^2 + 0.0011t + 1$$

We apply the summation 5.52

$$v(0.5) = \sum_{i=1}^{4} (5-i)a_{5-i}0.5^{4-i}$$
$$= 4(0.104)0.5^3 + 3(-0.1169)0.5^2 + 2(0.4879)0.5 + 0.0011$$
$$= 0.453325$$

$$a(0.5) = \sum_{i=1}^{4} (5-i)(4-i)a_{5-i}0.5^{4-i-1} = 0.9371$$

The computation from the exact form of the first derivative is

$$v(0.5) = d'(0.5) = \frac{[(1+\sin 0.5) - \cos 0.5]e^{0.5}}{(1+\sin 0.5)^2} = 0.453361$$

The exact value of the second derivative is

$$a(0.5) = d''(0.5) = 0.9377$$

Therefore, for the first derivative, the accuracy is better than 10^{-4}, while for the second derivative, the accuracy decreases to 10^{-3}. The reader certainly knows that for the second derivative, the omitted exact form is very complicated.

Another way to calculate the derivatives at a given point is to use the Lagrange interpolation formula with equidistance points to formulate an expression for calculation. To simplify the derivation of such formulas, let us consider three equidistance points $x_0, x_1, x_2,$ and their corresponding functional values $f(x_0), f(x_1), f(x_2)$ that represent data for a second-order polynomial approximation, then substituting in Lagrange formula yields

$$p_2(x) = \frac{(x-x_1)(x-x_2)}{(x_0-x_1)(x_0-x_2)}f(x_0) + \frac{(x-x_0)(x-x_2)}{(x_1-x_0)(x_1-x_2)}f(x_1) + \frac{(x-x_0)(x-x_1)}{(x_2-x_0)(x_2-x_1)}f(x_2)$$

$$p_2(x) = \frac{x^2-(x_1+x_2)x+x_1x_2}{(-h)(-2h)}f(x_0) + \frac{x^2-(x_0+x_2)x+x_0x_2}{(h)(-h)}f(x_1) + \frac{x^2-(x_0+x_1)x+x_0x_1}{(2h)(h)}f(x_2) \qquad (5.56)$$

The first derivative is

$$p_1'(x) = \frac{2x-(x_1+x_2)}{2h^2}f(x_0) - \frac{2x-(x_0+x_2)}{h^2}f(x_1) + \frac{2x-(x_0+x_1)}{2h^2}f(x_2) \qquad (5.57)$$

Now let us assume that it is required to calculate the $f'(x_0)$ or approximately the $p_1'(x_0)$ then substituting in the expression (5.57), we obtain

$$p_1'(x_0) = \frac{2x_0-(x_1+x_2)}{2h^2}f(x_0) - \frac{2x_0-(x_0+x_2)}{h^2}f(x_1) + \frac{2x_0-(x_0+x_1)}{2h^2}f(x_2)$$

$$= \frac{(x_0-x_1)+(x_0-x_2)}{2h^2}f(x_0) - \frac{(x_0-x_2)}{h^2}f(x_1) + \frac{(x_0-x_1)}{2h^2}f(x_2)$$

$$= \frac{(-h)+(-2h)}{2h^2}f(x_0) - \frac{(-2h)}{h^2}f(x_1) + \frac{(-h)}{2h^2}f(x_2)$$

$$p_1'(x_0) = -\frac{3}{2h}f(x_0) + \frac{2}{h}f(x_1) - \frac{1}{2h}f(x_2) \qquad (5.58)$$

In formula (5.58), let us denote $f_0, f_1,$ and f_2 instead of $f(x_0), f(x_1),$ and $f(x_2)$, respectively and rearrange to get a three points formula of calculating the first derivative at the first point.

$$p_1'(x_0) = \frac{1}{2h}[-3f_0 + 4f_1 - f_2] \tag{5.59}$$

Similarly, we can obtain formulas to calculate the first derivative at the other two points as

$$p_1'(x_1) = \frac{1}{2h}[-f_0 + 0f_1 + f_2] \tag{5.60}$$

$$p_1'(x_2) = \frac{1}{2h}[f_0 - 4f_1 + 3f_2] \tag{5.61}$$

Instead of the Lagrange interpolation formula, finite divided differences can also be used for calculating the derivatives of a given function at a given point. After choosing a suitable interval that includes the considered point, the evaluation can be done in two steps, first to determine the approximated polynomial and then to use formulas (5.54 and 5.55). The number of considered points should not be large but small equidistance is recommended.

Example 5.15.

Use three function values to evaluate $[x \sin\sqrt{x}]'$ at the point $x = 0.3$.
Before we do the numerical calculation, the exact first derivative is equal to

$$f'(x) = \frac{1}{2}\sqrt{x}\cos\sqrt{x} + \sin\sqrt{x};\ f'(0.3) = 0.7545$$

Let us first select the following three points $x_0 = 0.3, x_1 = 0.6, x_2 = 0.9$, i.e., $h = 0.3$ then the corresponding function values are $f(x_0) = 0.1562, f(x_1) = 0.4197$, and $f(x_2) = 0.7314$. To evaluate the first derivative at $x_0 = 0.3$, we use the formula (5.59).

$$[x \sin\sqrt{x}]' |_{x=0.3} \cong p_1'(0.3) = \frac{1}{2(0.3)}[-3(0.1562) + 4(0.4197) - (0.7314)] = 0.798$$

The absolute error of the numerical calculation is 0.0435. This error is a function of the distance h between the considered points, and it will be reduced significantly as h decreases. For example, with $h = 0.1$, the absolute error will be reduced to 8.5×10^{-3}.

Next, let us select another three points as $x_0 = 0, x_1 = 0.3, x_2 = 0.6$, then using formula (5.60) gives

$$[x \sin\sqrt{x}]' |_{x=0.3} \cong p_1'(0.3) = \frac{1}{2(0.3)}[-(0) + 0(0.1562) + (0.4197)] = 0.6995$$

For this selection of points, the absolute error is 0.055. Therefore, an accurate result can be obtained by decreasing the distance h and selecting the point at which the derivative is calculated, i.e., the calculation formula.

The accuracy also increases as the number of the considered points increases. There are derived formulas in the literature to calculate the derivatives when five, seven, etc. points are considered. Moreover, the formulas are extended to include higher derivatives calculations and at every considered point. Here, we provide as an example a formula that uses five points to calculate the first derivative at the middle point (like the formula (5.60), but it is derived from the center finite divided difference technique); it is

$$f'(x = x_2) \cong p_3'(x_2) = \frac{1}{12h}[f_0 - 8f_1 + 0f_2 + 8f_3 - f_4] \tag{5.62}$$

Example 5.16.

Let us evaluate the derivative of the previous example again by considering five points with $h = 0.1$. Then the data in Table 5.13 is computed for the function $f(x) = x \sin\sqrt{x}$.

Table 5.13. Computed data for example 5.16

x_i	0.1	0.2	0.3	0.4	0.5
$f(x_i)$	0.0865	0.1199	–	0.1952	0.2365

Substituting in formula (5.62) gives

$$[x \sin\sqrt{x}]' |_{x=0.3} \cong p_3'(0.3) = \frac{1}{12(0.1)}[0.0865 - 8(0.1199) + 8(0.1952) - 0.2365]$$
$$= 0.7552$$

The absolute error is 7×10^{-4}; therefore, the accuracy has been improved.

Example 5.17.

Use the technique of five points finite differences to evaluate $f'(x)$ at the point $x = 1.5$, where

$$f(x) = \frac{3\cos(\ln x)}{\sqrt{x^2+5}}$$

Let us select the approximation interval to be $[1,2]$; therefore, for the five points, the equidistance will be $h = 0.25$. Table 5.14 gives the finite difference table. Using Newton's interpolation formula, the following fourth-order polynomial that approximates the given function can be found as

Table 5.14. Differences table for example 5.17

x_i	$f(x_i)$	δ_1	δ_2	δ_3	δ_4
1	1.2247				
		$D_{11} = -0.3308$			
1.25	1.1420		$D_{12} = -0.2841$		
		$D_{21} = -0.4728$		$D_{13} = 0.2695$	
1.5	1.0238		$D_{22} = -0.0820$		$D_{14} = -0.1354$
		$D_{31} = -0.5138$		$D_{23} = 0.1314$	
1.75	0.8954		$D_{32} = 0.0186$		
		$D_{41} = -0.5045$			
2	0.7692				

$$p_4(x) = -0.1354x^4 + 1.0142x^3 - 2.8095x^2 + 2.9046x + 0.2508$$

Finally, formula (5.54) to obtain

$$f'(1.5) \cong \sum_{i=1}^{4}(5-i)a_{5-i}1.5^{4-i}$$

$$= 4(-0.1354)1.5^3 + 3(1.0142)1.5^2 + 2(-2.8095)1.5 + 2.9046$$

$$f'(1.5) \cong -0.50595$$

If we use formula (5.62), then we get

$$f'(x=1.5) \cong p_3'(1.5) = \frac{1}{12(.25)}[1.2247 - 8(1.142) + 8(0.8954) - 0.7692]$$

$$f'(x=1.5) \cong p_3'(1.5) = -0.50577$$

The two results are almost the same.

5.6. Numerical Integration

It is known that not all integrals have an explicit form of solutions, and hence, numerical evaluation is the alternative way. In such cases, the evaluation will be an approximate value that depends on the used numerical method. A general definition of numerical integration is stated as

$$\int_a^b f(x)\,dx \cong \sum_{i=0}^{n} w_i f(x_i) \tag{5.63}$$

where, w_i are properly chosen weights, and $f(x_i)$ are the function values at a specific number of points x_i lying between the integration limits.

The first direct numeric approach for calculating an integral is to determine an approximated polynomial of the integrand $f(x)$ and then use for each term of the polynomial the following known rule.

$$\int_a^b x^n\,dx = \frac{x^{n+1}}{(n+1)}\Big|_a^b \tag{5.64}$$

For this purpose, we can apply any of the previously discussed methods of polynomial approximations. The accuracy of this approach is a result of the most fitting polynomial to the integrand $f(x)$. This section will consider this approach in the given examples to compare different ways of calculation.

Another approach depends on the way of approximating the area under the curve of the integrand. Let us first consider a more straightforward way, which resembles the whole area between the limits. Figure 5.3 illustrates possible cases of approximations. Figure 5.3a shows the total area under the curve that is equal exactly to the integration value, while in Figure 5.3b, the integration is approximated by a rectangular area, or

$$\int_a^b f(x)\,dx \cong h[f(a) - f(b)];\ h = b - a \tag{5.65}$$

Figure 5.3c illustrates another method of computation named the midpoint rule, which approximates the area by a rectangle with height equal to the function value at the midpoint, i.e.,

$$\int_a^b f(x)\,dx \cong h\left[f\left(a + \frac{h}{2}\right)\right] \tag{5.66}$$

This category's last method is to approximate by a trapezoidal area, as shown in Figure 5.2d. The computation based on this trapezoidal rule is given by

$$\int_a^b f(x)\,dx \cong \frac{h}{2}[f(a) + f(b)] \tag{5.67}$$

Figure 5.3 shows that the rectangular rule is a bad approximation, while the integration of the middle point rule is a good approximation. For the trapezoidal, the rule is a moderated approximation. It is not a strict rule for measuring these methods' accuracy; in other cases, this accuracy ranking order may be altered. However, unless the integration step h is small enough, these rules' accuracy is always questionable. Moreover, in some examples, the mentioned method gives serious errors. The problem of measuring computation accuracy becomes more tedious when the integration has no explicit form of solution.

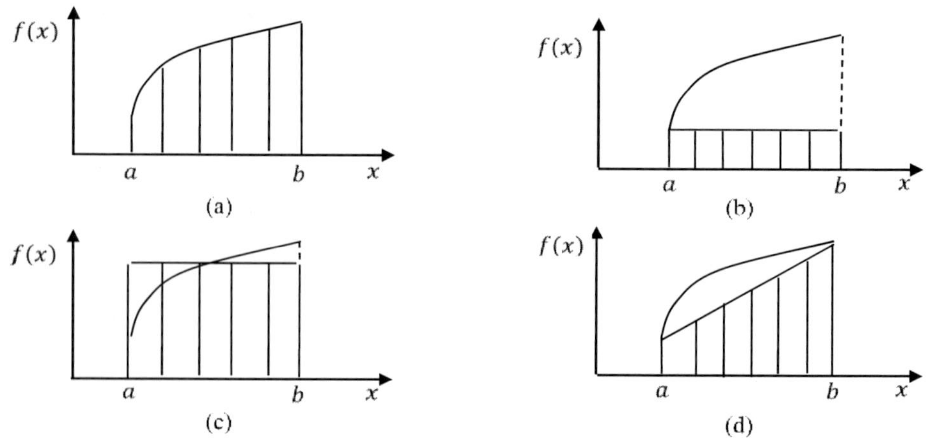

Figure 5.3. Methods of integration.

Example 5.18.

Use the mentioned integration rules to evaluate the integral

$$\int_{-1}^{1} \frac{1}{1+x^2}\, dx$$

The integral has an explicit solution form, and for the given limit, the value of integration is $\pi/2$.

Over the given limit, Figure 5.4 shows a plot of the integrand.

We have $a = -1, b = 1, h = 2, f(a) = f(b) = 0.5, f(0) = 1$. The three methods give the following results:

- The rectangular rule gives: $2\ (0.5 - 0.5) = 0$
- Midpoint rule gives: $2\ f(0) = 2 \times 1 = 2$
- Trapezoidal rule give: $1\ (0.5 + 0.5) = 1$

Therefore, all three methods give seriously erroneous results. This conclusion is reached because we know the exact value of integration. What to do if the exact value is not known? We must refer to an approximated value obtained by an approximated polynomial (i.e., using an approximation technique and equation (5.64)).

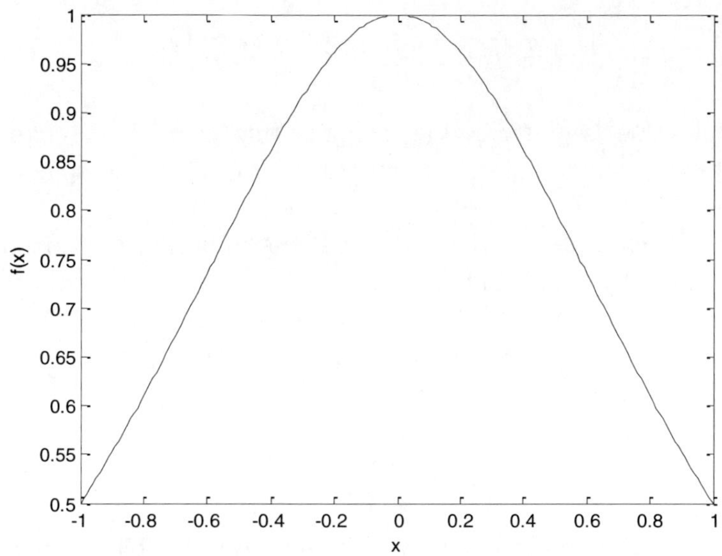

Figure 5.4. The integrand function (Example 5.18).

Example 5.19.

Evaluate the integral

$$\int_0^1 \frac{e^x}{1+\sin x}\,dx$$

This integral has no explicit form of solution.

We have $a = 0, b = 1, h = 1, f(a) = 1, f(b) = 1.4761, f(0.5) = 1.1144$. The three methods give

- Rectangular gives: $1(0 - 1.4761) = -1.4761$
- Midpoint rule gives: $1 \times f(0.5) = 1.1144$
- The trapezoidal rule gives: $0.5(1 + 1.4761) = 1.23805$

The question is which of these values we should take. Next, let us approximate the function by a 4th – order polynomial, using any previously mentioned methods. If we use the simple interpolation technique (not discussed throughout this section), then we obtain

$$p_4(x) = 0.104x^4 - 0.1169x^3 + 0.4879x^2 + 0.0011x + 1$$

As shown in Figure 5.5, the approximated 4th – order polynomial is sufficiently accurate for calculating the area under the $f(x)$ curve.

Applying equation (5.64) yields

$$\int_0^1 \frac{e^x}{1+\sin x}\,dx \cong \int_0^1 p_4(x)\,dx = 1.1548$$

The nearest value is given by the midpoint rule with an absolute error equal to 0.04, then by the trapezoidal rule of absolute error of 0.08. The result of the rectangular rule is rejected.

Another two famous methods of numerical integration are Simpson's method and what is called the three-eight rule. These two methods are based on quadratic and cubic interpolation polynomials, respectively. Simpson's method used three equidistance points and the function values at these points, i.e.,

$$\int_a^{a+2h} f(x)\,dx \cong \frac{h}{3}[f(a) + 4f(a+h) + f(a+2h)] \tag{5.68}$$

For the three-eight rule, we calculate similarly from the following formula

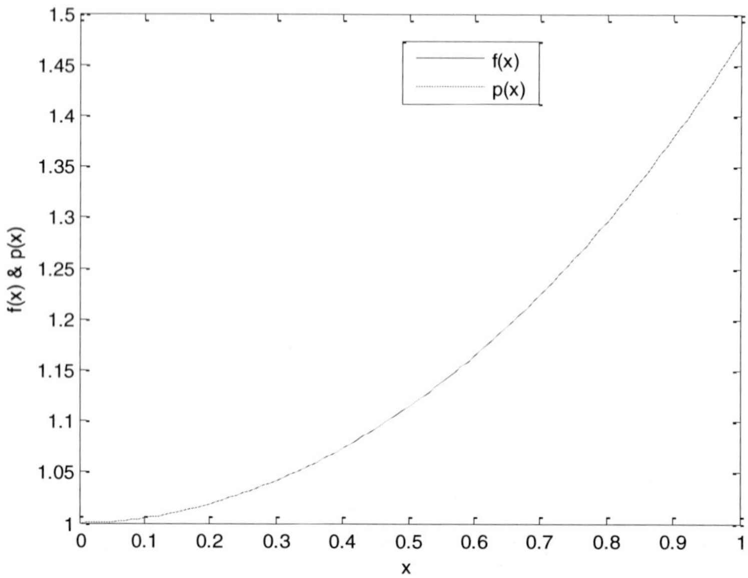

Figure 5.5. Integration solutions (Example 5.19).

$$\int_a^{a+3h} f(x)\,dx \cong \frac{3h}{8}[f(a) + 3f(a+h) + 3f(a+2h) + f(a+3h)] \tag{5.69}$$

For engineers, integration evaluation is one of the most frequent needs. There is a wide range of applications such as evaluating the distance of moving bodies, area of surfaces or volumes, a center of mass of beams or plates, energy and power, electric current or voltage, etc.

Example 5.20.

It is to find the center of mass of a beam of one-meter length and, which has a density function $\rho(x) = x\cos x$. The center of mass is calculated by

$$\bar{x} = \frac{\text{the moment of the beam around zero}}{\text{the total mass of the beam}} = \frac{\int_{x_0}^{x_1} x\rho(x)dx}{\int_{x_0}^{x_1} \rho(x)dx}$$

If the beam's total mass is equal to 0.3818 Kilograms, Use the three-eight rule to calculate the beam's moment around zero.

In our example, $x_0 = 0$, and $x_1 = 1$. Then, we must calculate the integral $\int_0^1 x^2 \cos x \ dx$

This integral has an explicit solution form, and for the given limit, the value of integration is 0.2391; however, we calculate numerically. For the three-eight rule, we consider the three points $x_0 = 0, x_1 = 1/3, x_2 = 2/3, x_3 = 1$, and their function values $f(0) = 0, f(1/3) = 0.105, f(2/3) = 0.3493, f(1) = 0.5403$, and then applying formula (5.69) gives

$$\int_0^1 x^2 \cos x \ dx \cong 3\frac{1/3}{8}[0 + 3(0.105) + 3(0.3493) + 0.5403] = 0.2379$$

Therefore, the center of mass of the beam is

$$\bar{x} = \frac{0.2379}{0.3818} = 0.6231$$

This result means that the center of mass is 62.31 centimeters from the left of the beam. In other words, if a fulcrum is located at this position, then the beam will be balanced horizontally.

One basic idea of improving the numerical calculation of an integral is to divide the given interval (limit of integration) into a finite number of subintervals. Then calculate individually for each subinterval using an applicable rule, and finally sum the results. In this category of methods, we introduce two techniques for numerical calculation. The first one is called the multiple-trapezoidal rule MTR, and the second one is the multiple Simpsons one-third method MS$\frac{1}{3}$M.

For the MTR, the interval of the limit $[a, b]$ is divided into a finite number of equal subintervals (segments n) such that

$$x_i - x_{i-1} = h; \ i = 1,2 \dots n, h = \frac{b-a}{n} \tag{5.70}$$

where n is any positive integer number. Next, we apply the trapezoidal rule to each segment. The result can be obtained from the following formula.

$$\int_a^b f(x)\,dx \cong \frac{h}{2}\left[f_0 + 2\sum_{i=1}^{n-1} f_i + f_n\right] \tag{5.71}$$

where, $f_i = f(x_i)$ are the function values at the points x_i .

Similarly, for the MS $\frac{1}{3}$ M, however, the number of segments n should be a positive even integer number. The following formula can obtain the result of the numerical integration.

$$\int_a^b f(x)\,dx \cong \frac{h}{2}\left[f_0 + 4\sum_{i=1,3,5..}^{n-1} f_i + 2\sum_{i=2,4,6..}^{n-2} f_i + f_n\right] \tag{5.72}$$

Simplified estimates expressions for the local truncation error of the MTR and MS $\frac{1}{3}$ M is found in the literature as

$$e_{MTR} = -\frac{(b-a)^3}{12n^2} f'(x)|_a^b \tag{5.73}$$

$$e_{MS\frac{1}{3}M} = -\frac{(b-a)^4}{180n^4} f'''(x)|_a^b \tag{5.74}$$

Despite the computation burden of the error estimation, the expressions (5.73 and 5.74) are useful, especially when the integration has no explicit solution. Therefore, a comparison between the two numerical solutions becomes available.

Example 5.21.

Use both MTR and MS $\frac{1}{3}$ M to evaluate the integral

$$\int_0^6 \frac{x^2 \sin x}{4 + \cos x}\,dx$$

The integral has no explicit form of solution. Estimate the local truncation errors for both methods.

To evaluate using both methods, we select an even number of segments $n = 8$; thus, the segments are of width $h = (6-0)/8 = 0.75$. Therefore, we can prepare the value Table 5.15 of the integrand function for the given integration limits.

Table 5.15. Nine computed data of example 5.21

x_i	0	0.75	1.5	2.25	3	3.75	4.5	5.25	6
$f(x_i)$	0	0.081	0.5513	1.1682	0.422	−2.528	−5.224	−5.2469	−2.0279

To have a computation reference, let us approximate the integrand by an 8th – order polynomial using Lagrange polynomial. The result is

$$p_8(x) = 0.0035x^8 - 0.0835x^7 + 0.7961x^6 - 3.8651x^5 + 10.157x^4 - 14.5272x^3 + 10.8585x^2 - 3.1008x$$

Figure 5.6 shows the plot of the given integrand function $f(x)$ and the eighth-order polynomial. The area under the polynomial curve differs from that under the integrand function. However, we will not calculate the area under the curve but only apply the integration's numerical methods. The integral value of the approximate polynomial is equal to -9.3701. Next, using formulas (5.71 and 5.72) gives respectively:

$$\text{MTR:} \cong \frac{0.75}{2}[0.081 + 2(0.5513 + 1.1682 + 0.422 - 2.528 - 5.224 - 5.2469) - 2.0279] = -8.8428$$

$$\text{MS}\frac{1}{3}\text{M:} \cong \frac{0.75}{3}[0.081 + 4(0.5513 + 0.422 - 5.224) + 2(1.1682 - 2.528 - 5.2469) - 2.0279] = -9.1580$$

Now, using the error expressions (5.73 and 5.74) to estimate the truncation error (although it is tedious, we can compute analytically both $f'(x)|_a^b$ and $f'''(x)|_a^b$).

$$e_{MTR} = -\frac{(6-0)^3}{12(8^2)}1.0678 = 0.3003$$

$$e_{MS\frac{1}{3}M} = -\frac{(6-0)^4}{180(8)^4}(-0.3822) = 0.004$$

Since the estimate of the truncation error obtained by MS$\frac{1}{3}$M is less than that obtained by MTR; we will consider that the integration value -9.158 is the more accurate. This value is also closer to the value obtained by integrating the approximated polynomial. In fact, with a larger number of segments, say 60, the MTR gives an integration value is -9.13875. Most software packages use the MTR with a large number of segments.

The previous ways of computing integrals use constant segment length, i.e., $h =$ constant. However, when the given n data of function values is of unequal segment lengths, $h_i, i = 1,2..n-1$, we have to compute according to the following two cases:

1. If all segments are of non-equal lengths, then we use the trapezoidal rule for each segment and add the results of the $n-1$ segments, i.e.,

$$\int_a^b f(x)\,dx \cong \frac{1}{2}\sum_{i=0}^{n-1} h_i[f(x_i) + f(x_{i+1})];\ h_i = x_{i+1} - x_i;\ \sum_{i=1}^{n-1} h_i = b - a \qquad (5.75)$$

2. When there are two or three adjacent segments of equal length distributed between some non-equal segments, we compute using Simpson's rule formula (5.68) or formula (5.69) for every two or three equal segments, respectively, and trapezoidal rule for others. The individual computations are added to give the area's numerical value under the given function value curve.

Note that the computation of case 1 can be used even if we have a data table described in case 2; however, computation accuracy will be lower.

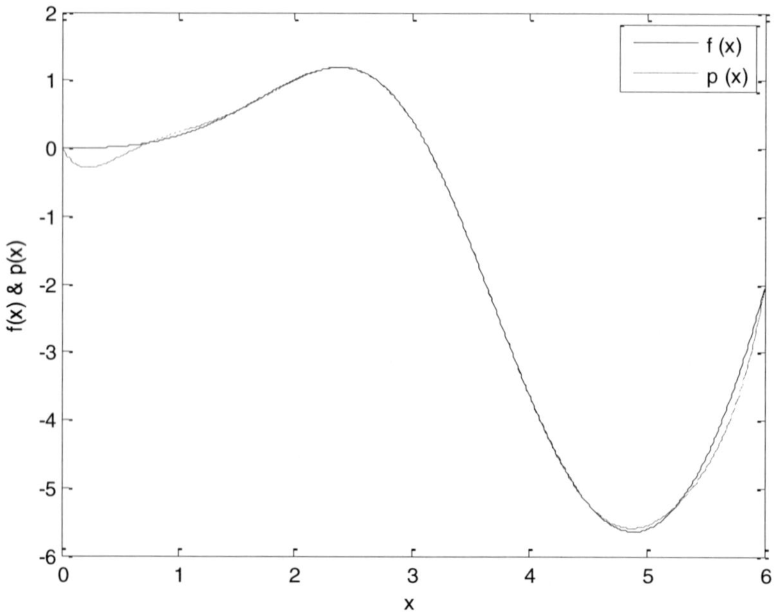

Figure 5.6. Approximation of the integrand (Example 5.21).

Example 5.22.

For a definite integration, its exact value is equal to 1.6405. If the integrand function values are given in Table 5.16, evaluate the integration value numerically.

Table 5.16. Integrand function values for example 5.22

x_i	0	0.12	0.22	0.32	0.36	0.4	0.44	0.54	0.64	0.7	0.8
$f(x_i)$	0.2	1.3097	1.3052	1.7434	2.0749	2.456	2.8429	3.5073	3.1819	2.363	0.232

First, we will use the formula (5.75) without regarding the existence of some equidistance segments.

$$\int = 0.12\frac{0.2 + 1.3097}{2} + 0.1\frac{1.3097 + 1.3052}{2} + \cdots + 0.1\frac{2.363 + 0.282}{2} = 1.5948$$

Next, we will consider the equidistance segments as follows:

- First segment

$$I_1 = 0.12\frac{0.2 + 1.3097}{2} = 0.0905$$

- $h_2 = h_3 = 0.1$, so we apply the formula (5.68)

$$I_2 = \frac{0.1}{3}[1.3079 + 4(1.3052) + 1.7434] = 0.2758$$

- $h_4 = h_5 = h_6 = 0.04$, so we apply the formula (5.69)

$$I_3 = \frac{3 \times 0.04}{8}[1.7434 + 3(2.0749) + 3(2.456) + 2.8429] = 0.2727$$

- $h_7 = h_8 = 0.1$, so apply the formula (5.68) again

$$I_4 = \frac{0.1}{4}[2.8429 + 4(3.5073) + 3.1819] = 0.6685$$

- $h_9 = 0.7 - 0.64 = 0.06$, we apply formula 5.67)

$$I_5 = \frac{0.06}{2}[3.1819 + 2.363] = 0.1663$$

- $h_{10} = 0.8 - 0.7 = 0.1$, we apply the formula (5.67) again

$$I_6 = \frac{0.1}{2}[2.363 + 0.232] = 0.1298$$

$$\int = \sum_{i=1}^{6} I_i = 0.0905 + 0.2758 + \cdots + 0.1298 = 1.6036$$

Therefore, the calculation, according to the second case, is closer to the exact value.

Example 5.23.

In an RLC electric circuit, the circuit current is recorded with time, as shown in Table 5.17. Determine the voltage across the capacitance, which has a value of 2 microfarads.

Table 5.17. RLC-circuit current of example 5.23

t (sec)	0	0.1	0.2	0.3	0.6	0.7	0.8	1.1	1.3	1.5
I (Amp)	0	0.0643	0.1377	0.0927	−0.1702	−0.0231	0.128	−0.1245	−0.0525	0.1683

The voltage across the capacitance is given by

$$V_c = \frac{1}{C}\int_0^{1.5} i(t)\,dt$$

For the evaluation of the integral, we will use the rule of multiple trapezoidal to obtain

$$V_c = \frac{10^{-6}}{2}\frac{1}{2}[0.1(0.0643) + 0.1(0.0643 + 0.1377) + 0.1(0.1377 + 0.0927) + 0.3(0.0927 - 0.1702)]$$

$$+0.3(-0.12702 - 0.0231) + 0.1(-0.0231 + 0.128) + 0.3(0.128 - 0.1245)$$

$$+0.2(-0.1245 - 0.0525) + 0.2(-0.0525 + 0.1683) = -\ 4.829 \text{ m volt}$$

In MATLAB, we can invoke the function *trapz* to calculate numerically the integration of two equal length vectors (table data) using a trapezoidal rule with equal spacing between data. If an integrand function is given, then the first step is to obtain the two equal-length vectors before applying this function. More accurate results can be obtained by writing a MATLAB code for numerical integration, where the programmer can control the step of integration and the method of calculation.

5.7. Numerical Methods for Solving Ordinary Differential Equations

As we have learned in chapters two and three, not all ordinary differential equations have an exact explicit form of solutions, especially nonlinear equations. The more crucial problem could be described by a set of equations that do not have an exact solution. For example, a projectile motion in space, considering all aerodynamic effects and the atmospheric conditions, is described by a highly nonlinear system of ordinary differential equations that can be solved only numerically. From a practical point of view, it is sometimes more efficient to find a numerical solution for specific initial or final conditions than to search for an exact solution. The accuracy of these numerical methods can always be controlled, and the desired solution can also be obtained.

Several numerical methods are based on different concepts and have different accuracy degrees, memory requirements, convergence behavior, and numerical stability. Since differential equations are of different natures, then numerical methods can be classified accordingly. However, we will be content with some basic numerical methods that illustrate the subject due to the text limitation.

It is convenient to start with first-order differential equations, but all numerical methods used to solve differential equations are developed essentially for first-order equations. These methods can be implemented for higher orders as well by introducing the concepts of reduction to first-order equations and utilizing the matrix formulation. The considered first-order equation has the form

$$y'(x) = f(x,y), y(x_0) = y_0 \tag{5.76}$$

where f is any scalar function, x_0 and y_0 are given initial values. It is required to find

$$y(x) = \int_{x_0=a}^{x_m=b} f(x,y)\, dx \tag{5.77}$$

When the limit value b is not given, then the initial-value problem is what we consider; however, numerically, the value $y(x_m)$ is what is required.

Let us assume a constant integration step h, which is usually of small value compared to the value of $(b - a)$, for example, one-tenth. Then, the integral (5.77) is approximated in each step of integration by following a specific numerical method; it is not necessarily the same method in each step. In this section, we will introduce only three fundamental methods. The reader is encouraged to learn more about these methods found in the specialist books of numerical methods.

5.7.1. Euler's Method

This method is the simplest illustrative method that is based on approximating the derivative of the function $f(x,y)$ by a straight line during the constant step of integration h. Figure 5.7 illustrates the concept of the method. A tangential straight line that represents the slope of the function $f(x,y)$ at a point x_i is drawn. The computation is carried out for one segment of the straight line between the point (x_i, y_i) and a next point $(x_{i+1,} y_{i+1,})$, where $h = x_{i+1} - x_i$; thus, we obtain the value y_{i+1}.

$$y_{i+1} = y_i + hf(x_i, y_i);\ i = 0,1,2 \ldots \tag{5.78}$$

where, $(x_0,\ y_0)$ is a given initial point.

The computation is performed iteratively to obtain $y_1, y_2, \ldots$, to reach the required value y_m. Now, in the interval $[x_0, x_m]$, the plot of the values $y_0, y_1, y_2, \ldots\ y_m$ gives the numerical solution. The numerical error is defined as

$$e_j = y_{act_j} - y_{com_j};\ j = 1,2 \ldots \tag{5.79}$$

where y_{act_j} and y_{com_j} are the actual and numerically computed values, respectively.

Euler's method obvious advantage is the simplicity of repeating the calculation; the programming of formula (5.78) is an easy task. However, the method has the characteristic that the error is accumulated as we move toward the required point. Therefore, this method is only suitable for small intervals and when the actual solution is almost linear. It gives unsatisfactory results when the interval $[a, b]$ is an unnecessary big. It is a crucial problem, especially when we do not have the exact solution, and hence the method is not recommended in practice. Generally, reducing the step of integration h will decrease the computation error; however, this reduction must be so significant in some problems to minimize the error. In the following two examples, we will consider two differential equations with exact solutions to grope for error.

Example 5.24.

Consider the following initial value problem

$$y'e^x - 2(x+1)y^2 = 0, y(0) = \frac{1}{6}$$

This first-order nonlinear differential equation has the following exact solution.

$$y(x) = \frac{1}{2xe^{-x} + 4e^{-x} + 2}$$

It is required to obtain numerically the value $y(0.5)$.

Let us take a step of integration h equal to $1/6$; thus, we must calculate three times to reach $x = 0.5$. Also,

$$y' = f(x,y) = 2(x+1)e^{-x}y^2, x_0 = 0, y_0 = 1/6$$

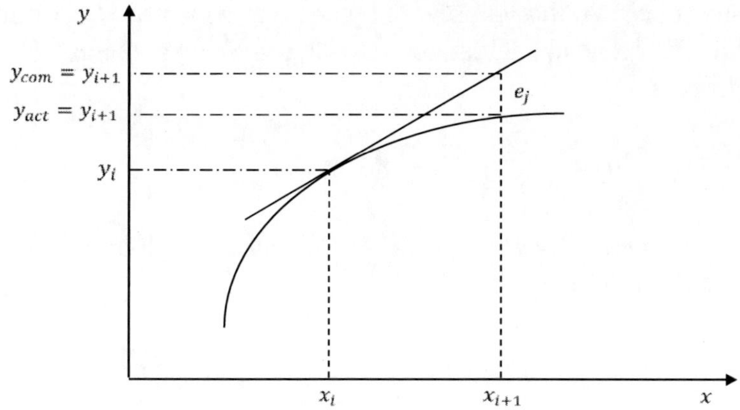

Figure 5.7. Euler's method concept.

For our example, the iterative formula (5.78)

$$y_{i+1} = y_i + \frac{1}{6}[(x_i + 1)e^{-x_i}y_i^2], i = 0,1,2$$

Then we calculate

$$y_1 = y_0 + \frac{1}{6}[(x_0 + 1)e^{-x_0}y_0^2] = \frac{1}{6} + \frac{1}{6}\left[2(0 + 1)e^0\left(\frac{1}{6}\right)^2\right] = 0.1759$$

$$y_2 = y_1 + \frac{1}{6}[(x_1 + 1)e^{-x_1}y_1^2] = 0.1759 + \frac{1}{6}\left[2\left(\frac{1}{6} + 1\right)e^{-\frac{1}{6}}(0.1759)^2\right] = 0.1861$$

$$y_3 = y(0.5) = y_2 + \frac{1}{6}[(x_2 + 1)e^{-x_2}y_2^2] = 0.1861 + \frac{1}{6}\left[2\left(\frac{2}{6} + 1\right)e^{-\frac{2}{6}}(0.1861)^2\right] = 0.1971$$

If we substitute $x = 0.5$ in the exact solution, then we will obtain $y(0.5) = 0.1987$; therefore, the numerical error is 0.0016. The accumulated error at $x = 1$ will be doubled, i.e., 0.0033. On the other hand, if we decrease the step of integration by 24 times, i.e., $h = 1/144$, then after 144 times of calculation (using a computer for that), we will obtain an accumulated error of only 10^{-4} at $x = 1$. This result is a time-cost.

Example 5.25.
Consider the following initial value problem

$$(1 + y + 2x)y' - (1 - 2y - 4x) = 0, y(0) = 1$$

This first-order nonlinear differential equation has the following exact solution

$$y(x) = -(1 + 2x) \pm \sqrt{6x + 4}$$

It is required to obtain the value $y(1)$ by calculating with $h = 0.1$ numerically.

Substituting the given initial value $x = 0$ into the exact solution yields two solutions of values as follows:

$$y(0) = -(1+0) + \sqrt{6(0)+4} = 1, \text{and } y(0) = -(1+0) - \sqrt{0+4} = -3$$

Therefore, the numerical solution will follow the first solution formula. For the 10-times calculation, we prepare

$$y_{i+1} = y_i + 0.1\frac{1 - 2y_i - 4x_i}{1 + y_i + 2x_i} \; ; \; i = 0,1,2 \dots 9$$

If we repeat the calculations up to $x = 1$, then we will obtain the numerical solution $y(1) = 0.1848$. To calculate the numerical error, we have to use the first solution, i.e.,

$$y(x) = -(1+2x) + \sqrt{6x+4}$$

$$y(1) = -(1+2)\sqrt{6+4} = 0.1623$$

The error is equal to $-\,0.0225$, which may be unaccepted in several practical cases. The reader is invited to solve the problem with the second solution assuming that the given initial is $y(0) = -3$, and he will discover that the numerical solution also follows the second exact solution approximately. Figure 5.8 illustrates, in the upper plots, both exact and numerical solutions and, in the lower plots, the solution errors, which are almost identical in absolute value. Although the decaying slops of the two solutions are different, the absolute error is nearly the same. As we see in the previous example, decreasing the step of integration will improve the numerical accuracy; for instance, for, $h = 0.001$ the numerical error $e(x = 1)$ will be -2×10^{-4}.

5.7.2. Heun's Method – Predictor Corrector Approach

This method performs the numerical solution by two formulas of calculation to improve the numerical accuracy. The idea is to calculate the average of two slops that are calculated by $f(x_i, y_i)$ and $f(x_{i+1}, \overline{y_i})$ in the ith – step of integration. The value $\overline{y_i}$ is a predicted value calculated by Euler's method (first formula). The final answer of the ith – step is corrected by a second formula.

Figure 5.9 illustrates the idea behind the method. The two formulas are

$$\text{Predictor: } \overline{y_{i+1}} = y_i + hf(x_i, y_i) \tag{5.80}$$

$$\text{Corrector: } y_{i+1} = y_i + \frac{h}{2}[f(x_i, y_i) + f(x_{i+1}, \overline{y_i})]; \; i = 0,1,2 \dots \tag{5.81}$$

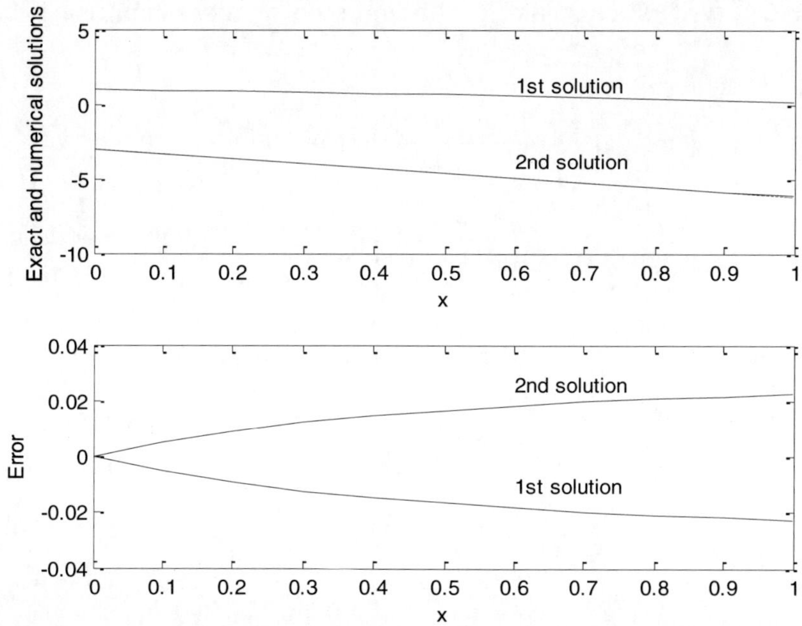

Figure 5.8. Numerical solutions (Example 5.25).

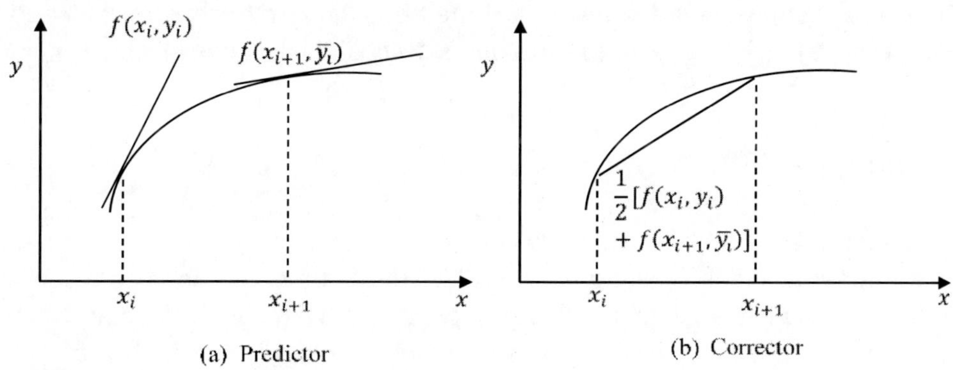

Figure 5.9. Heun's method concept.

Note that the first slop is calculated at the point x_i, while the second one must be calculated at the point x_{i+1}. Example 5.26 is considered again to see the efficiency of the method.

Example 5.26.
Solve the following first-order nonlinear differential equation

$$y' = \frac{1 - 2y - 4x}{1 + y + 2x};\ y(0) = 1$$

With $h = 0.1$, we first calculate for the first step of integration using the predictor formula

$$\bar{y}_1 = y_0 + h\frac{1 - 2y_0 - 4x_0}{1 + y_0 + 2x_0} = 1 + 0.1\frac{1 - 2}{1 + 1} = 0.95$$

$$y_1 = y_0 + \frac{h}{2}[f(x_0, y_0) + f(x_1, \bar{y}_1)] = 1 + \frac{0.1}{2}\left[\frac{1 - 2}{1 + 1} + \frac{1 - 2(0.95) - 4(0.1)}{1 + 0.95 + 2(0.1)}\right] = 0.9448$$

Next, we again calculate for the second step of integration

$$\bar{y}_2 = y_1 + h\frac{1 - 2y_1 - 4x_1}{1 + y_1 + 2x_1} = 0.9448 + 0.1\frac{1 - 2(0.9488) - 4(0.1)}{1 + -0.9448 + 2(0.1)} = 0.8846$$

$$y_2 = y_1 + \frac{h}{2}[f(x_1, y_1) + f(x_2, \bar{y}_2)]$$

$$y_2 = 1 + \frac{0.1}{2}\left[\frac{1 - 2(0.9488) - 4(0.1)}{1 + -0.9448 + 2(0.1)} + \frac{1 - 2(0.8846) - 4(0.2)}{1 + 0.8846 + 2(0.2)}\right] = 0.88036$$

Figure 5.10 shows in the interval $[0, 1]$ the absolute numerical error of Heun's and Euler's methods. Therefore, Heun's method significantly reduces errors.

5.7.3. Runge – Kutta Methods

These iterative numerical methods constitute a family of methods named after the German mathematicians C. Runge and M. W. Kutta. They are based on a similar idea for Heun's method in calculating in one step of integration multiple slops and averaging the calculations' results. We will consider the most famous one from this family, which is named the fourth-order method. In one constant step of integration, this method calculates the first derivative function $y' = f(x_p, y_q)$ four times as follows.

In one constant step of integration, this method calculates the first derivative function $y' = f(x_p, y_q)$ four times as follows:

$$y_{i+1} = y_i + \frac{1}{6}[k_1 + 2k_2 + 2k_3 + k_4];\ i = 0,1,2 \ldots \qquad (5.82)$$

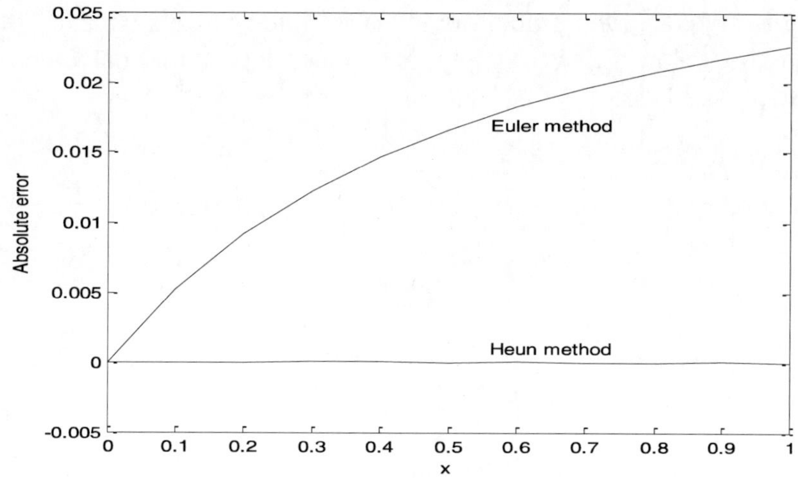

Figure 5.10. Absolute errors of Euler's and Heun's methods (Example 5.27).

where k_j are named gains, and they are given by

$$k_1 = hf(x_i, y_i) \tag{5.83a}$$

$$k_2 = hf\left(x_i + \frac{1}{2}h, y_i + \frac{1}{2}k_1\right) \tag{5.83b}$$

$$k_3 = hf\left(x_i + \frac{1}{2}h, y_i + \frac{1}{2}k_2\right) \tag{5.83c}$$

$$k_4 = hf(x_i + h, y_i + k_3) \tag{5.83d}$$

The method is well suited for computers because it needs no particular starting procedure, makes weak demand on storage, uses the same straightforward computational procedure, and is numerically stable.

Besides, in simple cases, we can derive one calculating formula that depends only on x_i, and y_i. For this method, there is a simple error estimation formula, which is based on computing the numerical solution twice with a step of integrations h and $2h$ to obtain

$$E_a = \frac{1}{15}[\tilde{y} - \tilde{\tilde{y}}] \tag{5.84}$$

where $\tilde{y}$ and $\tilde{\tilde{y}}$ are the solutions with h and $2h$ respectively. This estimate can be used to improve the result with h.

where $\tilde{y}$ and $\tilde{\tilde{y}}$ are the solutions with h and $2h$ respectively. This estimate can be used to improve the result with h.

The fourth-order Runge Kutta method's accuracy is improved significantly compared to Euler's or Heun's methods. Even for a relatively large integration step, the accuracy is high, and the stable numerical solution always satisfies the practical demands. In the following two simple examples, the numerical calculations are demonstrated in detail.

Example 5.27.
Consider the first-order non-homogeneous linear differential equation

$$y' - y = x, y(0) = 0$$

It has an exact solution

$$y(x) = e^x - x - 1$$

It is required to obtain $y(1)$ in five steps, i.e., with $h = 0.2$.
We have $f(x, y) = x + y$, two iterations are performed as follows:

First, iteration with $i = 0$, gives

$$k_1 = hf(x_0, y_0) = 0.2(0 + 0) = 0$$

$$k_2 = hf\left(x_0 + \frac{1}{2}h, y_0 + \frac{1}{2}k_1\right) = 0.2(0.1 + 0) = 0.02$$

$$k_3 = hf\left(x_0 + \frac{1}{2}h, y_0 + \frac{1}{2}k_2\right) = 0.2(0.1 + 0.1) = 0.022$$

$$k_4 = hf(x_0 + h, y_0 + k_3) = 0.2(0.2 + 0.011) = 0.0444$$

$$y_1 = y_0 + \frac{1}{6}[k_1 + 2k_2 + 2k_3 + k_4] = 0 + \frac{1}{6}[0 + 0.04 + 0.044 + 0.0444] = 0.0214$$

Similarly, the second iteration with $i = 1$, gives

$$k_1 = 0.2(0.2 + 0.0214) = 0.04428$$

$$k_2 = 0.2\left(0.2 + 0.1 + 0.0214 + \frac{0.04428}{2}\right) = 0.068708$$

$$k_3 = 0.2\left(0.2 + 0.1 + 0.0214 + \frac{0.068708}{2}\right) = 0.0711508$$

$$k_4 = 0.2(0.2 + 0.2 + 0.0214 + 0.0711508) = 0.0985101$$

$$y_2 = 0.0214 + \frac{1}{6}[0.04428 + 2(0.068708) + 2(0.0711508) + 0.0985101] = 0.091818$$

The results from the rest calculation are shown in Table 5.18. The last column of the table shows the error of the numerical calculation, which is of order 10^{-5}, or better. However, errors will be accumulated as we move forward in the variable x.

Table 5.18. Numerical solution for example 5.27

i	x_i	y_{axact}	$y_{numerical}$	$Error$
0	0	0	0	–
1	0.2	0.0214027	0.0214	$\cong 3 \times 10^{-6}$
2	0.4	0.0918246	0.091818	$\cong 7 \times 10^{-6}$
3	0.6	0.2221188	0.222106	$\cong 11 \times 10^{-6}$
4	0.8	0.425541	0.425521	$\cong 20 \times 10^{-6}$
5	1	0.718282	0.718251	$\cong 30 \times 10^{-6}$

Example 5.28.

Consider the first-order non-homogeneous linear differential equation

$$y' + \frac{1}{2}y = 4\, e^{0.8x}, y(0) = 2$$

It is given that the exact value $y(2) = 14.84392$; calculate this value numerically with $h = 1$ and $h = 2$. Use the estimated error to improve the result of the calculation.

We have $f(x, y) = x + y$, two iterations are performed with $h = 1$.

First iteration: with $i = 0, x_0 = 0, y_0 = 2$, using the gain formulas (5.83) yields

$$k_1 = 3, k_2 = 4.21729, k_3 = 3.91297, k_4 = 5.94567$$

Then,

$$y(1) = 2 + \frac{1}{6}[3 + 2(4.21729 + 3.91297) + 5.49567] = 6.20103$$

Second iteration with $i = 1, x_1 = 1, y_1 = 6.20103$, using the gain formulas (5.83) yields

$$k_1 = 5.80164, k_2 = 8.72954, k_3 = 7.99756, k_4 = 12.71283$$

Then,

$$\tilde{y} = y(2) = 6.20103 + \frac{1}{6}[5.80164 + 2(8.72954 + 7.99756) + 12.71283] = 14.86247$$

Next, we let $h = 2$; hence, we calculate only once to reach $y(2)$. With $x_0 = 0, y_0 = 2$, using the gain formulas (5.83) yields

$$k_1 = 6, k_2 = 12.80432, k_3 = 9.40216, k_4 = 28.22209$$

Finally, we calculate

$$\tilde{\tilde{y}} = y(2) = 2 + \frac{1}{6}[6 + 2(12.80432 + 9.40216) + 28.22209] = 15.10584$$

The error estimation is

$$E_a = \frac{1}{15}[14.86247 - 15.10584] = -0.01622$$

Therefore, a more accurate result is

$$y(2) = 14.86247 - 0.01622 = 14.84625$$

The absolute numerical error is equal to $14.84626 - 14.84392 = 0.00234$.

5.7.4. Numerical Solution of Higher-Order Differential Equations

Since all numerical methods are used for first-order differential equations, higher-order equations must be reduced to a system of first-order equations. With the aid of matrix notations, these methods can be modified to give a numerical solution. The technique of order reduction was thoroughly discussed in chapter two. The reduction is brief and mainly based on changing the single dependent variable to several dependent variables (sometimes called states). The number of variables is equal to the order of the given equation. There are different ways of obtaining a reduced system; however, the most popular way is the canonical form.

We shall first introduce the reduction technique for numerical solution purposes using a second-order equation. The fourth-order Runge Kutta method is used to illustrate and deliver a numerical solution of the second-order differential equation. The next example explains this technique.

Example 5.29.
Consider the following initial value problem.

$$y'' + 3y' + 2y = e^x; \; y(0) = 1, y'(0) = -1$$

For this second-order linear nonhomogeneous differential equation, the exact solution is

$$y(x) = \frac{1}{6}e^x + \frac{1}{3}e^{-2x} + \frac{1}{2}e^{-x}$$

It is required to calculate the value numerically, $y(0.1)$, using the fourth-order Runge Kutta method. To shorten the calculation set $h = 0.1$; thus, only a one-step calculation is needed.

Let us choose two states, z_1 and z_2 to denote the dependent variable and its first derivative, y and y' respectively, i.e.,

$$z_1 = y, \text{and } z_2 = y'$$

Then, we can write

$$z_1' = y' = z_2$$

$$z_2' = y'' = -2y - 3y' + e^x = -2z_1 - 3z_2 + e^x$$

The original initial conditions should be stated in terms of the new two dependent variables, i.e.,

$$z_1(0) = y(0) = 1, \text{and } z_2(0) = y'(0) = -1$$

We write the two equations system as

$$z_1' = z_2;\ z_1(0) = 1$$

$$z_2' = -2z_1 - 3z_2 + e^x;\ z_2(0) = -1$$

Now, we define $f_1(x, z_1, z_2)$ and $f_2(x, z_1, z_2)$ to be z_2 and $-2z_1 - 3z_2 + e^x$, respectively, i.e.,

$$z_1' = f_1(x, z_1, z_2);\ z_1(0) = 1$$

$$z_2' = f_2(x, z_1, z_2);\ z_2(0) = -1$$

For each iteration, the four gains of the fourth-order Runge-Kutta will be denoted by two indices. The first index is for the number of the gain, and the second index is for the number of the function f. Therefore, in our second-order example, we have to calculate the functions eight times to obtain the eight gains, i.e.,

$$k_{11} = hf_1\big(x_0, z_1(x_0), z_2(x_0)\big) = 0.1f_1(0,1,-1) = -0.1$$

$$k_{12} = hf_2\big(x_0, z_1(x_0), z_2(x_0)\big) = 0.1f_2(0,1,-1) = 0.1(-2(1) - 3(-1) + e^0) = 0.2$$

$$k_{21} = hf_1\left(x_0 + \frac{1}{2}h, z_1(x_0) + \frac{1}{2}k_{11}, z_2(x_0) + \frac{1}{2}k_{12}\right) = 0.1f_1(0.05, 0.95, -0.9) = -0.9$$

$$k_{22} = hf_2\left(x_0 + \frac{1}{2}h, z_1(x_0) + \frac{1}{2}k_{11}, z_2(x_0) + \frac{1}{2}k_{12}\right) = 0.1f_2(0.05, 0.95, -0.9) = 0.18513$$

$$k_{31} = hf_1\left(x_0 + \frac{1}{2}h, z_1(x_0) + \frac{1}{2}k_{21}, z_2(x_0) + \frac{1}{2}k_{22}\right) = 0.1f_1(0.05, 0.955, -0.90744)$$
$$= -0.090744$$

$$k_{32} = hf_2\left(x_0 + \frac{1}{2}h, z_1(x_0) + \frac{1}{2}k_{21}, z_2(x_0) + \frac{1}{2}k_{22}\right) = 0.1f_2(0.05, 0.955, -0.90744)$$
$$= 0.18636$$

$$k_{41} = hf_1(x_0 + h, z_1(x_0) + k_{31}, z_2(x_0) + k_{32}) = 0.1f_1(0.1, 0.90926, -0.81364)$$
$$= -0.081364$$

$$k_{42} = hf_2(x_0 + h, z_1(x_0) + k_{31}, z_2(x_0) + k_{32}) = 0.1f_2(0.1, 0.90926, -0.81364)$$
$$= 0.17275$$

Now, we calculate twice

$$z_1 = 1 + \frac{1}{6}[-0.1 + 2(-0.09 - 0.090744) - 0.081364] = 0.90952$$

$$z_2 = -1 + \frac{1}{6}[0.2 + 2(0.18513 + 0.18636) + 0.17275] = -0.81404$$

In terms of the original dependent variable, we assign

$$y(0.1) = 0.90952, \text{and } y'(0.1) = 0.81404$$

In the exact solution, we have

$$y(0.1) = \frac{1}{6}e^{0.1} + \frac{1}{3}e^{-0.2} + \frac{1}{2}e^{-0.1} = 0.909524$$

$$y'(0.1) = \frac{1}{6}e^{0.1} - \frac{2}{3}e^{-0.2} - \frac{1}{2}e^{-0.1} = -0.814044$$

Therefore, the numerical accuracy is better than 10^{-4}.

5.8. Numerical Solution of Boundary Value Problem

We realize that there are one general solution for differential equations and an infinite number of particular solutions obtained for every set of initial (or/and final) conditions. Readers also know that these conditions should be equal to the equation for getting a particular solution. In some practical problems like the heat transfer in a long thin rod, the differential equation solution is desired for a given initial value and a final value of the dependent variable (or/and their derivatives), i.e., to find a specific particular solution. The problem is named the boundary value problem.

The boundary value problem is not difficult whenever an exact solution exists; i.e., the constants of integration can be found by substituting the given conditions and solving the resultant algebraic system of equations. However, the situation becomes more complicated when the differential equation has no exact solution. In this section, the boundary-value problem will be considered numerically only for one case, for a definite class of second-order equations with a given initial and final values of the dependent variable.

Consider a second-order equation with boundary conditions of the form

$$y'' = f(x, y, y');\; y(x_0) = c_1, y(x_1) = c_2,\; c_1, c_2 \in \Re \tag{5.85}$$

where, c_1, and c_2 are given real constants. As it is noted easily, the second boundary condition $y(x_1) = c_2$ replaces the initial condition $y'(x_0)$ in the initial value problem. There are two methods for solving the boundary value problem, the shooting method and the finite difference method numerically.

5.8.1. Shooting Method

A method makes use of the usual numerical solution of an initial value problem with some approximation technique. The numerical solution of equation (5.85) is as in the following points:

1. Assume two different first derivative conditions, says $y'(x_0) = m_1$ and $y'(x_0) = m_2$. The selection of these values depends on the problem itself; unreliable selection could result in inaccurate results or slow convergence.
2. Solve first the initial value problem with the initial conditions $y(x_0) = m_0$, $y'(x_0) = m_1$, and then solve with $y(x_0) = m_0, y'(x_0) = m_2$ to obtain two solutions y_1 and y_2 respectively. Rename these two solutions in such a way as to have y_1 with a small value and y_2 with the larger value. It is essential to use the same numerical method that should be used with the same minimum accuracy demands.
3. If either of y_1 or y_2 satisfies the given boundary condition, then consider that on and stop, else the next step is carried out.
4. Make the linear interpolation (or extrapolation) to get a third initial derivative condition, says m_3, using the formula

$$y'(x_0) = m_3 = m_1 + \frac{m_2 - m_1}{y_2 - y_1}(y(x_1) - y_1);\ y_2 > y_1 \tag{5.86}$$

5. Check again that the boundary condition. Within the desired accuracy requirement, if the boundary condition is satisfied, the calculation is ended. Otherwise, repeat the procedure by considering the nearer of m_1, m_2, or m_3 to the boundary condition and a new assumed one, say m_4.

The following example illustrates the above numerical procedure and does not solve an utterly detailed problem.

Example 5.30.

Consider a long thin rod shown in Figure 5.11, where L and T_a are the rod length and temperature of the surrounding air respectively. The temperatures $T(0)$ and $T(L)$ are the rod temperatures at its ends, which are given by T_0 and T_L, respectively.

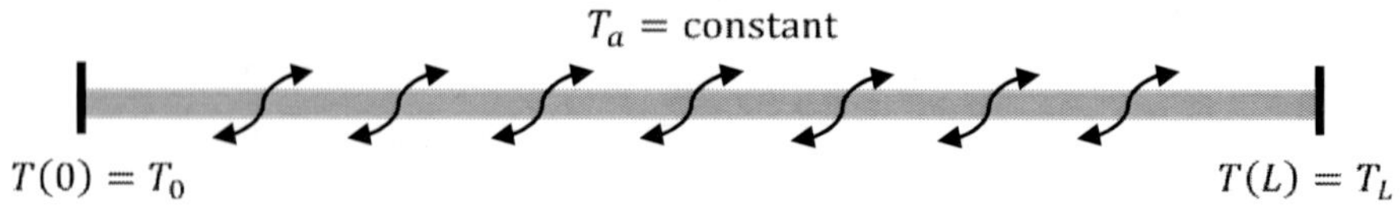

Figure 5.11. Heat problem of a long thin rod (Example 5.31).

The heat equation is

$$T'' + k(T_a - T) = 0, T' = \frac{dT}{dx}, x \in [0, L], T(x = 0) = T_0, T(x = L) = T_L$$

where k is the heat coefficient. Let $T_a = 20, k = 0.01, T_0 = 40, T_L = 200,$ and $L = 10\ m$ then we can write this boundary value problem as (for convenience, we replace T by y)

$$y'' - 0.01y = -0.2;\ y(0) = 40, y(10) = 200$$

1. We write the given equation as two first-order equations as follows.

$$x_1' = x_2;\ x_1(0) = y(0) = 40$$

$$x_2' = 0.01x_1 - 0.2, x_2(0) = y'(0) = \begin{cases} m_1 \\ m_2 \end{cases}$$

2. Assume the two values $m_1 = 10$ and $m_1 = 20$.

3. a. Use the fourth-order Runge-Kutta method with $h = 2 \in [0,10]$ to solve the problem

$$x_1' = x_2;\ x_1(0) = y(0) = 40$$

$$x_2' = 0.01x_1 - 0.2, x_2(0) = y'(0) = 10$$

We will find that $x_1(10) = y(10) = 168.3797$, which is not equal to the boundary condition $y(10) = 200$.

b) Use the fourth-order Runge-Kutta method with $h = 2 \in [0,10]$ to solve the problem

$$x_1' = x_2;\ x_1(0) = y(0) = 40$$

$$x_2' = 0.01x_1 - 0.2, x_2(0) = y'(0) = 20$$

We will find that $x_1(10) = y(10) = 285.898$, which is also not equal to the boundary condition $y(10) = 200$.

4. Name $y_1(10) = 168.3797,$ and $y_2(10) = 285.898$ and apply the linear interpolation formula (5.85) to obtain

$$y'(0) = 10 + \frac{20 - 10}{285.898 - 168.3797}(200 - 168.3797) = 12.6907$$

5. The following example illustrates the above numerical procedure and does not solve an utterly detailed problem).

$$x_1' = x_2;\ x_1(0) = y(0) = 40$$

$$x_2' = 0.01x_1 - 0.2, x_2(0) = y'(0) = 12.6907$$

We will find that $x_1(10) = y(10) = 200$, thus the boundary condition is satisfied.

6. This simple linear equation has an exact solution as

$$y(x) = c_1 e^{0.1x} + c_2 e^{-0.1x} + 20$$

Substituting the conditions $y(0) = 40,$ and $y(10) = 200$ to obtain

$$c_1 + c_2 + 20 = 40$$

$$c_1 e^{1} + c_2 e^{-1} + 20 = 200$$

Solving for the constants c_1, and c_2, we get $c_1 = 73.4523$, and $c_2 = -53.4523$. Now let us determine these constants using the condition $y(0) = 40$ and the numerically found condition $y'(0) = 12{,}6907$. The above two algebraic equations become

$$c_1 + c_2 + 20 = 40$$

$$0.1c_1 - 0.1c_2 = 12.6907$$

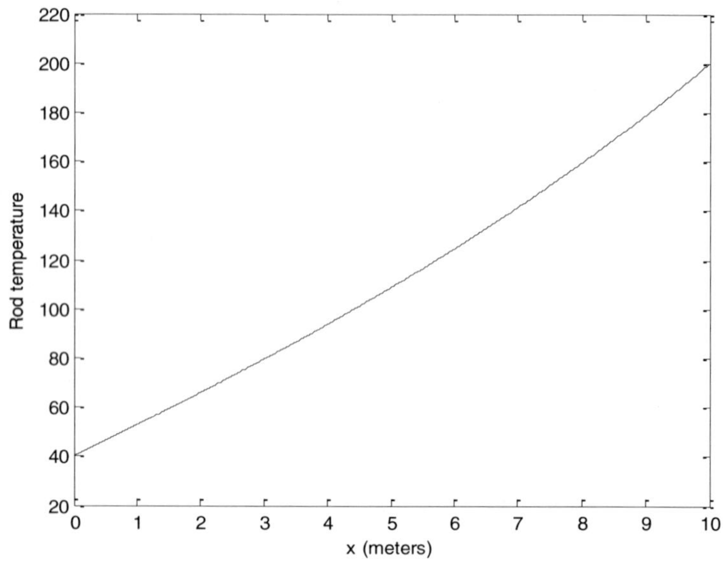

Figure 5.12. Solution of heat equation (example 5.31).

The solution of these last equations gives the same (at least for the four digits) values of the constants c_1, and c_2. These results prove the correctness of the numerical solution. Figure 5.12 shows the increase of the temperature to reach 200 at the end of the rod.

In other problems, especially with nonlinear equations, we may need a successive implementation for performing solutions and applying the linear interpolation to satisfy the accuracy demands. Furthermore, a quadratic polynomial fitting of three assumed second initial conditions m_1, m_2, and m_3 with three corresponding solutions y_1, y_2, and y_3 can be used to determine m_4 using the following equation.

$$m_4 = a + by_b + cy_b^2 \tag{5.87}$$

where, y_b is the given boundary condition and the constants $a, b, and\ c$ are found from

$$\begin{bmatrix} a \\ b \\ c \end{bmatrix} = \begin{bmatrix} 1 & y_1 & y_1^2 \\ 1 & y_2 & y_2^2 \\ 1 & y_3 & y_3^2 \end{bmatrix}^{-1} \begin{bmatrix} m_1 \\ m_2 \\ m_3 \end{bmatrix} \tag{5.88}$$

A higher-order fitting can be similarly implemented. However, this method's main problem is the correct and suitable selection of the second initial condition.

5.8.2. Finite Difference Method

This method is based on a different concept; the name of the method comes from the principle of using finite divided differences in determining the derivatives, as we have seen before. It is based on approximating the first and second derivatives (higher-order differential equations can be solved similarly). Let the initial and boundary conditions be given at points x_0, and x_b and the quantity (length, distance, etc.), $(x_b - x_0)$ is divided into n equal Δx segments by $(n - 1)$ nodes. The solutions at the end of each segment are denoted by $y_1, y_2 \dots y_n$, such that $y_0 = y(x_0)$, and $y_{n+1} = y(x_b)$. The first and second derivatives are approximated by three successive assumed solutions as

$$\frac{d^2y}{dx^2} = \frac{y_{i+1} - 2y_i + y_{i-1}}{(\Delta x)^2} \tag{5.89}$$

$$\frac{dy}{dx} = \frac{y_i - y_{i-1}}{\Delta x} \tag{5.90}$$

$$\Delta x = \frac{x_b - x_a}{n} \tag{5.91}$$

Then equation (5.85) becomes

$$\frac{y_{i+1} - 2y_i + y_{i-1}}{(\Delta x)^2} = f\left(x, y, \frac{y_i - y_{i-1}}{\Delta x}\right); \; i = 1,2 \dots n - 1 \tag{5.92}$$

The application of equation (5.92) will result in a definite system of n algebraic equations whose variables are the solutions $y_1, y_2 \dots y_n$. Therefore, this method's accuracy is directly proportional to the number of segments; a large number of segments will increase the accuracy. To illustrate, we again consider the problem of the heat equation of a long thin rod.

Example 5.31.

Let the rod length L is divided into five equal segments of length $\Delta x = (10 - 0)/5 = 2$; thus, the heat equation can be written approximately by

$$\frac{y_{i+1} - 2y_i + y_{i-1}}{(2)^2} = 0.01y_i - 0.2;\ i = 1,2,3,4$$

For $i = 1$, we obtain

$$\frac{y_2 - 2y_1 + y_0}{(2)^2} = 0.01y_1 - 0.2$$

Substituting $y_0 = y(x_0) = 40$ and arranging yields

$$2.04y_1 - y_2 = 40.8$$

For $i = 2,$ and $i = 3$, similarly, we obtain the following two equations

$$-y_1 + 2.04y_2 - y_3 = 0.8$$

$$-y_2 + 2.04y_3 - y_4 = 0.8$$

Finally, for $i = 5$, we obtain

$$\frac{y_5 - 2y_4 + y_3}{(2)^2} = 0.01y_1 - 0.2$$

Substituting $y_5 = y(x_b) = 200$ and arranging gives the fourth equation as

$$-y_3 + 2.04y_4 = 200.8$$

Putting these equations in a matrix form as

$$\begin{bmatrix} 2.04 & -1 & 0 & 0 \\ -1 & 2.04 & -1 & 0 \\ 0 & -1 & 2.04 & -1 \\ 0 & 0 & -1 & 2.04 \end{bmatrix} \begin{bmatrix} y_1 \\ y_2 \\ y_3 \\ y_4 \end{bmatrix} = \begin{bmatrix} 40.8 \\ 0.8 \\ 0.8 \\ 200.8 \end{bmatrix}$$

The coefficient matrix is a tridiagonal Toeplitz one; it is always like that. Solving this system gives

$$\begin{bmatrix} y_1 \\ y_2 \\ y_3 \\ y_4 \end{bmatrix} = \begin{bmatrix} 2.04 & -1 & 0 & 0 \\ -1 & 2.04 & -1 & 0 \\ 0 & -1 & 2.04 & -1 \\ 0 & 0 & -1 & 2.04 \end{bmatrix}^{-1} \begin{bmatrix} 40.8 \\ 0.8 \\ 0.8 \\ 200.8 \end{bmatrix} = \begin{bmatrix} 65.9698 \\ 93.7785 \\ 124.5382 \\ 159.4795 \end{bmatrix}$$

It means that we have

$$y(2) = 65.9698, y(4) = 93.7785, y(6) = 124.5382, y(8) = 159.4795$$

The corresponding values that are obtained from the exact solution are:

$$y(2) = 65.9518, y(4) = 93.7478, y(6) = 124.5036, y(8) = 159.4534$$

More accurate results can be found by reducing Δx to one meter; in this case, the linear system will have nine equations.

We notice that this method does not ask for solving the differential equation numerically but for solving a linear algebraic system. As compared to the shooting method, this needs to solve the differential equation several times and assume suitable shoots. The solution of only one linear algebraic system without any assumption represents a great advantage. Therefore, using the finite difference method reduces the computation burden significantly. Furthermore, the method can be generalized easily for programming purposes. Finally, the finite difference method's accuracy can be controlled more easily, only by increasing the number of the considered segments or/and by considering a higher order of derivatives approximations in complicated problems.

The variable-coefficient and nonhomogeneous equations can be solved without any difficulty. Despite these advantages, for nonlinear differential equations, the resulted algebraic system will be nonlinear too, and consequently, the solution will be a more difficult task. However, in some simple nonlinear cases, a nonlinear algebraic system's exact solution can be obtained. Alternatively, a numerical method must be invoked, for example, the method given in section 5.3.3.

Example 5.32.

Use the finite differences method to solve the boundary value problem using five segments.

$$y'' + Ay' + By = Cf(x);\ y(x_0) = m_1, y(x_b) = m_2$$

Then, apply the result to the following initial value problem.

$$y'' + 3y' + 2y = e^x;\ y(0) = 1, y(1) = 0.7$$

For five segments,

$$\Delta x = \frac{x_b - x_a}{5};\ \Delta x = \frac{1 - 0}{5} = 0.2$$

The given equation is approximated using equations 5.89 and 5.90, as

$$\frac{y_{i+1} - 2y_i + y_{i-1}}{(\Delta x)^2} + A\frac{y_i - y_{i-1}}{\Delta x} + By_i = Cf(i\Delta x);\ i = 1,2,3,4$$

$$\frac{y_{i+1}-2y_i+y_{i-1}}{(0.2)^2}+3\frac{y_i-y_{i-1}}{0.2}+2y_i=e^{0.2i};\ i=1,2,3,4$$

First, let us arrange the above equation to the form

$$(1-A\Delta x\,)y_{i-1}+(-2+A\Delta x+B(\Delta x)^2)y_i+y_{i+1}=(\Delta x)^2Cf(i\Delta x);\ i=1,2,3,4$$

$$(1-3\times 0.2)y_{i-1}+(-2+3\times 0.2+2\times 0.04)y_i+y_{i+1}=0.04e^{0.2i};\ i=1,2,3,4$$

If we rewrite

$$ay_{i-1}+by_i+cy_{i+1}=(\Delta x)^2Cf(i\Delta x);\ i=1,2,3,4$$

where,

$$a=(1-A\Delta x\,),b=(-2+A\Delta x+B(\Delta x)^2),c=1$$

As we did in the previous example, the above equation gives four linear algebraic equations, which are arranged in matrix form as

$$\begin{bmatrix} b & c & 0 & 0\\ a & b & c & 0\\ 0 & a & b & c\\ 0 & 0 & a & b\end{bmatrix}\begin{bmatrix} y_1\\ y_2\\ y_3\\ y_4\end{bmatrix}=\begin{bmatrix} (\Delta x)^2Cf(\Delta x)-am_1\\ (\Delta x)^2Cf(2\Delta x)\\ (\Delta x)^2Cf(3\Delta x)\\ (\Delta x)^2Cf(4\Delta x)-cm_2\end{bmatrix}$$

Therefore, we do not need to calculate the detailed equations but to have the required system directly from the above symbolic system as

$$\begin{bmatrix} -1.32 & 1 & 0 & 0\\ 0.4 & -1.32 & 1 & 0\\ 0 & 0.4 & -1.32 & 1\\ 0 & 0 & 0.4 & -1.32\end{bmatrix}\begin{bmatrix} y_1\\ y_2\\ y_3\\ y_4\end{bmatrix}=\begin{bmatrix} -0.3511\\ 0.0597\\ 0.0729\\ -0.611\end{bmatrix}$$

Then, we will have the solutions

$$\begin{bmatrix} y_1\\ y_2\\ y_3\\ y_4\end{bmatrix}=\begin{bmatrix} -1.32 & 1 & 0 & 0\\ 0.4 & -1.32 & 1 & 0\\ 0 & 0.4 & -1.32 & 1\\ 0 & 0 & 0.4 & -1.32\end{bmatrix}^{-1}\begin{bmatrix} -0.3511\\ 0.0597\\ 0.0729\\ -0.611\end{bmatrix}=\begin{bmatrix} 0.7885\\ 0.6898\\ 0.6548\\ 0.6613\end{bmatrix}$$

Therefore,

$$y(0.2)=0.7885,y(0.4)=0.6898,y(0.6)=0.6548,y(0.8)=0.6613$$

The corresponding exact values

$$y(0.2) = 0.8477, y(0.4) = 0.7506, y(0.6) = 0.6975, y(0.8) = 0.6819$$

The differences between numerical and exact values are significant. The number of points is increased to increase the accuracy of the numerical computation. However, in practice, the acceptance of the obtained accuracy depends on the problem at hand.

EXERCISES

Section 5.3

5.1. Use the simple iteration method to find one real root lying in the given interval of the following nonlinear equations. The desired accuracy is to be better than 10^{-3}:

a. $f(x) = x + \ln(x) = 0;\ I =< 0,1 >$

b. $f(t) = \sin t - t\cos t - \frac{\pi}{2} = 0;\ I =< 1,2 >$

c. $f(x) = \tan x + 2x\ln x - \frac{\pi}{4} = 0;\ I =< \pi/4, \pi/2 >$

5.2. Use the bisection method to find one real root lying in the given interval of the following nonlinear equations. The desired accuracy is to be better than 10^{-4}:

a. $f(x) = 5\cos^2 x + 10e^{-x} - 3 = 0;\ I =< \pi/4, \pi/2 >$

b. $f(x) = \sin^2 x - 2x - \sqrt{2x} + 5 = 0;\ I =< 1,2 >$

c. $f(x) = x^2 e^x - 1 = 0;\ I =< 0,1 >$

5.3. Use the secant method to find one real root lying into the given interval of the following nonlinear equations. The desired accuracy is to be better than 10^{-5}:

a. $f(x) = \sqrt{x}\tan x + x - 1 = 0;\ I =< 0, \pi/2 >$

b. $\tan x - 3e^x = 0;\ I =< 0, \pi/2 >$

5.4. Use the Newton method to find one real root, which is less than $\pi/2$ of the following nonlinear equation. The desired accuracy is to be better than 10^{-6}:

$$f(x) = \sqrt{x}\tan x + x - 2 = 0$$

5.5. Use Newton's method to find one real positive root of the following nonlinear equation. The desired accuracy is to be better than 10^{-6}:

$$f(x) = x^2\sqrt{2\sin x + 1} - 5\cos 2x - 3 = 0$$

5.6. Solve the following nonlinear systems with accuracy better than 10^{-3}:

a. $x^2 + y^2 - x = 0, x^2 - y^2 - y = 0;\ x_0 = 0.8, y_0 = 0.4$

b. $x + 3\ln(x) - y^2 = 0, 2x^2 - xy - 5x + 1 = 0;\ x_0 = 3.8, y_0 = 2.8$

c. $3x_1^2 - x_2^2 = 0, 3x_1x_2^2 - x_1^3 = 1;\ X_0 = [1 \quad 1]^T$

5.7. Consider the R-L load of the single-phase half-rectifier circuit shown in Figure 5.13. Due to the inductance, the circuit will have a deadlock angle β less than π, which can be determined from the transcendental equation

$$\sin(\beta - \varphi) + \sin(-\varphi)e^{-\frac{\beta}{\tan\varphi}} = 0\ ;\ \varphi = \tan^{-1}\left(\frac{wL}{R}\right), w = 2\pi \times 6o \text{ Hz}$$

Then, we can compute the average value of the voltage across the load from

$$V_{L\,ave} = \frac{\sqrt{2}}{2\pi}V_0(1 - \cos\beta)$$

Compute $V_{L\,ave}$ if the peak source voltage $V_0 = 220$ volts.

5.8. A flat plate of one-kilogram mass is freely falling in air with a velocity v. The plate is subjected to a downward gravitation force and upward frictional drag force due to air. The drag force F_D (Kg. m/ sec²) is given by

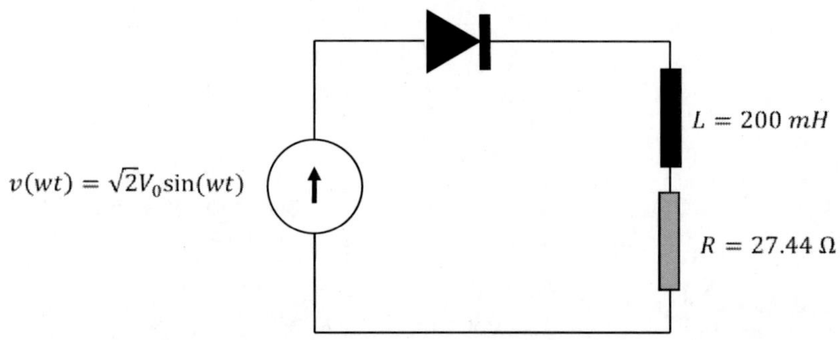

Figure 5.13. single-phase half-rectifier circuit (exercise 5.7).

$$F_D(v) = \frac{0.3\, v^2}{500 + (\ln v)^3} - 0.02\, v$$

The terminal velocity v_f is reached when the drag force is equal to the gravitational force, i.e., when

$$F_D(v_f) = mg;\ g = 9.81\, m/sec^2$$

Find the terminal velocity.

5.9. A necessary condition of a stable linear time-invariant dynamic system is that its characteristic equation has the same sign for all coefficients. In turn, it indicates that the characteristic equation has one or more positive roots. Thus, the system is unstable $s^3 + 5s^2 - 15s + 9 = 0$. Determine the unstable root(s).

5.10. Given that the following fifth-order polynomial equation has one complex conjugate pair $-1 \pm j$, determine the other 3 roots.

$$x^5 + 3x^5 + 3x^5 - 10x^5 - 22x - 20 = 0$$

Section 5.4

5.11. In the interval $[0, \pi/2]$, prepare a numerical form of equations 5.29 to approximate the function $f(x)$ by fourth-order polynomial, where

$$f(x) = \frac{\pi}{2} + x \cos x$$

Divide the given interval by five equidistance points.

5.12. In the interval $[0, 2]$, use the least square error technique to approximate, by a fourth-order polynomial, the function

$$f(x) = xe^x - 1$$

Determine the accuracy in the middle of the interval.

5.13. In the interval $[0, \pi/4]$, use Lagrange's interpolation formula to determine a third-order polynomial that approximates the function

$$f(x) = x + \cos x - 1$$

5.14. Consider the following experimental results

x_i	0	1.6408	2.1909	3.8439	4.1739	5.1708	6.9943	7.4398	8.34	9.3142
y_i	3.4536	5.2529	5.8601	7.6757	8.0382	9.1332	11.1361	11..6255	12.6143	13.6843

Assume a linear regression $y = ax + b$, to fit the data. Determine the parameters a and b.

5.15. A laboratory experiment of charging a battery records the following data table.

t_i (hours)	0	0.125	0.25	0.375	0.5	0.625	0.75	0.875	1
$v(t_i)$ (volts)	1.5	1.515	1.527	1.537	1.545	1.552	1.558	1.563	1.568

It is assumed that the function $v(t)$ has the following nonlinear expression

$$v(t) = k_1 e^{\left(0.2 - \frac{k_2}{t+1}\right)}$$

Use a linear regression technique to determine the parameters k_1 and k_2, then determine the time required to charge the battery to 1.6 volts.

5.16. Consider the following data table; note that some values of the independent variable are repeated.

x_i	0	3	5	8	8	9	10	12	12	16	16
$f(x_i)$	7.1	1.8	12.2	59.2	51.7	59.0	77.6	109.1	120.2	206.6	214.6

Assume a quadratic regression $y = ax^2 + bx + c$, to fit the data. Determine the parameters a, b, and c.

5.17. Use the interpolation formula of Newton divided differences to determine a 4th – order polynomial, represented by the following data.

x_i	−2	0	1	3	4	7
$f(x_i)$	−25	−51	−43	45	209	2189

5.18. In Figure 5.14, it is shown a four-bar linkage, where the relation between the angles α and φ is given by nonlinear equation

$$\cos(\alpha - \varphi) - \frac{5}{3}\cos\alpha + \frac{5}{2}\cos\varphi = \frac{11}{6}$$

The nonlinear equation solution is performed utilizing Newton's method to obtain the data in the following table.

α	0	10	20	30	40	50	60	70	80	90
φ	0	8.06934	16.1132	24.1049	32.0152	39.8104	47.4508	54.8887	62.0599	68.8887

Show that the values of φ can be obtained by the solution of the given nonlinear equation for the assigned values of α in the table.

Use the interpolation formula of Newton divided differences to approximate the function $\varphi = f(\alpha)$ by a third-order polynomial

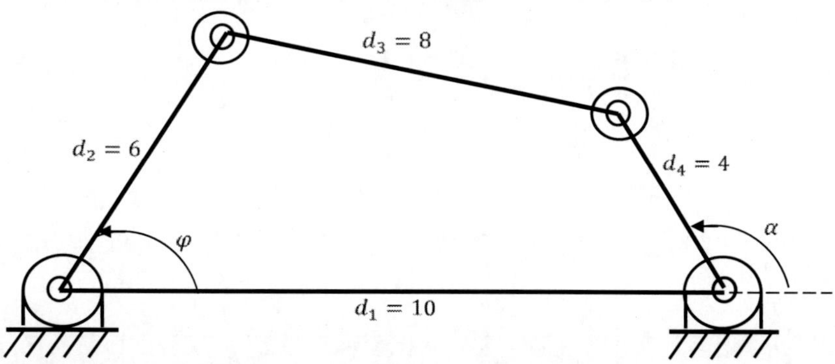

Figure 5.14. Four-bar linkages (exercise 5.18).

5.19. In some piping systems, it is assumed that the outcome flow rate Q_0 (m^3/minute) is related to the movement of the valve opening (mm) by

$$Q_0 = e^{\sqrt{\alpha x + \beta}}$$

The following table gives experimental data that describes this relationship.

x	2.54	3.09	3.55	4.12	4.75	5.21	5.95	6.47
Q_0	3.69	3.87	4.021	4.212	4.426	4.584	4.843	5.028

Use a linear regression technique to determine the force's value developed by the 4.5 mm movement of the screw.

Sections 5.5-5.6

5.20. Use Lagrange's interpolation formula to derive a formula that calculates the first-order derivative when four points (formula (5.62)) are used.

5.21. Consider the function

$$f(x) = \frac{x^2 \sin x}{4 + \cos x}$$

Use the formula (5.62) to calculate the first derivative at the point $x = 2.5$.

5.22. The following table describes the measurements of the electric current $i(t)$ through a coil of inductance L equals 12 mH in an AC electric circuit. Calculate the voltage V_L across this coil at the time $t = 1.5$.

i (m Amp)	107.8219	105.6868	103.5941	101.5428	99.5321	97.5612	95.6294
t (in sec.)	1	2	3	4	5	6	7

Hint: the voltage across the coil is given by

$$V_L(t) = L\frac{di}{dt}$$

5.23. The following table describes an airplane range R, which is approaching a tracking radar.

R (km)	29.399	25.338	21.354	17.493	13.856
t (in sec.)	0	10	20	30	40

Determine the speed of approaching the radar station after 6 and 7 seconds. Is the airplane in acceleration or deceleration mode?

5.24. Using the mentioned numerical integration method to evaluate the definite integral

$$I = \int_0^1 \frac{dx}{1 + x^3}$$

Compare the numerical solutions to the exact value given by

$$I = \frac{1}{6}\ln\left|\frac{(1+x)^2}{1 - x + x^2)}\right| + \frac{1}{\sqrt{3}}\tan^{-1}\left(\frac{2x-1}{\sqrt{3}}\right)$$

5.25. Consider the curve area that is given by the integral

$$A = \int_{\pi/2}^{\pi} \frac{x}{\sin(2x)}\, dx$$

The integral has no explicit solution form; thus, evaluate the area under the curve numerically using both MTR and MS$\frac{1}{3}$M rules. Also, an estimate of the truncation error for both rules

5.26. In Figure 5.15, a lake in the *x*- *y* plane is shown, where the coordinates (in kilometers) of select points on the lake boundary are also shown. Evaluate the lake area.

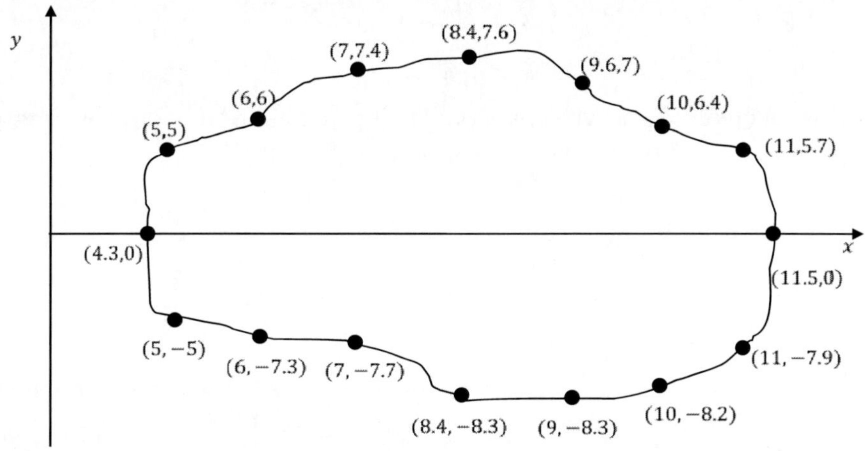

Figure 5.15. Lake Area in *x* – *y* plane (exercise 5.26).

5.27. Compute the area of the surface formed when the function $f(x) = x\sin 2x$ between $x = 0$ and $x = \pi$ is rotated around the y-axis. It is known that such an area is calculated from the integral

$$A_S = \int_{x=a}^{x=b} 2\pi x\sqrt{1 + f'(x)^2}\, dx$$

Use MTR and MS $\frac{1}{3}$M rules to compute the surface area.

5.28. Some indefinite integrations can be exactly computed by theories in a complex variable (see, example 4.35 Volume II); it is found that

$$\int_{-\infty}^{\infty} \frac{\cos 2x}{(x^2 - 4x + 13)} dx = -0.00169669$$

In practice, it might be required to compute similar but definite integration (which does not have an exact analytic solution), such as

$$\int_{-a}^{a} \frac{\cos 2t}{(t^2 - 4t + 13)} dt$$

For this integrand, it is desired to know the range of value of a that makes the value -0.00169669 accepted within an accuracy of 10^{-3}.

5.29. Use MTR and MS$\frac{1}{3}$M rules to compute the area under the curve of the function between $x_0 = 3$ and $x_1 = 5$.

$$f(x) = \int_{x_0}^{x_1} \sqrt{x} \ln x \; dx$$

5.30. For calculating the volume $V$of rotating a bounded area described by $A(x)$ around the *x*-axis, the disc method is carried out via the calculation of the following integral.

$$V = \pi \int_a^b f(x)^2 dx$$

For $A(x) = 1 - x^2, a = 0, b = 1$, the exact volume is equal to $8\pi/15$. Select one of the numerical methods to calculate this volume as accurately as possible.

Sections 5.7-5.8

5.31. Use Euler's method to find $y(1)$ by solving the following differential equation.

$$y' = \frac{x - y\cos x}{y + \sin x}; \; y(0) = 2$$

Use a step of integration $h = 0.1$. Compute the values $y(1), y(2)$.

5.32. Consider the following first-order nonlinear Riccati differential equation

$$y' + y^2 + \sqrt{8}\, y + 1 = 0; \; y(0) = 0$$

In the range $[0,1]$, use Heun's method to determine the numerical solution.

5.33. Make a table of comparison between the solutions obtained by applying Euler's and Heun's methods and the exact solution for the following linear first-order equation. Use a step of integration $h = 0.2$. For the interval $[0, 2]$, compare the computed value $y(2)$ with the exact value obtained from example 2.18.

$$\frac{dy}{dx} + \frac{2}{x + 1} y = 3x; \; y(0) = 1$$

5.34. Use Heun's method to solve Newton's first differential equation; use $h = 0.1$ to compute $y(1)$.

$$y' - (x+1)y = x^2 - 3x + 1;\ y(0) = 0$$

5.35. A room heating radiator is heated up to 353 Kelvin before it is switched off to cool. The following differential equation models this temperature gradient (Kelvin/seconds) problem

$$\frac{dT}{dt} = 2.98743 \times 10^{-3} - 9.4536 \times 10^{-12}\, T^4$$

Make two steps of calculating the fourth-order Runge-Kutta method to determine the radiator's temperature after 10 minutes.

5.36. A spherical projectile of a diameter $d = 12$ cm and a mass $m = 200$ gm is projected with an initial velocity $v_0 = 50$ m/sec, and an initial angle $\theta_0 = 60^o$. The mathematical model for calculating the velocity profile is given by

$$\frac{dv_y(t)}{dt} = g - \frac{1}{2m} A C_d \rho_0\, v(t) v_y(t);\ v_y(0) = v_0 \sin\theta_0$$

$$\frac{dv_x(t)}{dt} = -\frac{1}{2m} A C_d \rho_0\, v(t) v_y(t);\ v_x(0) = v_0 \cos\theta_0$$

$$v(t) = \sqrt{v_x^2(t) + v_y^2(t)}$$

where g is the gravitational acceleration ($9.81\ \text{m/sec}^2$), A is the reference area of the spherical projectile ($0.25\,\pi d^2$), C_d is the drag coefficient (0.47), and ρ_0 is the air density ($1.225\ \text{kg/m}^3$). Use four steps of the fourth-order Runge-Kutta method to determine the projectile velocity after 2.8 seconds (see example 5.30), thus calculating the coordinates $x(2.8)$ and $y(2.8)$.

5.37. Resolve the long thin rod problem given in example 5.30 for the following numerical: $T_a = 10, k = 0.015,\ T_0 = 55, T_L = 330,$ and $L = 12\ m.$ Determine the corresponding value of the initial condition $T'(12)$.

5.38. Use finite differences method to solve the boundary value problem of the linear variable- coefficient differential equation,

$$A(x)y'' + B(x)y' + C(x)y = f(x);\ y(x_0) = m_1, y(x_b) = m_2$$

Consider five segments to show that the linear algebraic system is given as

$$\begin{bmatrix} b_1 & c_1 & 0 & 0 \\ a_2 & b_2 & c_2 & 0 \\ 0 & a_3 & b_3 & c_3 \\ 0 & 0 & a_4 & b_4 \end{bmatrix} \begin{bmatrix} y_1 \\ y_2 \\ y_3 \\ y_4 \end{bmatrix} = \begin{bmatrix} (\Delta x)^2 f(\Delta x) - a_1 m_1 \\ (\Delta x)^2 f(2\Delta x) \\ (\Delta x)^2 f(3\Delta x) \\ (\Delta x)^2 f(4\Delta x) - c_4 m_2 \end{bmatrix}$$

where

$$a_i = A(i\,\Delta x) - \Delta x\, B(i\,\Delta x)$$

$$b_i = -2A(i\,\Delta x) + \Delta x\, B(i\,\Delta x) + \Delta x^2 C(i\,\Delta x)$$

$$c_i = A(i\,\Delta x);\; i = 1,2,3,4$$

5.39. Use the results in exercise 5.37 to solve the following boundary value problem

$$x^2 y'' - x(4-x)y' + (6-2x)y = 0;\; y(1) = 1, y(2) = -1.057$$

Check your results with exact values given by

$$y(x) = -x^2 + 2\,e^1\,x^2\,e^{-x}$$

ANSWERS TO SELECTED EXERCISES

SECTIONS 1.1–1.5

1.1 iii. $\det(B) = 2,\ adj(B) = \begin{bmatrix} 6 & 11 & 21 \\ 0 & -1 & -1 \\ -2 & -3 & -7 \end{bmatrix} B^{-1} = \frac{1}{2}\begin{bmatrix} 6 & 11 & 21 \\ 0 & -1 & -1 \\ -2 & -3 & -7 \end{bmatrix}$

1.3 $B = \begin{bmatrix} 1 & 0 & 2 \\ 0 & -1 & 1 \\ 2 & 1 & 4 \end{bmatrix}, C = \begin{bmatrix} 0 & 2 & 3 \\ -2 & 0 & 1 \\ -3 & -1 & 0 \end{bmatrix}$

1.4 $\det(A) = 0, \det(B) = 0$

1.6 $A^{-1} = \begin{bmatrix} 5 & -2 & 0 \\ -2 & 10 & -3 \\ 0 & -3 & 1 \end{bmatrix}$

1.8 $P = \begin{bmatrix} 4 & 1 & 1 \\ -13 & 0 & -4 \\ -2 & -1 & 0 \end{bmatrix}$

1.9 $E = \begin{bmatrix} 1 & 0 & 1 & 0 & 9/11 \\ 0 & 1 & 1 & 1 & -25/11 \\ 0 & 0 & 0 & 1 & 27/11 \\ 0 & 0 & 0 & 0 & 0 \end{bmatrix}$

1.11 $\text{rank}\,(M) = 2, \text{rank}(N) = 2, \text{rank}(P) = 4$

1.13 $X = [1 \quad 2 \quad 4 \quad 5]^T$

1.14 $P \neq 16,\ P = 5$

1.16 $i. X = [a - 2b \quad -4a - 5b \quad b \quad a]^T,\ ii. X = [3c/2 \quad -7c/3 \quad c]^T$

1.17 $a = 2, b = 7, c = 6, d = 4$

1.18 Linearly independent

1.20 $\begin{bmatrix} x_{11} & x_{12} & x_{13} \\ x_{21} & x_{22} & x_{23} \\ x_{31} & x_{32} & x_{33} \end{bmatrix} = \begin{bmatrix} 3 & 2 & -2 \\ -5 & -2 & 7 \\ 3 & 1 & -3 \end{bmatrix}$

1.22 No

1.26 $Q = \begin{bmatrix} 1 & -1 & 0 & 0 \\ 0 & 1 & -2 & 1 \\ 0 & 0 & 1 & -1 \\ 0 & 0 & 0 & \frac{1}{3} \end{bmatrix}, P = \begin{bmatrix} 1 & 0 & 0 & 0 \\ 0 & 1 & 0 & 0 \\ 1 & 0 & 1 & -1 \\ 0 & 0 & 3 & -2 \end{bmatrix},$

1.27 $A^{-1} = \begin{bmatrix} 1 & -1 & 0 & 0 \\ -2 & 1 & 1 & 0 \\ 1 & 0 & -2 & 1 \\ 0 & 0 & 1 & -2/3 \end{bmatrix}$

Sections 1.6–1.7

1.28 Yes

1.29 Yes

1.30 $p = 0$

1.32 Yes

1.35 $P = \begin{bmatrix} 3/5 & 0 & 4/5 \\ 0 & 1 & 0 \\ 4/5 & 0 & -3/5 \end{bmatrix}$

1.36 $[41 \quad -3 \quad 84]^T$

1.38 $\delta = [3/4 \quad 1/4 \quad 3/2]^T$

Sections 1.8–1.9

1.40 $\text{eig}(A): \{\lambda_1 = 4, \lambda_2 = 9\};\ X_1 = [-1 \quad 1]^T, X_2 = [1 \quad 4]^T$

$\text{eig}(M): \{\lambda_1 = -3, \lambda_2 = 3, \lambda_3 = 9\};\ X_1 = \begin{bmatrix} -2 \\ 1 \\ -2 \end{bmatrix}, X_2 = \begin{bmatrix} 2 \\ 2 \\ -1 \end{bmatrix}, X_3 = \begin{bmatrix} -1 \\ 2 \\ 2 \end{bmatrix}$

1.44 $p(\lambda) = \lambda^4 - 6\lambda^3 + \lambda^2 - 12\lambda - 29$

1.45 $\text{eig}(A^3): \{\lambda_1 = -1, \lambda_2 = 343\};\ \text{eig}(A^{-1}): \{\lambda_1 = -1, \lambda_2 = 1/7\}$

1.47 $i. V_C = \begin{bmatrix} 1 & 1 & 1 \\ -1 & -2 & -3 \\ 1 & 4 & 9 \end{bmatrix}$;

$$ii. C^{-1} = -1/6[A^2 + 6A + 11I] = -1/6\begin{bmatrix} 11 & 6 & 1 \\ -6 & 0 & 0 \\ 0 & -6 & 0 \end{bmatrix}$$

1.49 $\text{eig}(S): \{\lambda_1 = 1, \lambda_2 = 2, \lambda_3 = 3\}$; $p(S) = \begin{bmatrix} 30 & 90 & 90 \\ 30 & 210 & 210 \\ -30 & -90 & -90 \end{bmatrix}$

1.50 $m_1(A) = A(A - I)(A - 6I)$; $m_2(B) = (B - I)(B - I)(B + I)$

1.51 $C = \begin{bmatrix} 0 & 1 & 0 \\ 0 & 0 & 1 \\ -67 & -5 & 14 \end{bmatrix}$

SECTION 1.10–1.12

1.52 $R = \left(\frac{1}{5\sqrt{26}}\right)\begin{bmatrix} 3\sqrt{26} & 4 & -20 \\ 0 & 25 & 5 \\ 4\sqrt{26} & -3 & 15 \end{bmatrix}$; $R^T AR = \begin{bmatrix} -2 & \frac{7\sqrt{26}}{26} & \frac{7\sqrt{26}}{130} \\ 0 & -1 & \frac{338}{65} \\ 0 & 0 & -1 \end{bmatrix}$

1.55 $\overline{\sigma}(A) = \sqrt{24}, \underline{\sigma}(A) = \sqrt{7}$

1.57 $\overline{\sigma}(A) = 4.5807, \underline{\sigma}(A) = 1.579$

1.59 $||A||_1 = 4, ||A||_2 = 5.8286, ||A||_\infty = 5, ||A||_{Fr} = 7.2801$

1.60 $\kappa(A) = 126.3858, \kappa(B) = 2.846$

1.61 $U_{Ch}(A) = \begin{bmatrix} \sqrt{2} & \sqrt{12.5} & \sqrt{18} \\ 0 & \frac{1}{\sqrt{2}} & \sqrt{32} \\ 0 & 0 & 0 \end{bmatrix}, L_{LU}(A) = \begin{bmatrix} 1 & 0 & 0 \\ 2/3 & 1 & 0 \\ 1/3 & 4/5 & 1 \end{bmatrix},$

$$U_{LU}(A) = \begin{bmatrix} 6 & 27 & 50 \\ 0 & -5 & -43/3 \\ 0 & 0 & 4/5 \end{bmatrix}$$

1.63 $\begin{bmatrix} 14.6969 & 37.5588 \\ 0 & 32.0832 \end{bmatrix} X = \begin{bmatrix} 8.165 \\ 42.182 \end{bmatrix} \Rightarrow X = \begin{bmatrix} -2.8044 \\ 1.3148 \end{bmatrix}$

SECTION 2.1–2.4

2.2 i. $CR_1\,i_2'(t) - CR_1\,i_1'(t) + i_1 = CE'(t)$, and $L\,i_2'(t) + (R_2 - R_1)\,i_2(t) + R_1 i_1 = 0$

iii. $V'' + 2\xi w_n V' + w_n^2 V = \frac{1}{C} I', \xi w_n = \frac{1}{2RC}, w_n = \frac{1}{\sqrt{LC}},$
where V is the voltage across the circuit

2.3 $i_2'' + 10^3 i_2' + 198 \times 10^6\, i_2 = 15 \cos 0.075\, t;\ i_2(0) = i_2'(0) = 0.$

2.5 $\frac{dS}{dt} = 2.5 - \frac{S}{250+t};\ S(0) = 20$

2.7 i. $y = c + 0.25(2x - 1)e^{2x}$ ii. $y = 1 - \frac{1}{2}x - \frac{1}{4}\sin 2x$

iii. $w = \left(\frac{1}{25}\right)(5t \sin 5t + \cos 5t + 499)$, iv. $y = \ln(c\sqrt{x^2 + 2x})$

SECTION 2.5

2.9 14.286 grams

2.10 $S(60) = 4700/31$ kg

2.11 i. $y = 2/(x^2 + 1)^2$

ii. $3x^2 + 6x + 2y^3 + 12y = 24$

iii. $y = (c - 3\sqrt{x})^{2/3}$,

iv. $y = \frac{13}{5}\ln(x - 4) - \frac{8}{5}\ln((x + 1)/6) + 10$

2.12 ii. $6y^2x^2 + 3x^4 + 4x^3 = c$

2.13 iii. $y \cos x + x \sin y = c$

2.14 i. $y = \frac{1}{2}(-x + x\sqrt{8x + 1})$

2.15 $y \sin x + y^2/\cos x = c$

2.16 i. $y^4 = 4x^4 \ln(c/x)$

ii. $y^2 = x^2 - x^3$

v. $x + \tan^{-1}(x/y) = c$

vi. $y = 7 \cos x - 2 \cos^2 x$

vii. y= $x = 0.2\,(y^2 - y^{-3})$

ix. $y = \frac{1}{2}t(t^2 + 5) - t \ln t$

xii. $y = e^x + c_1 e^{-2x} + c_2$

SECTION 2.6

2.17 i. $y = e^t[c_1\cos(\sqrt{6}\,t)+c_2\cos(\sqrt{6}\,t)] + \frac{1}{7}t^2 + \frac{17}{49}t + \frac{197}{343}$

iii. $y = \frac{3}{4}\,e^{-x} + \frac{5}{12}e^{-5x} + \frac{5}{6}e^{x}$

v. $y = \;e^{-2x}[c_1\cos(3x) + c_2\sin(3x)] + \frac{1}{40}[3\cos\,(3x) + 9\sin(3x)] + \frac{1}{9}xe^{-2x}$

vii. $y = \cos t - \frac{1}{2}\sin t - \frac{1}{2}t\cos t$

2.18 i. $y = c_1\sin x + c_2\cos x - x\cos x + \sin(\ln(\sin t))$

ii. $y = c_1e^{-4x} + c_2 + \frac{1}{20}(2\sin 2x - \cos 2x)$

SECTIONS 2.7-2.8

2.19 i. $y = c_1e^{-x} + e^{x/2}\left[c_2\cos\left(\frac{\sqrt{3}}{2}x\right) + c_3\sin\left(\frac{\sqrt{3}}{2}x\right)\right] + 1 + \cos x + \sin x$

iii. $y = c_1\cos(2x) + c_2\sin(2x) - \frac{1}{4}c_3 + \frac{1}{5}e^{x} + \frac{1}{3}\sin x$

2.20 i. $y = c_1e^{-4x} + c_2e^{-t} + c_3 - \frac{1}{3}x\,e^{-x} + \frac{1}{9}e^{-x}$

iii. $y = \frac{17}{36}\,e^{-2x}\cos(\sqrt{5}x) - \frac{\sqrt{5}}{90}e^{-2x}\sin(\sqrt{5}x) - \frac{1}{36}\cos(3x) + \frac{14}{9}$

2.21 $\begin{bmatrix} y_1 \\ y_2 \\ y_3 \end{bmatrix} = c_1\begin{bmatrix} 2 \\ 1 \\ 3 \end{bmatrix}e^{t} + c_2\begin{bmatrix} 3 \\ 1 \\ 5 \end{bmatrix}e^{2t} + c_3\begin{bmatrix} 1 \\ -1 \\ 2 \end{bmatrix}e^{3t}$

SECTION 3.2

3.1 i. $y(x) = c_1\,x + c_2x^2 + c_3x^3 + 10\,x\ln|x|$

ii. $y(x) = x^3 + x^{-2}$,

iii. $y(x) = c_1x + c_2x^{-1} + c_3x^{-1}\ln x$

3.4 $y(x) = -0.5x^{-2} + x^{-1} + \ln x - 1.5$

3.5 i. All points are ordinary points except $x = \pm 1$, which are regular singular points.

ii. All points are ordinary points except $x = 0, x = 1$, which are regular singular and irregular singular points respectively.

3.7

$$a_{k+1} = \frac{2-(k+r)}{(k+r+1)(k+r-4)+6} a_k;\ a_0 = 1, k = 0,1,2 \ldots$$

3.9 $y(x) = c_1\left(1 + 9x^2 + \frac{15}{2}x^4 - \frac{7}{2}x^6 \ldots\right) + c_2\left(x + \frac{8}{3}x^3\right);\ \forall\,|x| < \frac{1}{\sqrt{2}}$

3.12

i. $y(x) = c_1 x^{-1} + c_2 x\left(1 + \frac{1}{4}x + \frac{1}{24}x^3 + \frac{1}{196}x^5 + \cdots\right)$

ii. $y(x) = c_1 x^2 + c_2 x\left(1 - x^2 - \frac{1}{3}x^4 - \frac{1}{5}x^6 - \cdots\right)$

iii. $y(x) = c_1 x^{\frac{1}{3}} e^{-x} + c_2 x^{-\frac{1}{3}}\left(1 - 3x + \frac{9}{4}x^4 - \frac{27}{28}x^3 + \frac{81}{280}x^4 - \cdots\right)$

Section 3.3

3.15 $y(x) = c_1 x^2 + c_2[x^{-1}(1 + 3x + 6x^2 - 4x^4 - \cdots) - 4x^2 \ln x]$

3.16 $\lambda = 1, n = 1,\ y(x) = P_1(x) = x$

3.18

$$y_1(x) = F\left(\frac{1,1,2}{3}, x\right) = 1 + \frac{3}{2}x + \frac{9}{5}x^2 + \frac{81}{40}x^3 + \frac{243}{110}x^4 + \cdots$$

$$y_2(x) = G(a,b,c,x) = Ax^{\frac{1}{3}}F\left(\frac{4}{3},\frac{4}{3},\frac{4}{3},x\right); A = \frac{\Gamma\left(\frac{4}{3}\right)\Gamma\left(\frac{4}{3}\right)\Gamma\left(-\frac{1}{3}\right)}{\Gamma\left(\frac{1}{3}\right)}$$

3.19

i. $y(x) = c_1 I_3(4\sqrt{x}) + c_2 K_3(4\sqrt{x})$

ii. $y(x) = x^{-1}[c_1 J_0(2\sqrt{x}) + c_2 Y_0(2\sqrt{x})]$

iii. Irreducible

iv. $y(x) = \sqrt{x}\left[c_1 J_{\frac{5}{4}}\left(\frac{1}{2}x^2\right) + c_2 J_{-\frac{5}{4}}\left(\frac{1}{2}x^2\right)\right]$

v. $y(x) = x\sqrt{x}e^{-x/2}[c_1 I_{0.5}(0.5x) + c_2 I_{-0.5}(0.5x)]$

vi. $y(x) = (x)^{-\frac{1}{2}}\left[c_1 J_{\frac{3}{2}}\left(\frac{\sqrt{5}x}{2}\right) + c_2 J_{-\frac{3}{2}}\left(\frac{\sqrt{5}x}{2}\right)\right]$

vii. $y(x) = x[c_1 I_0(x) + c_2 K_0(x)]$

viii. $y(x) = c_1 J_0(x) + c_2 Y_0(x)$

ix. $y(x) = x[c_1 J_2(2\sqrt{2x}) + c_2 Y_2(2\sqrt{2x}) + c_3 I_2(2\sqrt{2x}) + c_4 K_2(2\sqrt{2x})]$

x. $y(x) = x^{-1}[c_1 J_2(4\sqrt{x}) + c_2 Y_2(4\sqrt{x}) + c_3 I_2(4\sqrt{x}) + c_4 K_2(4\sqrt{x})]$

3.21

$$y(x) = \frac{x^6 - 2.5}{x(x^6 + 0.5)}$$

3.22 ii. $y(x) = c_1 y_1 + c_2 y_2$, where

$$y_1 = \frac{d}{dx}[\ln(1 - \frac{1}{2}x^3 + \frac{1}{2\times 5}x^6 - \frac{1}{2\times 5\times 8}x^9 + \cdots)$$

$$y_2 = \frac{d}{dx}[\ln(x - \frac{1}{3}x^4 + \frac{1}{3\times 6}x^7 - \frac{1}{3\times 6\times 9}x^{10} + \cdots)$$

iv. $y(x) = (e^{5x} - 4c)/(e^{5x} + c)$

3.24 The Bernoulli's equation is $u' - 4u + u^2 = 0$'

$$y(x) = 2 + 3x + \frac{4e^{4x}}{c - e^{4x}}$$

3.26

i. $\frac{y^3}{3} + \frac{3y^2}{2} + 2y = c_1 x + c_2$

ii. $y = -\frac{2}{3}\tan^{-1}(c_1 x) + c_2$

iii. $y = (3e^{2x} - 2)/(2 - e^{2x})$

iv. $y = c_1 - \sqrt{1 - (x - c_2)^2}$

v. $y = x^2$

3.27

$$y(x) = \frac{1}{2}(c_1 - x)^2 - \cos(c_1 - x) - 2c_2(c_1 - x) + c_3$$

3.28 $(y - c_2)^3 = (x - c_1)^2$

3.31 i. $y(t) = \mathrm{cn}(3t);\ k = 1/\sqrt{6}$

ii. $4p \cong 2.38$

SECTION 4.2

4.2 $F(s) = (1 + e^{-s} + e^{-2s})/s$

4.3 $F(s) = [(s+1)(1 - e^{-s}) + 1]/s$

4.4 $F(s) = (s^2 + 4)/(s^2 - 4)^2$

4.10

$$\text{ii. } F(s) = \frac{2\,s^6 + 48\,s^2 w^4 + 64\,w^6}{s^3\,(s^2\; +\; 4\,w^2)^3}, \text{iii. } F(s) = \frac{10(3s^2 + 12s - 13)}{(s^2 + 4s + 29)^3}$$

$$\text{vi. } F(s) = \tan^{-1}\left(\frac{2}{s+3}\right) - \tan^{-1}\left(\frac{4}{s+1}\right), \text{vii. } F(s) = \frac{s+3}{s^2 + 6s + 5}$$

SECTION 4.3–4.4

4.11 $f(t) = \frac{3}{2}\,e^{-t} - 2\,e^{-2t} + \frac{1}{2}\,e^{-3t}$

4.12 $k_1 = \frac{3}{122}, k_2 = \frac{5}{61}, k_3 = -\frac{3}{122}, k_4 = \frac{4}{61}$

4.15 $c(t) = \frac{3}{25}\,u(t) - \frac{1}{25}\,e^{-3t}(3\cos\,4t + 4\,\sin\,4t)$

4.17

$$\text{i. } f(t) = \frac{1}{2}e^t - \frac{1}{17}\,e^{-4t} - \frac{15}{34}\cos t - \frac{25}{34}\sin t$$

$$\text{iii. } f(t) = 3\,e^{-3t} + 2\,t\,e^{-t}, \text{vi. } f(t) = \frac{2}{t}(\cos t - \cosh t)$$

$$\text{vii. } f(t) = (2t^2 + t)e^{-t}, \text{ix. } f(t) = \cos\left(\sqrt{3}t\right) + \cos(2t)$$

4.18

$$\text{i. } y(t) = \frac{3}{4}e^{-2t} + \frac{1}{4}\,e^{-4t} + \frac{3}{2}\,t\,e^{-2t}, \text{ii. } y(t) = \cos 2t + \sin 2t\; + \frac{3}{4}t\sin 2t$$

$$\text{v. } y(t) = \frac{1}{9}\,te^{-2t} + \frac{25}{27}e^t - \frac{8}{9}\,te^t + \frac{2}{27}e^{-2t}$$

vi. $y(t) = 34\cos(\sqrt{3}t) + 3\sqrt{3}\,\sin(\sqrt{3}t) - e^t(34\cos t +$
$14\sin t) + 39\,t\,e^t\cos t +$

$$\frac{26}{169}\,te^{-t}\sin t$$

$$\text{viii. } y(t) = \frac{1}{4}\left((\cos^2(t) - \cos\left(\sqrt{2}t\right)\right)$$

4.20

i. $y(t) = c_1\, t + c_2\, t \ln t$

iii. $y(t) = c_1 t + c_2 t^2 + c_3 t^3$

iv. $y(t) = c_1 t^3 + c_2 \cos(\sqrt{2}\ln(t)) + c_3 \sin(\sqrt{2}\ln(t))$

4.21 $y(t) = C(sI - A)^{-1}(BU(s) + X_0)$

4.22 a. $f(t) = [u(t) - u(t-1)]t + [u(t-1) - u(t-2)] + [u(t-2) - u(t-3)](t-1)$

$$+[u(t-3) - u(t-4)](-2t+8)$$

b. $f(t) = \frac{1}{2}[u(t) - u(t-2)]t^2$

4.23

i. $F(s) = e^{-2s}\mathcal{L}\{(t+2)^2 + 3(t+2) + 2\} - e^{-3s}\mathcal{L}\{(t+3)^2 + 3(t+3) + 2\}$

iii. $F(s) = \frac{s}{s^2+1}(1 - e^{-2\pi s})$

iv. $F(s) = \frac{(e^{-\pi s} + 1)}{(s^2+1)(1 - e^{-\pi s})}$

vi. $F(s) = \frac{2 - 3e^{-s} + e^{-3s}}{2s^2(1 - e^{-3s})}$

4.24

ii. $\frac{1}{8}\sin 2t - \frac{1}{4}t\, e^{-2t} - \left(\frac{1}{8}\sin 2t - \frac{1}{4}(t - 2\pi)e^{-2(t-2\pi)}\right)u(t - 2\pi)$

4.25

$$y(t) = \frac{1}{6}\sin^3 t + \frac{1}{6}\sum_{m=1}^{\infty} \sin^3(t - (m+1)\pi)\, u(-(m+1)\pi)$$

4.27

i. $\ln\left(\frac{5}{2}\right)$, ii. $\frac{\pi}{5}$ iii. $\tan^{-1}\left(\frac{2}{3}\right)$ iv. $\frac{7}{121}$

4.28

$$F(s) = \frac{2s+4}{(s^2+4s+8)^2} - \frac{1}{(s+2)(s^2+4s+8)} + \frac{1}{(s+2)^2}$$

4.30 $c_{ss} = 34.67$

4.32

$$\frac{V_0(s)}{I(s)} = \frac{R(1 + LCs^2)}{1 + RC\, s + LC\, s^2}$$

4.34

$$T_1(s) = \frac{U(s) + 0.5\, T_0}{s^2 + 2.75\, s + 0.5}$$

SECTION 5.3

5.1 $a = 0.5671, b = 1.9056, c = 0.8303$

5.2 $a = 1.31594, b = 1.94627, c = 0.70346$

5.3 $a = 0.548070, b = 1.496281$

5.4 0.8762533

5.5 0.9438497

5.6 $a. x = 3.7568, y = 2.7798$ $b. x = 0.7718, y = 0.4196$

5.7 $V_{L\,ave} = 54.6408$ volt

5.8 $v_f = 166.65$ m/sec

5.9 There are two unstable roots $r_1 = 2.53, r_2 = 0.68$

5.10 $r_1 = 2, r_{2,3} = -1.5 \pm 1.66j$

SECTION 5.4

5.11 The norm system is

$$\begin{bmatrix}\sum_{k=1}^{5} w & \sum_{k=2}^{6} w & \cdots & \cdots & \sum_{k=5}^{9} w\end{bmatrix}^T \begin{bmatrix} a_1 \\ a_2 \\ a_3 \\ a_4 \\ a_5 \end{bmatrix} = \begin{bmatrix} \frac{2\pi + \pi^2}{4} - 1 \\ \frac{4\pi^2 + \pi^3}{16} - 2 \\ \frac{-144\pi + 6\pi^3 + \pi^4}{48} + 6 \\ \frac{-384\pi^2 + 8\pi^4 + \pi^5}{128} + 24 \\ 60\pi - \frac{5}{2}\pi^2 + \frac{\pi^5}{32} + \frac{\pi^6}{320} - 120 \end{bmatrix} ;; w$$

$$= \frac{1}{k}\left(\frac{\pi}{2}\right)^k$$

5.12 $f(x) \cong 0.6032x^4 - 0.4451x^3 + 1.7809x^2 + 0.7634x - 0.9836,$ middle accuracy is better than 10^{-4}.

5.13 $p_3(x) = 0.0632x^3 - 0.5296x^2 + 1.0041x$

5.14 $a = 1.0984, b = 3.4536$

5.15 $k_1 = 1.341, k_2 = 0.0876, t = 2.75$ hours

5.16 $a = 0.9167, b = -1.711, c = 3.5616$

5.17 $p_3(x) = x^4 - x^3 + 3x^2 + 5x - 51$

5.18 $p_3(\varphi) = -7.0925 \times 10^{-6}\, \alpha^3 + 2.1667 \times 10^{-4}\, \alpha^2 + 0.80338\, \alpha$

5.19 $\alpha = 0.23, \beta = 1.12, Q_0(x = 4.5) = 4.34\ \mathrm{m}^3/\mathrm{minute}$

Section 5.5–5.6

5.21 $f'(x) = -0.41511$

5.22 $V_L(1.5) = -24.365\ \mu\ \mathrm{volt}$

5.23 $321.73\ \mathrm{m/sec}, 316.67\ \mathrm{m/sec}$, in deceleration mode

5.24 $I_{num} = 0.835659, I_{exact} = 0.835648$

5.25 $\mathrm{MTR}:\ -5.105088, \mathrm{Ms}\frac{1}{3}\mathrm{M}:\ -5.759586$, $\mathrm{error}_{\mathrm{est}}: 0.0135, -0.000056$

5.26 93.25 km^2

5.27 eight segments: $\mathrm{MTR}: A_S =\ 94.1601, \mathrm{Ms}\frac{1}{3}\mathrm{M}: A_S = 93.2806$

5.28 $a \in [5.8, 9.9]$

5.29 eight segments: $\mathrm{MTR}: \mathrm{Area} =\ 5.5302, \mathrm{Ms}\frac{1}{3}\mathrm{M}: \mathrm{Aera} = 5.5307$

5.30 $\mathrm{Ms}\frac{1}{3}\mathrm{M}: V = 1.6756$

Section 5.7–5.8

5.31 $y(1) = 1.499, y(2) = 1.983$

5.32 $y(0.1) = -0.14731, y(1) = -0.36191$

5.33 $y_{Heun}(2) = 3.906, y_{exact}(2) = 3.8889$

5.34 $y(0.1) = 0.091, y(1) = 0.19099$

5.35 $T(5) = 318.2427, T(10) = 294.316$

5.36 $x(2.8) \cong 47.75\ m, y(2.8) \cong 46.25\ m$

5.37 $T'(12) = 12.907$

References

[1] Johnston, Nathaniel, *Introduction to Linear algebra and Matrix theory*, Springer-Verlag, New York, 2021.

[2] Boyd, Stephen, *Introduction to Applied Linear Algebra, Vectors, Matrices, and Least Squares*, Cambridge University Press, 2018.

[3] Kalbaugh, David V., *Differential Equations for Engineers*, Routledge CRC Press, New York, 2018.

[4] Axler, Sheldon, *Linear Algebra Done Right*, 3rd ed., Springer-Verlag, New York, 2019.

[5] Shores, Thomas S., *Applied Linear Algebra and Matrix Analysis*, Springer-Verlag, New York, 2007.

[6] Strang, Gilbert, *Introduction to linear Algebra*, 5th ed, Cambridge University Press, 2016.

[7] Meckes, Elizabeth S., and Mark W. Meckes, *Linear Algebra*, Cambridge Mathematical Textbooks, 2018.

[8] Kyrchei, Ivan, *Advances in Linear Algebra Research*, Nova Science Publishers Inc., New York, 2015.

[9] Charles G. Gullen, *Linear Algebra & Differential Equations,* 2nd ed, Prindle Weber & Schmidt, Boston, 1980

[10] Lin, R. M., J. E. Mottershead, and T. Y. Ng, A state-of-the-art review on theory and engineering applications of Eigenvalue and Eigenvector derivatives, *Journal of Mechanical Systems and Signal Processing*, Vol. 138, April 2020,106536.

[11] Compton, Elizabeth A., and Stacey L. Ernstberger, Singular Value Decomposition: Applications to Image Processing, *Citations Journal of Undergraduate Research* 2020, Volume 17.

[12] Ahmad, Shair and Antonio Ambrosetti, *Differential Equations*, De Gruyter Textbook, 2019.

[13] Polyanin, Andrei D., Valeatin Zaitsev, and Inna K. Shingareva, *Handbook of Ordinary Differential Equations: Exact Solutions, Methods, and Problem*, Routledge CRC Press, New York, 2017.

[14] Keskin, Ali Omit, *Ordinary Differential Equations for Engineers*, Springer-Verlag, New York, 2019.

[15] N'Guerekata, Gaston M., *Differential Equations Application Systems and Functions*, Nova Science, New York Publishers Inc., 2012.

[16] Cengel, Yunus A., and William J. Palm, *Differential Equations for Engineers and Scientists*, McGraw Hill, New York, 2019.

[17] Xie, Wei-Chau, *Differential Equations for Engineers*, Academic-Cambridge, 2010.

[18] Sparks, Seth and Bill Willis, *Mathematical Modeling- Methods, Applications and Research*, Nova Science Publishers Inc., New York, 2018.

[19] Li, Dongfang, Jinming Wen, and Jiwei Zhang, Recent Advances in Numerical Methods and Analysis for Nonlinear Differential Equations, *International Journal of Differential Equations* Vol. 2019, Article ID 3243510, Hindawi.

[20] Roberts, Charles, *Elementary Differential Equations Applications, Models, and Computing*, 2nd ed., Routledge CRC Press, New York, 2019.

[21] Cohen, Harold and Daniel Gallup, *An Introduction to Differential Equations with Applications*, World Scientific, September 2020.

[22] Apelblat, Alexander, *Laplace Transforms & their Applications*, Nova Science Publishers Inc., New York, 2012.

[23] Stroud, K. A. and Dexter J. Booth, *Advanced Engineering Mathematics*, 6th ed., Macmillan international Higher Education, 2020.

[24] Ghil, Byungmoon and Hwajoon Kim, The Solution of Euler-Cauchy Equation Using Laplace Transform, *International Journal of Mathematical Analysis*, Vol. 9 No. 53, 2015.

[25] Seacrest, Tyler, *Mathematical Models of Image Processing*, Harvey Mudd College, Department of Mathematics, 2006.

[26] Harold Klee, *Simulation of Dynamic Systems with MATLAB and Simulink*, CRC Press, New York, NY, 2007.

[27] Kraniauskas, Peter, *Transforms in Signals and Systems*, Addison Wesley, Wokingham England, 1992.

[28] Schiff, Joel L, *The Laplace Transform Theory and Application*, Springer-Verlag, New York 1999.

[29] Shaikh, Zameer, *Laplace Transform*, independently published, 2019.

[30] Hsu, Tai-Ran, *Applied Engineering Analysis*, John Wiley & Sons, 2018.

[31] C. Ray Wyle, Louis C. Barrett, *Advanced Engineering Mathematics,* 5th ed, McGraw-Hill International Book Company, New Delhi, 1982

[32] Dubner, H., and J. Abate, Numerical Inversion of Laplace Transforms by Relating Them to the Finite Fourier Cosine Transform, *Journal of AMC*, Vol. 15, No. 1, 1968.

[33] Hupper, Bruno and Eli Pollak, Numerical inversion of the Laplace transform, *AIP The Journal of Chemical Physics*, Vol. 110, Issue 23, 1999.

[34] Ungar, Abraham, The Laplace Transform Inversion by Inspection, *The American Mathematical Monthly*, Vol. 93, No. 10, 1986.

[35] Schiff, Joel L., *The Laplace Transform Theory and Applications*, Springer-Verlag, New York, 1999.

[36] McNamee, J. M., *Numerical Methods for Polynomial-Part I*, Volume 14, 1st ed., Elsevier Sciences 2007.

[37] Hoffman, Joe D., *Numerical Methods for Engineers and Scientists*, Marcel Dekker, INC., New York, 2001.

[38] Davis, Mark E., *Numerical Methods and Modeling for Chemical Engineers*, John Wiley &Sons, Ney York, 1984.

[39] Chapra, Steven C., Raymond P. Canale, *Numerical Methods for Engineers*, 6th ed., McGraw-Hill Higher Education, New York, 2014.

[40] Gupta, Rajesh Kumar, *Numerical Methods Fundamentals and Applications*, Cambridge University Press, 2019.

[41] Kharab, Abdelwahab, Ronald Guenther, *An Introduction to Numerical Methods A MATLAB Approach,* 4th ed, Routledge CRC Press, New York, 2019.

[42] Li, Dongfang, Jinming Wen, and Jiwei Zhang, Recent Advances in Numerical Methods and Analysis for Nonlinear Differential Equations, *International Journal of Differential Equations*, Volume 2019 |Article ID 3243510, Hindawi.

About the Author

Rami A. Maher, PhD
Professor in Engineering Project Management, MSc Program,
Isra University Amman, Jordan

Rami A. Maher, PhD in Engineering Cybernetics – Automatic Control branch of science in 1982 and 1986 from VAAZ – Brno – Czechoslovakia, awarded the title Professor in 2002. He is the author of the book *Optimal Control Engineering with MATLAB*.

Email: rami.maher@iu.edu.jo; dr.ramimaher@gmail.com

INDEX

A

B

C

D

E

F

G

H

I

J

K

L

M

N

O

P

Q

R

S

T

U

V

W